Predictive Analytics

Advanced Research in Reliability and System Assurance Engineering
Series Editor: Mangey Ram, Professor, Assistant Dean (International Affairs), Department of Mathematics, Graphic Era University, Dehradun, India

Reliability Engineering
Theory and Applications
Edited by Ilia Vonta and Mangey Ram

Modeling and Simulation Based Analysis in Reliability Engineering
Edited by Mangey Ram

For more information about this series, please visit: https://www.crcpress.com/Reliability-Engineering-Theory-and-Applications/Vonta-Ram/p/book/9780815355175

Predictive Analytics

Modeling and Optimization

Edited by

Vijay Kumar
and
Mangey Ram

CRC Press
Taylor & Francis Group
Boca Raton London New York

CRC Press is an imprint of the
Taylor & Francis Group, an **informa** business

First edition published 2021
by CRC Press
6000 Broken Sound Parkway NW, Suite 300, Boca Raton, FL 33487-2742

and by CRC Press
2 Park Square, Milton Park, Abingdon, Oxon, OX14 4RN

CRC Press is an imprint of Taylor & Francis Group, LLC

Library of Congress Cataloging-in-Publication Data
Names: Kumar, Vijay, editor. | Ram, Mangey, editor.
Title: Predictive analytics : modeling and optimization / edited by Vijay Kumar and Mangey Ram.
Description: First edition. | Boca Raton, FL : CRC Press/Taylor & Francis Group, LLC, 2021. | Series: Advanced research in reliability and system assurance engineering | Includes bibliographical references and index.
Identifiers: LCCN 2020038821 (print) | LCCN 2020038822 (ebook) | ISBN 9780367537463 (hardback) | ISBN 9781003083177 (ebook)
Subjects: LCSH: Engineering--Data processing. | Predictive analytics.
Classification: LCC TA340 .P726 2021 (print) | LCC TA340 (ebook) | DDC 003/.2--dc23

LC record available at https://lccn.loc.gov/2020038821
LC ebook record available at https://lccn.loc.gov/2020038822

ISBN: 978-0-367-53746-3 (hbk)
ISBN: 978-1-003-08317-7 (ebk)

Typeset in Times
by SPi Global, India

Contents

Preface

Technological innovation has created opportunities to monitor and measure the performance of engineering systems with broader enterprise operations, such as manufacturing operations, service logistics, maintenance management, and after-sales services. The technological revolution demands methodologies and solutions which are capable of analysing and modelling to support and facilitate optimal decision-making policies. Combined with state-of-the-art, real-time optimization techniques, nearly instantaneous decisions can be computed in fast-changing business environments unlocking significant cost-savings. Thus predictive analytics plays a vital role in technological innovation. Predictive analytics is the practice of extracting information from existing data sets in order to determine patterns and predict future outcomes and trends. Predictive analytic tools are capable of synthesizing and extracting information from existing systems to the forecast future performance of systems. Optimization models compute optimal decisions by leveraging the information embedded in the data. The development of modern methodologies allows for efficient updating when information changes, as well as automatic model calibration using techniques from machine learning, information theory, and statistics.

This book presents state-of-the-art predictive analytics and optimization in the areas of system reliability, quality, supply chain management, and logistics. This book helps companies overcome the challenges posed by increasingly complex systems in today's competitive marketplace. Examining both research on and practical aspects of products, the book features contributions written by active researchers and/or experienced practitioners in the field so as to effectively bridge the gap between theory and practice and address new research challenges in system reliability, quality, supply chain management, and logistics management in practice. The topic is also covered in most of the international scientific conferences, seminars, and workshops. These facts provide sufficient evidence for the increasing interest in this scientific area. All the chapters in the book are written by leading researchers and practitioners in their respective fields of expertise and present various innovative methods, approaches, and solutions not covered before in the literature.

The book would be useful to senior graduate and postgraduate students, as well as research scholars. Of course, it can also be used as a reference book by academicians, scientists, engineers, and managers working in different organizations.

Vijay Kumar
Department of Mathematics,
Amity Institute of Applied Sciences,
Amity University Uttar Pradesh, Noida-201313, India

Mangey Ram
Graphic Era Deemed to be University,
Dehradun, India

Acknowledgements

The editors acknowledge CRC Press for this opportunity and professional support. Also, we would like to thank all the chapter authors and reviewers for their availability for this work.

Contributors

Reshu Agarwal
Amity Institute of Information Technology
Amity University
Uttar Pradesh, Noida, India

A. Mohamed Ashik
Assistant Professor in Statistics
Department of Mathematics
Merit Arts and Science College
(Affiliated to Manonmaniam Sundaranar University)
Idaikal, Tirunelveli, Tamil Nadu, India

Astha Bathla
Guru Gobind Singh Indraprastha University
New Delhi, India

Murray V. Calichman
Retired Director
Department of Operational Research & Analytics
St. Francis Hospital
Roslyn, NY, USA
Catholic Health Services of Long Island (11 Years) and past President of GBC Consulting Corporation (35 Years)

Adarsh Dixit
Amity Institute of Information Technology
Amity University
Uttar Pradesh, Noida, India

Aakanshi Gupta
Guru Gobind Singh Indraprastha University
New Delhi, India

Saksham Jain
Amity School of Engineering and Technology
Delhi Campus, Noida, India
Data Scientist, GreenTech ITS

Mahesh Kumar Jayaswal
Banasthali Vidyapith
Jaipur, Rajasthan, India

Shylaja Vinaykumar Karatangi
G. L. Bajaj Institute of Technology and Management
Greater Noida, Uttar Pradesh, India

Maninder Jeet Kaur
Canadore College
Ontario, Canada

Faizan Ahmad Khan
University of Tabuk
Tabuk, Saudi Arabia, UAE

Digeshkumar B. Shah
L. D. College of Engineering
Ahmedabad, Gujarat, India

A. Loganathan
Professor
Department of Statistics
Manonmaniam Sundaranar University
Abishekapatti, Tirunelveli, Tamil Nadu, India

Rana Majumdar
MSIT, TIG
Kolkata, India

Mohamed Arezki Mellal
LMSS, Faculty of Technology
M'Hamed Bougara University
Boumerdes, Algeria
Center for Advanced Life Cycle Engineering (CALCE)
University of Maryland
College Park, Maryland, USA

Mandeep Mittal
AIAS, Amity University
Uttar Pradesh, Noida, India

Arif Mushtaq
City University College of Ajman
Ajman, UAE

R. Jeromia Muthuraj
Assistant Professor
Department of Statistics (DD&CE)
Manonmaniam Sundaranar University
Abishekapatti, Tirunelveli, Tamil Nadu, India

Subhrapratim Nath
Jadavpur University
Kolkata, West Bengal, India

Milan B. Patel
Department of Mathematics
Gujarat University
Ahmedabad, Gujarat, India

Dushyantkumar G. Patel
Government Polytechnic
Ahmedabad, Gujarat, India

Rajkumar Bhimgonda Patil
Center for Advanced Life Cycle Engineering (CALCE)
University of Maryland
College Park, Maryland, USA
Department of Mechanical Engineering
Annasaheb Dange College of Engineering & Technology
Ashta, India

Suyog S. Patil
Research Scholar
Department of Mechanical Engineering
Zeal College of Engineering
Savitribai Phule Pune University
Pune, Maharashtra, India
Department of Mechanical Engineering
Sharad Institute of Technology College of Engineering
Yadrav, Maharashtra, India

Anand K. Bewoor
Professor
Department of Mechanical Engineering
Cummins College of Engineering
Pune, Maharashtra, India

Akshit Pradhan
Kalinga Institute of Industrial Technology
Bhubaneswar, Odisha, India

Sadia Riaz
S.P. Jain School of Global Management
Dubai, UAE

Rita Yadav
Banasthali Vidyapith
Jaipur, Rajasthan, India

Isha Sangal
Banasthali Vidyapith
Jaipur, Rajasthan, India

Nita H. Shah
Department of Mathematics
Gujarat University
Ahmedabad, Gujarat, India

Meghna Sharma
Department of Computer Science and Engineering
The NorthCap University
Gurugram, Haryana, India

Chetansinh R. Vaghela
Marwadi University
Rajkot, Gujarat, India

Priyanka Vashisht
Department of Computer Science and Engineering
The NorthCap University
Gurugram, Haryana, India

Vijay Kumar
Department of Mathematics
AIAS, Amity University
Uttar Pradesh, Noida, India

Editors

Dr. Vijay Kumar received his MSc in applied mathematics and an MPhil in mathematics from the Indian Institute of Technology, Roorkee, India, in 1998 and 2000, respectively. He has completed his PhD in the Department of Operational Research at the University of Delhi. Currently, he is an associate professor in the Department of Mathematics, Amity Institute of Applied Sciences, Amity University, Noida, India. He has published more than 40 research papers in the areas of software reliability, mathematical modelling, and optimization in international journals and conferences of high repute. His current research interests include software reliability growth modelling, optimal control theory, and marketing models in the context of innovation diffusion theory. He has reviewed many papers for IEEE Transactions on Reliability, Soft Computing (Springer), IJRQSE, IJQRM, IJSAEM, and other reputed journals. He has edited special issues of the IJAMS and RIO journals. He is an editorial board member with IJMEMS and a lifetime member of the Society for Reliability Engineering, Quality and Operations Management (SREQOM).

Prof. Dr. Mangey Ram received his PhD in mathematics with a minor in computer science from G. B. Pant University of Agriculture and Technology, Pantnagar, India. He has been a faculty member for around 12 years and has taught several core courses in pure and applied mathematics at the undergraduate, postgraduate, and doctorate levels. He is currently a research professor at Graphic Era (deemed to be university), Dehradun, India. Before joining Graphic Era, he was a deputy manager (probationary officer) with Syndicate Bank for a short period. He is editor in chief of *International Journal of Mathematical, Engineering and Management Sciences*; book series editor with *Elsevier, CRC Press-A Taylor and Frances Group, De Gruyter Publisher Germany, River Publisher, USA*; and the guest editor and member of the editorial board of various journals. He has published 225 plus research publications with IEEE, Taylor & Francis, Springer, Elsevier, Emerald, World Scientific, and many other national and international journals and conferences. His fields of research are reliability theory and applied mathematics. Dr. Ram is a senior member of the IEEE and lifetime member of the Operational Research Society of India, SREQOM in India, and Indian Society of Industrial and Applied Mathematics. He has been a member of the organizing committee of a number of international and national conferences, seminars, and workshops. He was conferred with the "Young Scientist Award" by the Uttarakhand State Council for Science and Technology, Dehradun, in 2009. He was awarded the "Best Faculty Award" in 2011, "Research Excellence Award" in 2015, and "Outstanding Researcher Award" in 2018 for his significant contributions in academics and research at Graphic Era, Dehradun, India.

1 Role of MCDM in Software Reliability Engineering

Rana Majumdar
MSIT, TIG
Kolkata, India

Subhrapratim Nath
Jadavpur University
Kolkata, India

Vijay Kumar
Department of Mathematics, AIAS, Amity University
Uttar Pradesh, Noida, India

CONTENTS

1.1 INTRODUCTION

One of the fundamentals adhered to by business organizations in vigorous and ambitious environments is stability in providing reliable products. The maturation of software quality remains a traditional obstacle to any software industry. Enhancing testing efficiency and effectiveness will account for the production of high quality software.

As a result, orientation toward effective defect management is a requisite for an organization to achieve better efficiency and software reliability by incorporating the testing attribute that is facilitated through empirical investigation performed using testing attributes through a data collection mechanism. A tool called the analytical hierarchy process in association with requirement prioritization when dealing with the origin of defects enables the identification of the relative significance of attributes for getting better testing competence. This analysis examines testing efficiency during the process of software development from both the stakeholders' and users' perspectives. Accordingly, the study seeks to rectify the approaches to testing by practicing the analytical hierarchical process (AHP) by guiding the proper solutions to testing in lieu of old conventional approaches to static testing. This work proposes a theory called the 'efficacy premise,' which is used to assess testing efficiency from three perspectives: (1) software team leader, (2) developers/managers, and (3) testers. This effort also demonstrates the perspectives of the team leader, developer, and tester in assessing the functioning of each of the attributes and can serve as an important indicator for measuring testing efficiency. Efficient-testing practices impart reliability and eliminate software flaws during the foundation phase, which reduces the time and cost incurred in the rework, thereby reducing penalties; moreover, these practices make the software industry productive, sustainable, and efficient. Through experimental verification and numeric examples, the latter section of this chapter will introduce the following elements: impact of cost and penalty and the benefit of the software upgrade.

Today's organizational needs for survival and growth – whether through the globalization of markets, brain waves of a futurist entrepreneur, or even the needs of its stakeholders – are recurrently highlighted concerns within organizational literature and practice. Software companies are open systems in continuous interaction with their business environment. They go through adaptation processes and are likely to present organizational dysfunctions, which in turn reflect in their performance.

For the past several decades, researchers have been concerned with measuring the effectiveness of software testing techniques and processes. The standards are definite properties to which a potential solution must comply up to a certain extent. Measuring effectiveness is independent of any solution without specifying performance or criteria. Efficient testing techniques not only enhance software performance in an operational environment but also improve the productivity, quality, and sustainability of the software industry, thus reducing rework [1]. There is a need for endless checking of the testing practice that in turn establishes an admissible level of poise for the release of high-quality software products.

Here, our emphasis is to identify testing parameters from different users' perspectives based on their usage. The objective is not to get the computed results and draw comparisons but to establish their consistency and address their importance and scope for enhancing testing efficiency. A two-mode mechanism is introduced to model the efficiency measurement process that takes into account both management's and the testers' perspectives to analyze the effectiveness of the testing system based on the efficacy premise. The two-mode assessment framework tries to capture a thorough understanding of the tester/developer and team leader so that the effectiveness of a given technique, readiness for release, and customer satisfaction are emphasized.

In the subsequent sections, an altered approach is suggested by using AHP for weighing the testing parameters of interest and their relative importance from the team leader's, developer's, and tester's points of view. The idea of using AHP is far from new but through examples, it establishes a pattern that exhibits the importance of the relative measurement of testing attributes to enhance the performance of testing efficiency. These days, because of the importance of software application, specialized testing of software becomes a task of the utmost importance to counter problems, such as the costs and penalties associated with every phase of development. Software reliability is associated with software in many aspects, including the structure, operational environment, cost, penalty, and amount of risk involved during the development stage, which is described in the last section. In any software development, the quality of the software is very important for meeting customer satisfaction [2]. As previously mentioned, testing, like almost every other activity, must start with objectives and terminate with results. Finally, a managerial implication concludes this chapter.

1.2 NECESSITIES TO MEASURE TESTING EFFICIENCY

Over a period of time, the complexity of software products has increased unprecedentedly. There are elements that account for the complexity of software products: comprehensive product functionalities, advanced frameworks, and vastly supported configurations. Quality is free, while the lack of quality costs [3]. Defect discovery and removal are subsequently challenging to ensuring the reliability and quality of a product [4–6]. In construct validity, one must know that measuring an attribute involves a metric dependent on the value of the attribute [7]. This is important because the assessment of the effectiveness of the attributes of a system can't be done from a single point of view when the significance of various combinations is not in one's knowledge. Thus all metrics should be validated. Multidimensional analyses of attributes come out promising as a means of recording the quality of the attribute, thus improving efficiency [8]. Therefore, success in terms of testing efficiency depends on a structured approach, and so an effective testing practice is established for continuous enhancement [9, 10]. Thus it is important from the perspective of an organization to develop products of high quality to seek responsive services and earn customer's value in the market [9]. Fault finding and removal certainly play a major role in estimating the reliability and quality of products [4, 5, 9]. In construct soundness, one must know that measuring an attribute has a metric dependent on the value assigned to the attribute [11]. This is essential because the usefulness of the attributes of a system cannot be measured appropriately.

1.2.1 Process Adopted

Software testing techniques are meant to prevent and cure malfunctions, as well as demand ample performance. Improving testing efficiency is a boundless challenge in the industry. Unfortunately, until now, it was viewed as highly ad hoc, expensive, and randomly effective. The identification of key attributes is a crucial activity in practices of testing software that is done on the basis of analyzing the efficiency of software testing by using test metrics and test cases [12]. Indeed, software testing is an extensive term that expands up to a class of operations along the development cycle over and above that are aimed at different goals. Consequently, software testing research encounters many challenges. In a software project, efficacy theory acts as a key tool in assisting testers and managers with estimating overall efficacy and assessing the effect of testing efficiency in a given project. Further, on the grounds of the relative significance per attribute, an efficacy matrix is generated.

1.2.2 Case Study I

This section is a road map to the most applicable challenges addressed. The initial idea was determined by taking inputs from industry people to constitute initial testing attributes based on a questionnaire presented to various software engineers and team leaders/project managers. Based on their response, principle testing attributes were determined. To rank these attributes, they were scaled numerically in accordance with their importance with respect to testing efficiency and then ranked on a scale of 1 to 5. An attribute obtaining Rank 1 signifies the least importance in assessing testing efficiency. Further, ranks are marked with indices of 10, 8, 6, 4, and 2, respectively ranging from highest to lowest (5–1). The data collection process ensured confidentiality. It was assumed that collected data/opinions or expert views would possess a precise insight into testing practices.

Table 1.1 includes the key attributes as identified from the literature review and from industry experts providing the means of computing various efficiency features of the software testing process.

TABLE 1.1
Testing Attributes

Attributes	Definition
Budgetary Constraint (BC)	It's important to consider the allocated testing budget in prioritizing the features of a given release.
Release Target (RT)	This is the optimal release time of software subject to the system constraints.
Quality Control and Assurance (QCA)	This attribute ensures maximal performance in its operational usage.
Clear Requirements and Objectives (CLO)	This is a requirement specification/correctly specified functional and nonfunctional documentation.
Application Type (AT)	It is important to identify various types of systems based on their importance (e.g., real-time control system, scientific computation system for assigning priority for maximum functionality).
Dedicated Resources (DR)	Dedicated resources nurture individuals' resources more cost-efficiently.
Test Cases and Test Plans (TC&TP)	A test plan lowers the risk to a certain extent by offering a great repository to identify risk-prone areas that assist managers in employing resources.
Develop Environment (DI&R)	Here the emphasis is given to a methodology adopted and tools used to achieve goals.
Testing Methodology (TM)	This attribute is important because tests are derived from behavior models of software, so proper methodology needs to be applied for better sustainable results
Fault Tolerance (FT)	These techniques enable the system to tackle system failure.

1.3 AHP

AHP is a collective benchmark of decision-making approaches originally built up by Prof. Thomas L. Saaty (1977). It offers assistance to engineers in developing priorities with prototypes, redundancy, and measures of judgment consistency, and it simplifies preference ratings among decision criteria. AHP is aimed at solving collective benchmark decision problems with certainty in conjunction with the availability of information, but one needs to trade off many factors. AHP is an operative instrument for measuring consistency in the judgments. The judgments recorded determine the total weights of the selected attributes. On collecting the grades of discovered attributes that are intended to measure testing efficiency, an expert opinion has been recorded on the grounds of the comparative contribution of these attributes to each other. This section pursues the following steps by taking the scale of comparison into account:

> 1 – equal importance, 3 – moderate importance of one factor over another, 5 – high importance, 7 – ve ry high importance, 9 – extreme importance, and 2, 4, 6, 8 – values for inverse comparison.

- Ratings are developed for every decision alternative per criterion by designing a pair-wise comparison matrix per criterion (refer to Table 1.2).
- The resultant matrix is normalized (refer to Table 1.3).
- The values in each row are averaged to obtain a corresponding rating.
- The consistency ratio is figured and checked.

1.3.1 Efficacy Measure Using AHP

This section demonstrates the weighted average. Consequently, individual efficacy assessment is calculated for all of the chosen attributes. The aggregate efficacy measure (Table 1.3) is taken as a whole depending on the contribution to the total expected efficacy value. The aggregate measure of efficacy supports an organization

TABLE 1.2
Use of AHP on Selected Attributes (Sample Data for Experimental Purpose)

	BC	RT	QCA	CLO	AT	DR	TP&TC	DI&R	FT	TM
BC	1	0.33	0.2	0.33	2	3	0.33	0.25	0.33	0.2
RT	3	1	0.33	1	3	2	0.33	0.25	0.33	3
QCA	5	3	1	3	4	5	2	3	3	2
CLO	3	1	0.33	1	5	2	3	0.5	0.33	1
AT	0.5	0.33	0.25	0.2	1	3	0.25	2	0.33	2
DR	0.33	0.5	0.2	0.5	0.33	1	2	3	0.33	1
TP&TC	3	3	0.5	0.33	4	0.5	1	3	7	3
DI&R	4	4	0.33	2	0.5	0.33	0.33	1	7	5
FT	3	3	0.33	3	3	3	0.14	0.143	1	7
TM	5	0.33	0.5	1	0.5	1	0.33	0.2	0.14	1
Sum	27.83	16.5	3.98	12.37	23.3	20.8	9.73	13.34	19.81	25.2

TABLE 1.3
Testing Attributes with Normalized Values

	BC	RT	QCA	CLO	AT	DR	TP&TC	DI&R	FT	TM	Sum	Weights
BC	0.03	0.02	0.05	0.02	0.09	0.14	0.03	0.02	0.01	0.008	0.44	4.41
RT	0.10	0.06	0.08	0.08	0.13	0.1	0.03	0.02	0.01	0.119	0.74	7.46
QCA	0.18	0.18	0.25	0.24	0.17	0.24	0.21	0.22	0.15	0.079	1.92	19.28
CLO	0.10	0.06	0.08	0.08	0.21	0.1	0.31	0.04	0.01	0.04	1.04	10.46
AT	0.01	0.02	0.06	0.01	0.04	0.14	0.03	0.15	0.01	0.079	0.57	5.76
DR	0.01	0.03	0.05	0.04	0.01	0.05	0.21	0.22	0.01	0.04	0.68	6.82
TP&TC	0.10	0.18	0.12	0.02	0.17	0.02	0.1	0.22	0.35	0.119	1.43	14.38
DI&R	0.14	0.24	0.08	0.16	0.02	0.02	0.03	0.07	0.35	0.198	1.3	13.3
FT	0.10	0.18	0.08	0.24	0.13	0.14	0.01	0.01	0.05	0.278	1.24	12.42
TM	0.18	0.02	0.12	0.08	0.02	0.05	0.03	0.01	0.00	0.04	0.57	5.72

$\lambda_{\max} = 11.53$ $CR = 0.17$

in numerous ways by drawing its comparison among both favorable and unfavorable case synopses. When testers' perspectives are brought to effect, they encourage management to seek detailed analysis and replace existing attributes with the better ones.

The probability of nature exhibited by testing attributes brought into a testing environment is recorded in Table 1.4. The likelihood is that more than *80% (0.8)* of respondents conceived that BC is very good, and *50% (0.5)* of respondents believed that QCA, AT, and DR have been managed effectively by the testers and organization.

The aggregate measure of utility consists of attributes' total effectiveness but is unable to showcase the criterion or level of acceptance of utility that is expected to emerge from such attributes. For this purpose, the 'acceptable' level of utility is not derived from any specific attribute. The two scenarios are presented next.

1.3.1.1 Ideal Efficacy Measure

This is computed by taking the claims of every tester and claiming the high effectiveness of attributes used by the organization. As a result, such attributes are rated as '10.' The likelihood of all instruments being of high effectiveness is weighted as '1,' resulting in the weight of the highest/ideal utility for every single attribute mentioned in Table 1.3 that eventually results in the highest possible weight of utility shown in Table 1.3 obtained from any given class of attributes.

1.3.1.2 Worst Efficacy Measure

This measure, on the other hand, is computed by taking the claims of every single respondent into account. Claiming attributes used by the organization were the least effective and thereby such attributes are rated as '2.' The likelihood of all instruments being of least effectiveness is weighted as '1,' resulting in the weight

TABLE 1.4
Overall Efficacy Measures

Testing Attributes	Weights (W_i)	Levels: Very Good 10	Good 8	Average 6	Poor 4	Very Poor 2	Expected Level Weight	Contribution to Total Expected Utility (U_i)
BC	4.41	0.8	0.1	0.1	0	0	9.4	41.454
RT	7.46	0.6	0.3	0.1	0	0	9	67.14
QCA	19.28	0.5	0.1	0.2	0.2	0	7.8	150.384
CLO	10.46	0.5	0.2	0.2	0.1	0	8.2	85.772
AT	5.76	0.3	0.2	0.5	0	0	7.6	43.776
DR	6.82	0.3	0.1	0.2	0.2	0.2	6.2	42.284
TP&TC	14.38	0.2	0.2	0.2	0.3	0.1	6.2	89.156
DI&R	13.3	0.3	0.2	0.2	0.2	0.1	6.8	90.44
FT	12.42	0.4	0.2	0.3	0.1	0	7.8	96.876
TM	5.72	0.2	0.1	0.2	0.4	0.1	5.8	33.176
								740.458

of least utility for every single attribute that eventually results in the lowest possible weight of utility, as explained in Table 1.6 and obtained from any given class of attributes.

To govern the outcome of the AHP method, the consistency ratio of each of the matrices and the overall inconsistency of the hierarchy were calculated. The measure of consistency is called the consistency index (CI):

$$CI = \frac{l_{\max} - n}{n-1} \tag{1.1}$$

The CR is used to directly estimate the consistency of pair-wise comparisons (Table 1.5). The CR is computed by dividing the CI by a value obtained from a table of the random consistency index (RI).

$$CR = \frac{CI}{RI} \tag{1.2}$$

If the CR is less than 0.10, the comparisons are acceptable. For this application, all CR inconsistency ratios values were less than 0.1 (CR < 0.1); therefore, all the judgments were consistent. Each alternative possesses a score on all criteria. The criteria scores were combined into an overall score. The overall score indicates the relative importance of each alternative.

TABLE 1.5
Pair-Wise Comparison between Instruments

	Instrument 1 BC					Instrument 2 RT			
Experts	**ES**	**VS**	**S**	**MS**	**EQ**	**MS**	**S**	**VS**	**ES**
A									√
B								√	
C									√
D									√
E									√

TABLE 1.6
Ideal and Worst-Case Scenarios

Testing Attributes	Weights	Levels					Expected Level Weight	Contribution to Total Expected Utility (U_i)
		Very Good	*Good*	*Average*	*Poor*	*Very Poor*		
	(W_i)	*10*	*8*	*6*	*4*	*2*		
BC	4.41	1	0	0	0	0	10	44.1
RT	7.46	1	0	0	0	0	10	74.6
QCA	19.28	1	0	0	0	0	10	190.2
CLO	10.46	1	0	0	0	0	10	100
AT	5.76	1	0	0	0	0	10	57.6
DR	6.82	1	0	0	0	0	10	68.2
TP&TC	14.38	1	0	0	0	0	10	140.3
DI&R	13.3	1	0	0	0	0	10	133
FT	12.42	1	0	0	0	0	10	124.2
TM	5.72	1	0	0	0	0	10	572
								10.2
							Total Utility	696
	(W_i)	*10*	*8*	*6*	*4*	*2*		
BC	4.41	0	0	0	0	1	2	8.825
RT	7.46	0	0	0	0	1	2	14.92
QCA	19.28	0	0	0	0	1	2	38.56
CLO	10.46	0	0	0	0	1	2	20.92
AT	5.76	0	0	0	0	1	2	11.52
DR	6.82	0	0	0	0	1	2	13.64
TP&TC	14.38	0	0	0	0	1	2	28.76
DI&R	13.3	0	0	0	0	1	2	26.6
FT	12.42	0	0	0	0	1	2	24.84
TM	5.72	0	0	0	0	1	2	11.44
							Total Efficacy	200.2

1.4 MEASURING TESTING EFFICIENCY BY MEANS OF ALTERNATING POLICY

It is well-known that software organizations keep trying to increase their quality standards within assessed cost and agendas with decent domain-specific knowledge in circulating approaches of quality checks, which include the process of reviewing, auditing, and testing. Therefore, success depends on a structured approach, and so the testing process is adopted for uninterrupted enhancement [9]. As a consequence, from the perspective of a developer, the production of high-quality products, earning value in the eyes of the customer by providing efficient services is significant for keeping an organization's place in the highly competitive market [10]. Here the emphasis is placed on identifying testing parameters from diverse perspectives based on software development and usage. The objective is to get the computed results and do comparisons by capturing the attention of the tester, developer, and team leader; determining the effectiveness of a given technique; and enabling promptness for release and customer fulfillment.

Software testing procedures are dedicated to check and device malfunctions of a system and to demand ample performance. As mentioned in Section 1.2.2, refining testing efficiency is a well-known task in engineering, but inappropriately, it is still implemented on an ad hoc basis, which only adds to the functioning and testing price and is arbitrarily effective. The different phases in the process of testing assists in the discovery of attributes that serve as mediums for computing the testing efficiency of software by implementing a class of test attributes. Subsequently, research conducted in the domain of software testing addresses many experiments for improving the efficiency of testing. The effect of the assessment of testing efficiency on a theory of efficacy constituting a software project is taken as a key tool for helping team leaders, developers, and testers in estimating overall utility during development phases.

1.5 CASE STUDY II

The approach addressed in this section worked on primary data collected from information technology companies and obtained a significant result that offers far greater insights into testing than existing processes. Based on users' interactions and customer comments, industry feedback attributes were identified, and attributed weights were calculated. Both qualitative and quantitative information can be compared using informed judgments to derive weights and priorities. They are classified and placed based on their virtues/facts to check their relative priority weights and consistency using the AHP form of individual perceptions. This section demonstrates separate resolution and patterned CI on individuals' assertiveness.

Here the computed results were compared with the previous result. A comparison matrix A is said to be consistent if $A_{ij} * A_{jk} = A_{ik}$ for all i, j, and k. However, consistency can't be forced, as too much consistency is undesirable because it deals with human judgment. To be called consistent, the rank can be transitive, but the values of judgment are not necessarily forced into the multiplication formula $A_{ij} * A_{jk} = A_{ik}$. The largest Eigenvalue is equal to the number of comparisons, $l_{max} = n$.

In the past, most of the relative weights of system attribute effectiveness were known to be of equal weight. However, in practice, the importance of each attribute in overall effectiveness is not exactly the same. From the tabular results in Table 1.7, we see that the team leader has more indications toward BC, RT, and quality of the product, and it is consider a prime testing attribute among other identified attributes. Calculation, as demonstrated in Table 1.8, of AHP shows a higher priority for QCA and quality of the product. The team leaders' perspective is clear: quality is more important than BC and RT. Interestingly, the developer shows more emphasis on CLO, illustrated in Table 1.8, as it heads to the proper design of the software, thus it is easy to code by keeping in mind the other parameters, such as AT and availability of resources.

Testing teams' priority is on TC&TP, as they believe an effective plan and execution of test cases will help them to identify software bugs, and the removal of those bugs will improve the quality of the software described in Table 1.9. Once again, the emphasis here is on quality. The reliability of software is measured on the basis of the testing effort. In this tabular representation, the *CI* has values 0.019, 0.23, and 0.70, which ensures that the evaluation of attributes is consistent to establish the fact that QCA, CLO, and TC&TP are the prime attributes for the testing process.

TABLE 1.7
Team Leader Perspectives (Based on Experimental Data)

TL	BC	RT	QCA	Third root of product		PV
BC	1	0.33333	0.2	0.0666667	0.40548	0.104729
RT	3	1	0.3333	1	1	0.258285
QCA	5	3	1	15	2.466212	0.636986
Sum	9	4.33333	1.5333		3.871692	1
Sum*PV	0.942565	1.11923	0.9767		l_{max}= 3.03	
Now Calculating CI = (Lambda Max − n)/n − 1					0.019	ok

TABLE 1.8
Developer Perspectives (Based on Experimental Data)

Developer	CLO	AT	DR	Third root of product		PV
CLO	1	5	2	10	2.154435	0.607193
AT	0.2	1	3	0.6	0.843433	0.237708
DR	0.5	0.33333	1	0.166667	0.550321	0.155099
Sum	1.7	6.33333	6		3.548189	1
Sum*PV	1.032228	1.50548	0.9306		l_{max}= 3.46	
Now Calculating CI = (Lambda Max − n)/n − 1					0.23	ok

TABLE 1.9
Tester Leader Perspectives (Based on Experimental Data)

Tester	TP&TC	DI&R	FT	TM	Third root of the product		PV
TP&TC	1	3	7	3	63	3.979057	0.569917
DI&R	0.333333	1	7	5	11.66667	2.268031	0.324848
FT	0.142857	0.14286	1	7	0.142857	0.522758	0.014129
TM	0.333333	0.2	0.1429	1	0.009524	0.211968	0.03036
Sum	1.809524	4.34286	15.143	16		6.981814	0.939254
Sum*PV	1.031279	1.41077	0.2139	0.4857602		l_{max}= 3.14	
Now Calculating CI = (Lambda Max – n)/n – 1						0.0705	ok

1.6 EFFICIENCY MEASUREMENT FOR UPGRADED SOFTWARE

Software quality can be amplified by performing suitable testing during various stages of its development. Currently, because of the importance of software application, specialized testing of software has become a more important task to counter problems such as costs and penalties associated with every phase of development. Software reliability is associated with software in many aspects, including the structure, operational environment, cost, penalty, and amount of risk involved during the development stage [13]. This section demonstrates a more realistic approach to measuring testing efficiency and effectiveness, as well as the importance of introducing the concept of early testing at every phase of development once developers have sufficient domain-specific knowledge related to software architecture, customers' feedback based on its first release, total no of post-release faults and their severity, and, finally, in association with the testing team, they can uncover an untouched portion of the software to counter the maximum number of slipped faults at the various stages of the software development phase, which go unnoticed during the first release of the software because of its unknown usage pattern and behavior. Here the objective is to demonstrate the usefulness of software upgrades both from the release time point of view and its relative effect on overall cost, penalty, and benefit of developed/developing software. Partial specification addressed at the time of coding, the design phase, etc., results in residual defects in requirements up to a certain proportion.

1.6.1 Methodological Approach

The objective of any organization is to deliver a result on time and within certain constraints. They are, namely BC; quality and reliability should be quantified by the customer [14]. To understand this objective, the whole team should work competently and incoordination, and they must show advancement in the effort they put in [15]. Test efficiency and effectiveness are essential quality attributes the software development team possesses to carry out all testing activities in an efficient manner, saving cost and time to avoid any unwanted penalties [16]. It is known that faults identified in the testing phase were already introduced in the early phases of the development process. So testing should start in a timely manner to avoid the introduction of defects in the early phase.

Computer-based systems, aided with automated tools that offer testers a variety of testing methods, test cases, and, hence, enhance the probability of error detection within time and BC, speeding up the testing process and thus reducing the risk of failure of software penalty either in the form of warranty or gaining maximum benefits to serve organizational purpose [9]. Defect removal efficiency is another approach deployed by industry people to identify and analyze the root cause of various defects so that they can be removed to achieve a higher level of defect removal efficiency for the future release of the software.

1.6.2 Case Studies III

A software testing principle is *start testing early* in the software development life cycle, but in a conventional software development life cycle process, sometimes planned activities won't go smoothly because of inadequate information, less learning, and complex software architecture, so time taken to initiate the process could be more than estimated, which causes issues in the overall development process. Moreover, if the release date is fixed, then it has a direct effect on the quality of the software application irrespective of whether it is a new product or an upgraded version, as there is less time to test the complete application and fix the bugs, which if found at later stages are costlier than when discovered in the early stages. This results in a premature release of the software with a maximum number of post-release faults reported either by the customer or found by the testing team. A proven and accepted means of assessing the probability of finding defects in software is the percentage of code coverage, bad smells, and repetition of unused codes during testing: the higher the code coverage, the lower the probability of finding new defects [16, 17].

It is worth mentioning that if testing starts at a later stage, then only A% of testing efficiency will be attained by the development team based on the model that testing starts only at the testing phase of the software referred to in Table 1.10. Although if testing starts, initially, fault intensification will be less and, hence, higher B% efficiency will be achieved, as explained in Table 1.11. In any development process, testing often begins as soon as there is something to test to get started finding and fixing problems. The rate at which test cases can be run initially is high, but later stages may be limited by the rate at which bugs can be fixed. This clearly demonstrates the effect of testing on the performance of the software. Here it is assumed that at every stage of testing, a detected bug removed by the testing team is in association with the development team so that fault gathering for the next phase will be significantly less, this is presented by x and y and elaborated in

TABLE 1.10
Fault Found at Various Phases of Software Development (Base on Sample Data) [16]

		Defect Origin					
Defect found in phase		Req. Analysis	Design	Coding	Testing	Maintenance	Total
	Req. Analysis	1					***1***
	Design	12	6				***18***
	Coding	4	1	20			***25***
	Testing	1	2	6	10		***19***
	Maintenance	1	1	2	2	5	11
	Total	***19***	***10***	***28***	**12**	5	74

Notations used in Table 1.10:
X = Total number of identified faults: 12 *(refer to the fifth column final row)*
Y = Total number of faults propagated: 19 *(refer to the fifth row of the final column)*
Z = 19 + 10 + 28 − {1 + 18 + 25} = 13
Testing efficiency = Number of defects found/ (existing + injected) = A%

TABLE 1.11
Fault Found at Various Phases of Software Development [14]

		Defect Origin					
Defect found in phase		Req. Analysis	Design	Coding	Testing	Maintenance	Total
	Req. Analysis	**X(Prop of Faults Removed)**	-	-	-	-	*zz*
	Design	x	**Y(Prop of Faults Removed)**	-	-		*z*
	Coding	x	y	20			***20***
	Testing	x	y	6	10		**R**
	Maintenance	x	y	1	1	3	5
	Total	***xx***	***y***	***27***	**M**	3	41

Notations used in Table 1.11:
X = Total number of identified faults: R *(refer to the fifth column final row)*
Y = Total number of faults propagated: M *(refer to the fifth column row final col)*
Z = xx + y + 27 − {zz + z + 20} = 7
Testing efficiency = Number of defects found/ (existing + injected) = B%

Table 1.11 and thus improves testing efficiency. This is true for the upgraded version of the software; one will need to test only the added functionality, not the entire software, thus it is easier to maintain the efficiency level.

The requirements prioritization model [7] estimates the relative values of implementing specific features in software systems. During development, if developers agreed on a subjective judgment of the benefit being twice as important as the

TABLE 1.12
Software Development Phase Rated with Three Dimensions

Development Phase	Cost	Benefit	Penalty
First Version (Late Testing)	2	1	2
Second Version (Early Testing)	1	2	0.5
2nd Version (Early Testing)			

penalty, which is the same importance as cost, they use the weighting factors I, J, K, respectively, for getting results for removing faults at the injected phase and calculating the relative benefit, cost, and penalty from customers' points of view either to maximize benefits with reduced fines or to minimize cost, as shown in Table 1.12. The values entered for each column are taken from the survey from industry people with a condition that people will provide independent results. After entering the relative values for all the phases, the relative cost and cost percentage for each feature is calculated by considering the percentage of the weighted feature desirability (see Table 1.13).

This section shows the importance of identifying defects at the early stage for upgrades so that corrective and preventive steps can be taken for future release. Two scenarios were presented: first, early release of software, measuring testing efficiency and effectiveness in terms of relative values, and second, during upgrade by initiating the testing process at every stage of development on a relative scale. The observation is illustrated by taking a numerical example. From a developer's perspective, test effectiveness is defined in terms of the ability to find defects and isolate them from a product to ensure quality and conformance to product specification. As compared to effectiveness, efficiency is an attribute, which means to maximize the useful output for a given input by reducing waste or losses [16]. Testing efficiency depends on the identification and removal of faults; at the same time, care should be taken so that the removal of faults will not introduce any new faults [2, 18]. If faults are injected at the early stages and not noticed by the respective team members, they will affect overall testing efficiency and effectiveness. During the testing phase, 10 faults were identified and 19 were propagated from previous stages. Two parameters for measuring testing efficiency were considered, but the faults that were not identified were prime factors for judgmental purposes. Clearly, efficiency becomes 76% based on the illustrations in Tables 1.11 and 1.12. If faults are identified early, testing efficiency will improve; now it becomes 88% in the absence of faults in phases such as the requirement analysis design (see Table 1.13). On the other hand, a released product will have a lower relative risk percentage and cost in terms of penalty, and relative benefit will be maximized (see Table 1.13).

1.7 CONCLUSIONS

There are many attributes that don't capture the essence of what they are supposed to do. Some of them don't even recognize the importance of defining the AT, measuring the development environment, and identifying the relative importance of testing

TABLE 1.13
Relative Benefit, Penalty, Cost Values for Software Upgrades

	2V	1V	2V	1V	2V	1V	2V	1V	2V	1V	2V	1V
Relative Weights:	2	1	0.5	2					1	2		
Feature	Relative Benefit	Relative Benefit	Relative Penalty	Relative Penalty	Total Value	Total Value	Value %	Value %	R. Cost	R. Cost	Cost %	Cost %
Req. Analysis	8	3	4	7	18	17	23.7	23.6	3	4	25.0	16.7
Design	7	4	3	5	15.5	14	20.4	19.4	3	4	25.0	13.3
Coding	5	3	3	4	11.5	11	15.1	15.3	2	5	16.7	16.7
Testing	9	3	4	7	20	17	26.3	23.6	3	9	**P**	**Q**
Maintenance	5	5	2	4	11	13	14.5	18.1	1	7	**R**	**S**
	34	18	16	27	76	72	100	100	12	29	100	100.0

efficiency and thus the effectiveness. This chapter set forth the use of the efficacy of attributes alone and taken together, and it compared the results with AHP to validate the importance of testing attributes to enhance testing efficiency. The relative weights assigned to each of the testing attributes highlight the user's perspective regarding the importance of an attribute and provides better insight into the overall estimation of the efficacy value. One of the major contributions lies in identifying testing attributes that require immediate attention based on the efficacy derived from each of them. In addition, the technique can also be used for benchmarking system standards based on organizational goals and objectives. Measurement of testing efficiency and effectiveness is a very powerful risk management tool. The relative importance assigned to each of the development phases based on three parameters highlights both the customers' and developers' perspectives toward the importance of and provides better insight into the overall estimation of the testing process and software performance. For software professionals, this study provides insight into the software process model and its effect on software release decisions; an early release decision may end up with high costs for fixing bugs or a newer version of the software in the form of upgrades. For software project managers, this study provides a methodology for allocating resources for developing reliable and cost-effective software. Testing must occur over a substantial portion of the useful life of the system to detect a substantial portion of the total errors, which will inevitably occur. Thus cost-effective testing should be incorporated into the maintenance of the system with user feedback loops to report on errors when they occur and to analyze and correct them rapidly.

REFERENCES

1. Kumar, V., P. Mathur, R. Sahni and M. Anand, 2016. Two-Dimensional Multi-Release Software Reliability Modeling for Fault Detection and Fault Correction Processes. *International Journal of Reliability, Quality and Safety Engineering*, vol. 23, no. 03, p. 1640002.
2. Kumar, V., P.K. Kapur, N. Taneja and R. Sahni, 2017. On allocation of resources during testing phase incorporating flexible software reliability growth model with testing effort under dynamic environment. *International Journal of Operational Research*, vol. 30, no. 4, pp. 523–539.
3. Kapur, P. K., S. K. Khatri, and S. Nagpal. 2013. ERP health assessment model commensurate with ERP success factor rate metric. *International Journal of Computer Applications—Special Issue*, pp. 25–29.
4. Suma, V., and T. R. Gopalakrishnan Nair. 2008. Effective defect prevention approach in software process for achieving better quality levels. *Fifth International Conference on Software Engineering*, vol. 32, pp. 2070–3740.
5. Suma, V., and T. R. Gopalakrishnan Nair. 2008. Enhanced approaches in defect detection and prevention strategies in small and medium scale industries. *The International Conference on Software Engineering Advances*, pp. 389–393. doi:10.1109/ICSEA.2008.79
6. Suma, V., and T. R. Gopalakrishnan Nair. 2012. *Defect management strategies in software development. Recent Advances in Technologies*, pp. 379–404, Vienna, Austria: Intec web Publisher.
7. Kaner, C. 2004. *Software engineering metrics: What do they measure and how do we know?* (10th ed.). *International Software Metrics Symposium Metrics*, Washington, DC: IEEE CS.

8. Majumdar, R., A.K. Shrivastava, P.K. Kapur, and S.K. Khatri. 2017. Release and testing stop time of a software using multi-attribute utility theory. *Life Cycle Reliability Safety Engineering*, vol. 6, pp. 47–55. Springer. doi:10.1007/s41872-017-0005-9.
9. Talib, F., and Z. Rahman. 2010. Studying the impact of total quality management in service industries. *International Journal of Productivity and Quality*, vol. 6, no. 2, pp. 249–268.
10. Spiewak, R., and K. McRitchie. 2008. Using software quality methods to reduce cost and prevent defects. *CROSSTALK: The Journal of Defense Software Engineering*, vol. 21, no. 12. pp. 23–27.
11. Majumdar, R., P. K. Kapur, and S. K. Khatri. 2015. Measuring testing efficiency: An alternative approach. *4th International Conference on Reliability, Infocom Technologies and Optimization*, pp. 1–6. doi:10.1109/ICRITO.2015.7359219.
12. Agarwal, M., R. Majumdar. 2103.Software maintainability and usability in agile environment. *International Journal of Computer Applications*, vol. 68, no. 4, pp. 30–36.
13. Sproles, N. 2001. The difficult problem of establishing measures of effectiveness for command and control: A systems engineering perspective. *Systems Engineering*, vol. 4, pp. 145–155.
14. Majumdar, R., P. K. Kapur, and S. K. Khatri. 2018. Assessing software upgradation attributes and optimal release planning using DEMATEL and MAUT. *International Journal of Industrial and Systems Engineering*, vol. 31, no. 1. doi:10.1504/IJISE.2019.096886.
15. Majumdar, R., R. Gupta, and A. Singh. 2018. Software performance measuring benchmarks. *International Conference on Wireless Intelligent and Distributed Environment for Communication*, pp. 21–129. Cham, Switzerland: Springer.
16. Majumdar, R., P. K. Kapur, and S. K. Khatri. 2016. Measuring testing efficiency & effectiveness for software upgradation and its impact on CBP. *International Conference on Innovation and Challenges in Cyber Security*, pp. 123–128. doi:10.1109/ICICCS.2016.7542347.
17. Gupta, A., B. Suri, V. Kumar, S. Misra, T. Blažauskas, and R. Damaševičius. 2018. Software Code Smell Prediction Model Using Shannon, Rényi and Tsallis Entropies. *Entropy*, vol. 20, p. 372.
18. Kumar, V. and R. Sahni, 2016. An effort allocation model considering different budgetary constraint on fault detection process and fault correction process. *Decision Science Letters*, vol. 5, no. 1, pp. 143–156.

2 Fault Tree Analysis of a Computerized Numerical Control Turning Center

Rajkumar Bhimgonda Patil
CALCE, University of Maryland
College Park, MD, USA
Annasaheb Dange College of Engineering & Technology
Ashta, India

Mohamed Arezki Mellal
M'Hamed Bougara University
Boumerdes, Algeria
CALCE, University of Maryland
College Park, MD, USA

CONTENTS

2.1 INTRODUCTION

Reliability estimation is crucial for the effectiveness and safety of any system (Mellal & Williams, 2018; Mellal & Zio, 2016, 2020; Patil, 2019; Patil, Kothavale, Waghmode, & Joshi, 2017). The fault tree analysis (FTA) method is a powerful tool used widely in system reliability and risk assessment. H. A. Watson developed the FTA method in 1961 at Bell Telephone Laboratories (1961). FTA uses the tree structure to decompose the system level faults into combinations of lower-level events (Amari, Dill, & Howald, 2003). FTA can help to prevent the occurrence of the failure by providing the data that show how and under what circumstances the failure could occur, allowing for alternative measures to prevent catastrophic failure. It represents graphically the logical interactions and probabilities of occurrence of component failures and other events in a system (Dugan, Sullivan, & Coppit, 2000). The most undesirable event (fault) is generally considered a 'top event,' and intermediate events and basic events are connected to it by using Boolean gates. The Boolean algebra is used to evaluate the fault tree (FT) to identify the qualitative and quantitative characteristics of the system (Haasl, Roberts, Vesely, & Goldberg, 1981; Lee, Grosh, Tillman, & Lie, 1985). The qualitative analysis of the FT determines the probability of system failure based on a single failure cause or common cause and the combination of component failures (minimal cut sets). The quantitative analysis of the FT focuses on the probability of the occurrence of the top event and intermediate events based on the probabilities of failure of the basic events. FTA has been widely used in industry, such as natural gas storage tanks (Yin et al., 2020), oil and natural gas pipelines (Badida, Balasubramaniam, & Jayaprakash, 2019), software testing technology (Li, Ren, & Wang, 2018), aircraft electrical systems (Nyström, Austrin, Ankarbäck, & Nilsson, 2006), wastewater treatment plants (Piadeh, Ahmadi, & Behzadian, 2018), and railway systems (Huang et al., 2020).

In this chapter, the reliability study of a computerized numerical control turning center (CNCTC) is presented using the FT method to identify critical faults. The remainder of the chapter is organized as follows. Section 2.2 presents the basic concepts of the FTA. Section 2.3 illustrates the FTA of a CNCTC and the estimation of its reliability. The last section concludes the chapter.

2.2 BASICS OF AN FTA

2.2.1 Symbols

The selection of effective symbols and their implementation plays an important role in the construction of the FT diagram. Different symbols are used while constructing the FT diagram. The basic symbols used in FTA are grouped as gates, events, and transfer symbols. Table 2.1 shows the various symbols used for constructing the FT diagram (Chen, Ho, & Mao, 2007; Volkanovski, Čepin, & Mavko, 2009).

2.2.2 Steps

In FTA, there are several steps to be followed. The first step is to increase the understanding level of the system and define the most undesirable event as the 'top event' (Aggarwal, 1979; Andow, 1980). The next step is to identify all the events (faults) responsible for the occurrence of the top event and construct an FT diagram by defining the boundaries of the system in the analysis. The symbols and gates are used to show the relationship between basic events, intermediate events, and top events. The qualitative evaluation provides information on the minimal cut sets for the top events (Allan, Rondiris, & Fryer, 1981; Kohda, 2006). The nature of the basic events and the number of basic events in the combined sets give important information about the occurrence of the top event. The quantitative evaluation gives not only the probability of the occurrence of the top event but also the dominant cut sets that contribute to the top event probability, as well as the quantitative importance of each basic event

TABLE 2.1
Symbols Used for Constructing an FT Diagram

Gate symbol		Event symbol		Transfer symbol	
Symbol	**Meaning**	**Symbol**	**Meaning**	**Symbol**	**Meaning**
	AND gate		Basic event		Transfer in
	OR gate	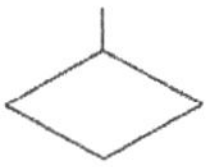	Incomplete event		
	Exclusive OR gate	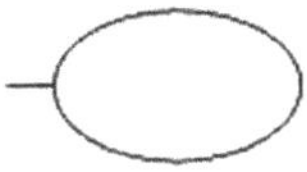	Conditional event		
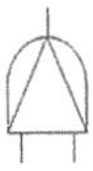	Priority AND gate		Normal event		Transfer out
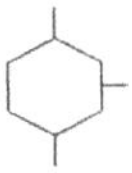	Inhibit gate	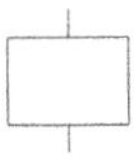	Intermediate event		

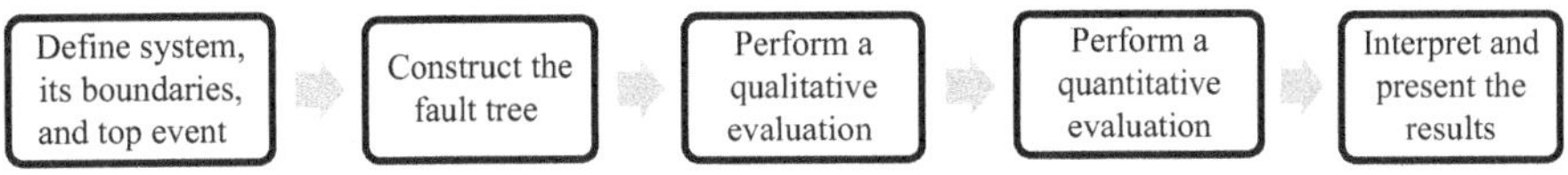

FIGURE 2.1 Steps Involved in FTA.

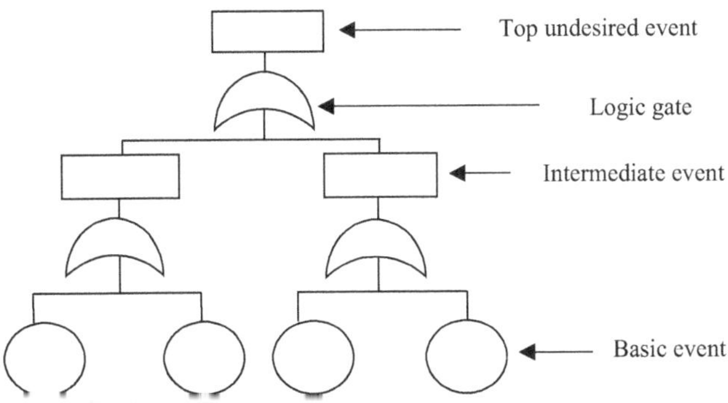

FIGURE 2.2 A Typical FT Diagram.

contributing to the top event (Bengiamin, Bowen, & Schenk, 1976; Dugan, Bavuso, & Boyd, 1992). Cut sets, in this case, are sorted by probability, and low-probability cut sets are truncated from the analysis. Emphasis is placed on the interpretation and not simply on the presentation. The results are interpreted to provide tangible implications, especially concerning the potential effect on the research objective (Bobbio & Raiteri, 2004; Meshkat, Dugan, & Andrews, 2002). Figures 2.1 and 2.2 illustrate the steps involved in FTA and a typical FT diagram, respectively.

2.3 CONSTRUCTION OF THE FT DIAGRAM FOR A CNCTC

In this section, the FT diagram for a CNCTC is constructed. The CNCTC is expected to manufacture the component or job with designed tolerances. The failure of the CNCTC owing to the failure of any of the components (or sub-systems) or the inaccuracy of the machining operation is considered the most undesirable event. Therefore, 'failure of CNCTC (F)' is considered the 'top event' of the FT diagram. The FT diagram in Figure 2.3 consists of a top event, 16 intermediate events, and 76 basic events. The first-level decomposition of the FT diagram consists of 12 intermediate events: CNC sub-system (CNCS), main transmission (MT), x- and z-axis sub-system (XZAS), hydraulic sub-system (HS), chuck sub-system (ChS), spindle sub-system (SS), electrical and electronic sub-system (EES), coolant sub-system (CS), turret sub-system (TS), lubrication sub-system (LS), swarf conveyor (SC), and other sub-systems (OS). The first level of decomposition is based on the main sub-systems of the CNCTC. The CNCTC may fail because of the failure of one of the 12 sub-systems and, therefore, these 12 intermediate events are connected to the top event by using the 'OR' gate.

Each intermediate event is further decomposed into unwanted causal events. The most complex intermediate event is CNCS. The failure of CNCS is due to one of six

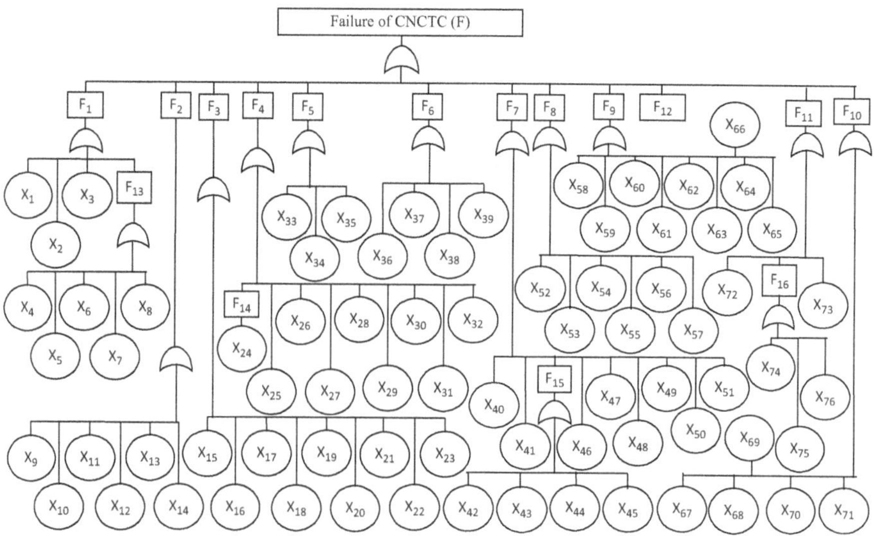

FIGURE 2.3 FT Diagram of CNCTC.

reasons: failure of drive cards, sensors, cables, encoders, or software. The encoder may fail because of its pulley, belt, shaft, or coupling. MT drives the spindle and may fail because of one of the six events: spindle motor, cooling fan, cooling fan motor, spindle belt, pulley, and misalignment. The failure cost of the spindle motor is very high, and the cause of the major failure is overheating and worn-out bearings. The failure of the spindle belt is mostly due to worn-out and misalignment.

The tool movement is controlled by XZAS. Nine events could be the failure cause of XZAS: x- and z-axis servomotors, servomotor couplings, drive belt, ball screw, ball screw bearing, timing belt, guideways, and backlash. Proper lubrication and continuous removal of chips from guideways can minimize the problem of jamming guideways. The faults associated with HS are hydraulic motor and its bearings, vane pumps, power packs, strainers, oil seals, hoses, cylinders, valves, and pressure switches. In the same manner, the faults associated with each intermediate event are identified.

The FT diagram of the CNCTC shows that the intermediate events and the basic events are directly or indirectly connected to the top event using logic gates. It is observed that the occurrence of a single event may lead to the occurrence of the top event and, therefore, all the intermediate events and basic events are connected to the top event by the OR gate.

2.3.1 Qualitative FTA

In the previous section, the FT diagram of the CNCTC was constructed. Qualitative analysis of the CNCTC gives the relationship between basic and intermediate events with the top event. Minimal cut sets (i.e., all the combinations of basic events when they happen simultaneously) leading to the system failure can also be obtained. For analysis purposes, the developed FT diagram is transformed into a reliability block diagram. All the basic events are connected to the top event directly or indirectly by the 'OR' gate, and the occurrence of a single event led to the occurrence of the top event. Therefore, all the basic events are connected in the series (series configuration), as shown in Figure 2.4.

In the case of series configuration, all events are considered to be critical events. Since reliability is a probability, the reliability of CNCTC (R_{CNCTC}) may be determined from the probability of non-occurrence of the events as follows:

$$\begin{aligned} X_1 &= \text{the event 1 does not occur} \\ X_2 &= \text{the event 2 does not occur} \\ &\vdots \\ X_{76} &= \text{the event 76 does not occur} \end{aligned} \tag{2.1}$$

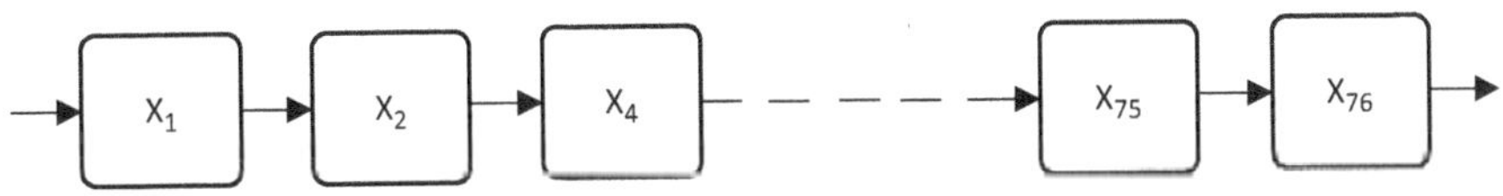

FIGURE 2.4 Reliability Block Diagram for CNCTC.

The probability of survival (reliability) of CNCTC (R_{CNCTC}) is estimated using the laws of probability as follows:

$$
\begin{aligned}
R_{\text{CNCTC}} = P\big(&X_1 \cap X_2 \cap X_3 \cap X_4 \cap X_5 \cap X_6 \cap X_7 \cap X_8 \cap X_9 \cap X_{10} \\
&\cap X_{11} \cap X_{12} \cap X_{13} \cap X_{14} \cap X_{15} \cap X_{16} \cap X_{17} \cap X_{18} \cap X_{19} \cap X_{20} \\
&\cap X_{21} \cap X_{22} \cap X_{23} \cap X_{24} \cap X_{25} \cap X_{26} \cap X_{27} \cap X_{28} \cap X_{29} \cap X_{30} \\
&\cap X_{31} \cap X_{32} \cap X_{33} \cap X_{34} \cap X_{35} \cap X_{36} \cap X_{37} \cap X_{38} \cap X_{39} \cap X_{40} \\
&\cap X_{41} \cap X_{42} \cap X_{43} \cap X_{44} \cap X_{45} \cap X_{46} \cap X_{47} \cap X_{48} \cap X_{49} \cap X_{50} \\
&\cap X_{51} \cap X_{52} \cap X_{53} \cap X_{54} \cap X_{55} \cap X_{56} \cap X_{57} \cap X_{58} \cap X_{59} \cap X_{60} \\
&\cap X_{61} \cap X_{62} \cap X_{63} \cap X_{64} \cap X_{65} \cap X_{66} \cap X_{67} \cap X_{68} \cap X_{69} \cap X_{70} \\
&\cap X_{71} \cap X_{72} \cap X_{73} \cap X_{74} \cap X_{75} \cap X_{76}\big)
\end{aligned}
\tag{2.2}
$$

$$
\begin{aligned}
R_{\text{CNCTC}} = {} & P(X_1) \cap P(X_2) \cap P(X_3) \cap P(X_4) \cap P(X_5) \cap P(X_6) \cap P(X_7) \cap P(X_8) \\
& \cap P(X_9) \cap P(X_{10}) \cap P(X_{11}) \cap P(X_{12}) \cap P(X_{13}) \cap P(X_{14}) \cap P(X_{15}) \cap P(X_{16}) \\
& \cap P(X_{17}) \cap P(X_{18}) \cap P(X_{19}) \cap P(X_{20}) \cap P(X_{21}) \cap P(X_{22}) \cap P(X_{23}) \cap P(X_{24}) \\
& \cap P(X_{25}) \cap P(X_{26}) \cap P(X_{27}) \cap P(X_{28}) \cap P(X_{29}) \cap P(X_{30}) \cap P(X_{31}) \cap P(X_{32}) \\
& \cap P(X_{33}) \cap P(X_{34}) \cap P(X_{35}) \cap P(X_{36}) \cap P(X_{37}) \cap P(X_{38}) \cap P(X_{39}) \cap P(X_{40}) \\
& \cap P(X_{41}) \cap P(X_{42}) \cap P(X_{43}) \cap P(X_{44}) \cap P(X_{45}) \cap P(X_{46}) \cap P(X_{47}) \cap P(X_{48}) \\
& \cap P(X_{49}) \cap P(X_{50}) \cap P(X_{51}) \cap P(X_{52}) \cap P(X_{53}) \cap P(X_{54}) \cap P(X_{55}) \cap P(X_{56}) \\
& \cap P(X_{57}) \cap P(X_{58}) \cap P(X_{59}) \cap P(X_{60}) \cap P(X_{61}) \cap P(X_{62}) \cap P(X_{63}) \cap P(X_{64}) \\
& \cap P(X_{65}) \cap P(X_{66}) \cap P(X_{67}) \cap P(X_{68}) \cap P(X_{69}) \cap P(X_{70}) \cap P(X_{71}) \cap P(X_{72}) \\
& \cap P(X_{73}) \cap P(X_{74}) \cap P(X_{75}) \cap P(X_{76})
\end{aligned}
\tag{2.3}
$$

However,

$$P(X_i) = R_i, (i = ,\ldots, 76) \tag{2.4}$$

where R_i ($i = 1,\ldots,76$) is the reliability or probability of the non-occurrence of event i.

Therefore,

$$
\begin{aligned}
R_{\text{CNCTC}} = {} & R_1 \times R_2 \times R_3 \times R_4 \times R_5 \times R_6 \times R_7 \times R_8 \times R_9 \times R_{10} \\
& \times R_{11} \times R_{12} \times R_{13} \times R_{14} \times R_{15} \times R_{16} \times R_{17} \times R_{18} \times R_{19} \times R_{20} \\
& \times R_{21} \times R_{22} \times R_{23} \times R_{24} \times R_{25} \times R_{26} \times R_{27} \times R_{28} \times R_{29} \times R_{30} \\
& \times R_{31} \times R_{32} \times R_{33} \times R_{34} \times R_{35} \times R_{36} \times R_{37} \times R_{38} \times R_{39} \times R_{40} \\
& \times R_{41} \times R_{42} \times R_{43} \times R_{44} \times R_{45} \times R_{46} \times R_{47} \times R_{48} \times R_{49} \times R_{50} \\
& \times R_{51} \times R_{52} \times R_{53} \times R_{54} \times R_{55} \times R_{56} \times R_{57} \times R_{58} \times R_{59} \times R_{60} \\
& \times R_{61} \times R_{62} \times R_{63} \times R_{64} \times R_{65} \times R_{66} \times R_{67} \times R_{68} \times R_{69} \times R_{70} \\
& \times R_{71} \times R_{72} \times R_{73} \times R_{74} \times R_{75} \times R_{76}
\end{aligned}
\tag{2.5}
$$

$$R_{\text{CNCTC}} = \sum_{i=1}^{76} R_i \tag{2.6}$$

Eq. (2.6) is the governing reliability model of the CNCTC. In this case, all the events of the CNCTC are assumed to be independent (i.e., the occurrence or non-occurrence of one event component does not affect the occurrence or non-occurrence of the other event). In other words, for the system to function, all 76 events should not occur. Furthermore, if the values of the probability of occurrence (F_i) for all 76 events are known, then the reliability of the CNCTC (R_{CNCTC}) is calculated as follows:

$$R_{\text{CNCTC}} = \sum_{i=1}^{76} (1 - F_i). \tag{2.7}$$

The reliability model for the first-level intermediate events can be estimated as follows:

$$R_{\text{CNCTC}} = R_1 \times R_2 \times R_3 \times R_4 \times R_5 \times R_6 \times R_7 \times R_8 \tag{2.8}$$

$$R_{\text{MT}} = R_9 \times R_{10} \times R_{11} \times R_{12} \times R_{13} \times R_{14} \tag{2.9}$$

$$R_{\text{XZAS}} = R_{15} \times R_{16} \times R_{17} \times R_{18} \times R_{19} \times R_{20} \times R_{21} \times R_{22} \times R_{23} \tag{2.10}$$

$$R_{\text{HS}} = R_{24} \times R_{25} \times R_{26} \times R_{27} \times R_{28} \times R_{29} \times R_{30} \times R_{31} \times R_{32} \tag{2.11}$$

$$R_{\text{ChS}} = R_{33} \times R_{34} \times R_{35} \tag{2.12}$$

$$R_{\text{SS}} = R_{36} \times R_{37} \times R_{38} \times R_{39} \tag{2.13}$$

$$R_{\text{EES}} = R_{40} \times R_{41} \times R_{42} \times R_{43} \times R_{44} \times R_{45} \times R_{46} \times R_{47} \times R_{48} \times R_{49} \times R_{50} \times R_{51} \tag{2.14}$$

$$R_{\text{CS}} = R_{52} \times R_{53} \times R_{54} \times R_{55} \times R_{56} \times R_{57} \tag{2.15}$$

$$R_{\text{TS}} = R_{58} \times R_{59} \times R_{60} \times R_{61} \times R_{62} \times R_{63} \times R_{64} \times R_{65} \times R_{66} \tag{2.16}$$

$$R_{\text{LS}} = R_{67} \times R_{68} \times R_{69} \times R_{70} \times R_{71} \tag{2.17}$$

$$R_{\text{SC}} = R_{72} \times R_{73} \times R_{74} \times R_{75} \times R_{76} \tag{2.18}$$

$$R_{\text{OS}} = R_{F12} \tag{2.19}$$

Eqs. (2.8)–(2.19) can be used to estimate the reliability of sub-systems or the probability of the non-occurrence of the intermediate events.

2.3.2 Quantitative FTA

In quantitative analysis, the probability of the occurrence of all 76 basic events is estimated using the laws of probability. The database required for the analysis has been collected from the users of the CNCTC. For some basic events, the failure probabilities are assumed or estimated using expert judgment. The experts included the employees engaged in maintenance activity, operators, service engineers, and design engineers of the CNCTC. The probability of the occurrence of all basic events is also cross-verified using expert judgments.

Table 2.2 summarizes the probability of occurrence and the reliability of all basic events. It shows that drive cards, spindle bearings, spindle belts, drawbar extension vane pumps, chucks, and encoder belts are the most critical events of the CNCTC with a reliability of less than 0.95. Furthermore, door rollers, drive batteries, alignment issues, screens/displays, cables, spindle motors, spindle motor cooling fans, gear pumps, ball screws, ball screw bearings, spindle axial plays, and coolant motors are critical basic events of the CNCTC with a reliability between 0.95 and 0.99. It should be noted that the $R_{F_{12}}$ is the reliability of the first-level intermediate event.

The reliability of all intermediate events (sub-systems) is estimated using Eqs. (2.8)–(2.19) and summarized in Table 2.3. It reveals that CNCS is the most critical sub-system with a reliability value of 0.8111 followed by ChS, SS, MT, EES, and HS, having a reliability of less than 0.91. The failure of SC is very rare and, therefore, its reliability of SC is taken as 1. Furthermore, the failure of SC does not affect the operation of the CNCTC. SC is the most reliable sub-system followed by LS, OS, TS, CS, and XZAS.

The reliability of the top event is calculated using Eq. (2.6) (with $R_{\mathrm{CNCTC}} \leq \min\{R_1, R_2, \ldots, R_{76}\}$) and estimated to be 0.3608. The most critical component of the CNCTC is the drive card with a reliability of 0.9048. However, the reliability of the CNCTC is less than the drive card. Therefore, it is concluded that the system reliability is less than the component with the least reliability in the case of series configuration. Thus it can be concluded that all components should have high reliability, especially if the system contains a large number of components.

2.4 CONCLUSIONS

FTA is an effectual tool that graphically models a system and estimates its reliability. Symbols to be used should be carefully selected, and the steps of the analysis should be followed. The goal of this chapter was to present the basics of the FTA and estimate the system reliability of a CNCTC involving 76 events. It allows the identification of the structure of the system and the events that may occur. Both qualitative and quantitative FTA have been investigated for this purpose. It has been observed that all the components of this system should be high in order to reach high system reliability as the sub-systems are connected in a series configuration. Further work could be devoted to the fuzzy FTA to deal with the uncertainties.

TABLE 2.2
Probability of Occurrence and Non-occurrence of the Basic Events

Event	Basic event	F_{X_i}	R_{X_i}	Event	Basic event	F_{X_i}	R_{X_i}
X_1	Drive cards	0.0952	0.9048	X_{39}	Spindle axial play	0.0166	0.9834
X_2	Sensors	0.0135	0.9865	X_{40}	Screen/ display	0.0238	0.9762
X_3	Cables	0.0207	0.9793	X_{41}	Power contactor	0.0124	0.9876
X_4	Encoder pulley	0.0052	0.9948	X_{42}	Foot switch	0.0083	0.9917
X_5	Encoder belt	0.0476	0.9524	X_{43}	Proximity switches	0.0052	0.9948
X_6	Encoder	0.0062	0.9938	X_{44}	Panel board switches	0.0031	0.9969
X_7	Encoder coupling	0.0104	0.9896	X_{45}	Other switches	0.0052	0.9948
X_8	Software	0.0041	0.9959	X_{46}	MCB	0.0031	0.9969
X_9	Spindle motor	0.0197	0.9803	X_{47}	Transformer	0.0021	0.9979
X_{10}	Spindle motor cooling fan	0.0197	0.9803	X_{48}	Connectors	0.0031	0.9969
X_{11}	Cooling fan motor	0.0062	0.9938	X_{49}	Control panel wiring	0.0031	0.9969
X_{12}	Spindle belt	0.0652	0.9348	X_{50}	Relays	0.0062	0.9938
X_{13}	Pulley	0.0010	0.9990	X_{51}	Drive battery	0.0300	0.9700
X_{14}	Misalignment	0.0041	0.9959	X_{52}	Coolant motor	0.0155	0.9845
X_{15}	X-axis servomotors	0.0062	0.9938	X_{53}	Coolant gear pump	0.0186	0.9814
X_{16}	Z-axis servomotors	0.0021	0.9979	X_{54}	Coolant hose	0.0104	0.9896
X_{17}	Servomotor coupling	0.0145	0.9855	X_{55}	Coolant strainer	0.0041	0.9959
X_{18}	Drive belt	0.0052	0.9948	X_{56}	Coolant ON/OFF coil	0.0062	0.9938
X_{19}	Ball screw	0.0052	0.9948	X_{57}	Coolant tank	0.0010	0.9990
X_{20}	Ball screw bearing	0.0176	0.9824	X_{58}	Turret indexing motor	0.0041	0.9959
X_{21}	Timing belt	0.0010	0.9990	X_{59}	Turret indexing plate	0.0093	0.9907
X_{22}	Guideways	0.0072	0.9928	X_{60}	Turret back side plate	0.0072	0.9928
X_{23}	Back lash	0.0083	0.9917	X_{61}	Turret thrust bearing	0.0083	0.9917
X_{24}	Hydraulic motor bearing	0.0031	0.9969	X_{62}	Turret ball bearing	0.0083	0.9917
X_{25}	Vane pump	0.0549	0.9451	X_{63}	Allan screws	0.0062	0.9938
X_{26}	Hydraulic power pack	0.0072	0.9928	X_{64}	Spring clip	0.0072	0.9928

(Continued)

TABLE 2.2 *(Continued)*

Event	Basic event	F_{X_i}	R_{X_i}	Event	Basic event	F_{X_i}	R_{X_i}
X_{27}	Hydraulic tank and strainer	0.0072	0.9928	X_{65}	Thread roller	0.0010	0.9990
X_{28}	Oil seal	0.0021	0.9979	X_{66}	Tool holder	0.0021	0.9979
X_{29}	Hydraulic hose	0.0135	0.9865	X_{67}	Lubrication motor	0.0031	0.9969
X_{30}	Cylinder	0.0031	0.9969	X_{68}	Lubrication pump	0.0114	0.9886
X_{31}	Valves	0.0041	0.9959	X_{69}	Lubrication hose	0.0021	0.9979
X_{32}	Pressure switch	0.0021	0.9979	X_{70}	Lubricant tank	0.0021	0.9979
X_{33}	Drawbar	0.0083	0.9917	X_{71}	Oil level indicator	0.0021	0.9979
X_{34}	Drawbar extension	0.0580	0.9420	X_{72}	Door roller	0.0321	0.9679
X_{35}	Chuck	0.0528	0.9472	X_{73}	Tube light	0.0021	0.9979
X_{36}	Spindle bearing	0.0694	0.9306	X_{74}	Control panel cooling fan	0.0031	0.9969
X_{37}	Alignment	0.0248	0.9752	X_{75}	Control panel cooling motor	0.0031	0.9969
X_{38}	Balancing	0.0083	0.9917	X_{76}	Short circuit	0.0021	0.9979

TABLE 2.3
Reliability of Intermediate Events (Sub-systems)

Event	Sub-system	R_{F_i}	F_{F_i}	Criticality Rank
F_1	CNCS	0.8111	0.1889	1
F_2	MT	0.8882	0.1118	4
F_3	XZAS	0.9346	0.0654	7
F_4	HS	0.9057	0.0943	6
F_5	ChS	0.8849	0.1151	2
F_6	SS	0.8851	0.1149	3
F_7	EES	0.8990	0.101	5
F_8	CS	0.9453	0.0547	8
F_9	TS	0.9474	0.0526	9
F_{10}	LS	0.9794	0.0206	11
F_{11}	SC	1.0000	0	12
F_{12}	OS	0.9579	0.0421	10

REFERENCES

Aggarwal, K. K. (1979). Comments on "an efficient simple algorithm for fault tree automatic synthesis from the reliability graph". *IEEE Transactions on Reliability*. doi:10.1109/TR.1979.5220615

Allan, R. N., Rondiris, I. L., & Fryer, D. M. (1981). An efficient computational technique for evaluating the cut/tie sets and common-cause failures of complex systems. *IEEE Transactions on Reliability*, R-30(2), 101–109. doi:10.1109/TR.1981.5220998

Amari, S., Dill, G., & Howald, E. (2003). A new approach to solve dynamic fault trees. *Proceedings of the Annual Reliability and Maintainability Symposium*. doi:10.1109/rams.2003.1182018

Andow, P. K. (1980). Difficulties in fault-tree synthesis for process plant. *IEEE Transactions on Reliability*, R-29(1), 2–9. doi:10.1109/TR.1980.5220684

Badida, P., Balasubramaniam, Y., & Jayaprakash, J. (2019). Risk evaluation of oil and natural gas pipelines due to natural hazards using fuzzy fault tree analysis. *Journal of Natural Gas Science and Engineering*, 66, 284–292. doi:10.1016/j.jngse.2019.04.010

Bell Telephone Laboratories. (1961). *Launch control safety study*. New Jersey.

Bengiamin, N. N., Bowen, B. A., & Schenk, K. F. (1976). An efficient algorithm for reducing the complexity of computation in fault tree analysis. *IEEE Transactions on Nuclear Science*, 23(5), 1442–1446. doi:10.1109/TNS.1976.4328508

Bobbio, A., & Raiteri, D. C. (2004). Parametric fault trees with dynamic gates and repair boxes. *Proceedings of the Annual Reliability and Maintainability Symposium*, 459–465. doi:10.1109/rams.2004.1285491

Chen, S. K., Ho, T. K., & Mao, B. H. (2007). Reliability evaluations of railway power supplies by fault-tree analysis. *IET Electric Power Applications*, 1(2), 161–172. doi:10.1049/iet-epa:20060244

Dugan, J. B., Bavuso, S. J., & Boyd, M. A. (1992). Dynamic fault-tree models for fault-tolerant computer systems. *IEEE Transactions on Reliability*, 41(3), 363–377. doi:10.1109/24.159800

Dugan, J. B., Sullivan, K. J., & Coppit, D. (2000). Developing a low-cost high-quality software tool for dynamic fault-tree analysis. *IEEE Transactions on Reliability*. doi:10.1109/24.855536

Haasl, D. F., Roberts, N. H., Vesely, W. E., & Goldberg, F. F. (1981). *Fault tree handbook*. U.S. Nuclear Regulatory Commission, Washington, DC.

Huang, W., Liu, Y., Zhang, Y., Zhang, R., Xu, M., De Dieu, G. J., ... & Shuai, B. (2020). Fault tree and fuzzy d-s evidential reasoning combined approach: An application in railway dangerous goods transportation system accident analysis. *Information Sciences*, 520, 117–129. doi:10.1016/j.ins.2019.12.089

Kohda, T. (2006). A simple method to derive minimal cut sets for a non-coherent fault tree. *International Journal of Automation and Computing*, 3, 151–156. doi:10.1007/s11633-006-0151-4

Lee, W. S., Grosh, D. L., Tillman, F. A., & Lie, C. H. (1985). Fault tree analysis, methods, and applications – a review. *IEEE Transactions on Reliability*, R-34(3), 194–203. doi:10.1109/TR.1985.5222114

Li, H. W., Ren, Y., & Wang, L. N. (2018). Research on software testing technology based on fault tree analysis. *Procedia Computer Science*, 154, 754–758. doi:10.1016/j.procs.2019.06.118

Mellal, M. A., & Williams, E. J. (2018). Large scale reliability-redundancy allocation optimization problem using three soft computing methods. In M. Ram (Ed.), *Modeling and simulation based analysis in reliability engineering* (pp. 199–214). Boca Raton, FL: CRC Press.

Mellal, M. A., & Zio, E. (2016). A penalty guided stochastic fractal search approach for system reliability optimization. *Reliability Engineering and System Safety*, 152, 213–227.

Mellal, M. A., & Zio, E. (2020). System reliability-redundancy optimization with cold-standby strategy by an enhanced nest cuckoo optimization algorithm. *Reliability Engineering & System Safety*, 201, 106973. doi:10.1016/j.ress.2020.106973

Meshkat, L., Dugan, J. B., & Andrews, J. D. (2002). Dependability analysis of systems with on-demand and active failure modes, using dynamic fault trees. *IEEE Transactions on Reliability*, 51(2), 240–251. doi:10.1109/TR.2002.1011531

Nyström, B., Austrin, L., Ankarbäck, N., & Nilsson, E. (2006). Fault tree analysis of an aircraft electric power supply system to electrical actuators. *2006 9th International Conference on Probabilistic Methods Applied to Power Systems, PMAPS*. doi:10.1109/PMAPS.2006.360325

Patil, R. B. (2019). Integrated reliability and maintainability analysis of computerized numerical control turning center considering the effects of human and organizational factors. *Journal of Quality in Maintenance Engineering*, 26(1), 87–103. doi:10.1108/JQME-08-2018-0063

Patil, R. B., Kothavale, B. S., Waghmode, L. Y., & Joshi, S. G. (2017). Reliability analysis of CNC turning center based on the assessment of trends in maintenance data: A case study. *International Journal of Quality and Reliability Management*, 34(9), 1616–1638. doi:10.1108/IJQRM-08-2016-0126

Piadeh, F., Ahmadi, M., & Behzadian, K. (2018). Reliability assessment for hybrid systems of advanced treatment units of industrial wastewater reuse using combined event tree and fuzzy fault tree analyses. *Journal of Cleaner Production*, 201, 958–973. doi:10.1016/j.jclepro.2018.08.052

Volkanovski, A., Čepin, M., & Mavko, B. (2009). Application of the fault tree analysis for assessment of power system reliability. *Reliability Engineering and System Safety*, 94(6), 1116–1127. doi:10.1016/j.ress.2009.01.004

Yin, H., Liu, C., Wu, W., Song, K., Liu, D., & Dan, Y. (2020). Safety assessment of natural gas storage tank using similarity aggregation method based fuzzy fault tree analysis (SAM-FFTA) approach. *Journal of Loss Prevention in the Process Industries*, 66, 104159. doi:10.1016/j.jlp.2020.104159

3 How to Schedule Elective Patients in Hospitals to Gain Full Utilization of Resources and Eliminate Patient Overcrowding

Murray V. Calichman
St. Francis Hospital
Roslyn, NY, USA

CONTENTS

INTRODUCTION

For the most part, hospitals allocate procedural time in their operating room (OR) suites, catheterization laboratories, electrophysiology (EPS) laboratories, and all other procedural areas by block time on a volume/seniority basis. A group of orthopedic surgeons, for example, may be allocated two rooms at 12 hours/room, three days each week, Monday, Wednesday, and Thursday. All other surgeons and surgeon groups are allocated block time in a similar manner, until all surgeons are assigned. Remaining time, if any, is allocated each day for emergent cases. Note that throughout this chapter, the OR will represent all procedural areas.

Knowing the amount of time each surgeon takes to perform his/her cases, including room turnaround time, the surgeon's office scheduler calls the OR in advance (days or even weeks) to schedule cases in the hospital. The OR scheduler keeps track of the time pre-scheduled in each room on each day and closes that room to further scheduling when room capacity is reached.

Although presented quite simplistically, the aforementioned explains how hospitals schedule patients for ORs and other procedural areas each day. And therein lies a major problem. Because there is no thought given to manage bed use (average length of stay could range from one day to twelve days, depending upon the type of surgery performed), hospitals do not manage, nor do they control, the demand of beds throughout the week. Consequently, they have more patients than beds on some days, usually Tuesdays–Thursdays and fewer patients than beds Fridays–Sundays.

The overutilization of beds Tuesdays–Thursdays is a cause of major concern as it leads to disruptions in patient flow, from minor delays to possible operational gridlock during the day. When all available floor beds become occupied and all Post-Anesthesia Care Unit (PACU) beds are in use (because there are no floor beds available for newly recovered patients to be transferred to), postsurgical patients cannot be sent to PACU to recover and so remain in the OR. Thus the following cases scheduled for that OR cannot be performed and are held in pre-op areas pending eventual patient movement. It may take hours to rectify this problem (resolved by discharging patients from the hospital to make beds available) and return to normal patient flow. In addition, because all floor beds are occupied, the (admitted) patients without assigned beds will be transferred to the emergency department to spend the night and, perhaps, many days, while disrupting patient flow in this service area also.

Hospitals do not have to operate in this fashion. However, most hospitals either fail to recognize the scheduling of elective surgical patients as the source of their patient flow problems or lack the knowledge to schedule their patients more effectively. The objectives of the daily/weekly elective patient schedule should be to comply with all operational constraints, e.g., demand for surgical service, OR time availability, and bed availability; while minimizing the need for those resources.

To achieve full utilization of the beds allocated for elective surgical patients and avoid patient overcrowding, hospitals have to adopt a more scientific means for scheduling those patients by:

1. Reviewing their historical data to obtain necessary statistics, and
2. Using linear programming, a methodology to achieve the best possible outcome in a mathematical model, to derive an optimum patient schedule.

SECTION 1: A REPEATING MONDAY–FRIDAY WEEKLY OPERATING ROOM (OR) SCHEDULE

Tables 3.1–3.3, below, illustrate the statistics that are required to derive an optimum weekly patient schedule Monday–Friday. Table 3.4, also below, is the Solution table.

A. Table 3.1 – Derivation Worksheet (Problem Setup): This table illustrates the problem setup leading to an optimum Monday–Friday patient schedule. As can be seen, the scheduling "problem" consists of 175 unknowns spread over the seven days of the week. Each unknown is the number of cases to be scheduled by surgical category each day of the week. Because the resulting schedule will

TABLE 3.1
Optimum Scheduling Derivation Worksheet for Monday–Friday Schedule

Hospital: Sample Hospital | **Schedule Option: MONDAY-FRIDAY** | **Prepared by: Murray V. Calichman**

SURGICAL CATEGORIES	AVE. CASE TIME (INCL. TRNRND)	OR TIME AVAILABILITY @ 87.5%							AVE. WKLY VOL.	MIN. CASES PER DAY M - Su	MAX. CASES PER DAY M - Sun
		M	T	W	TH	F	S	SU			
OHS - INPT	372	X_1	X_2	X_3	X_4	X_5	X_6	X_7	30	2	8
CARDIOLOGY - INPT	0	X_8	X_9	X_{10}	X_{11}	X_{12}	X_{13}	X_{14}	2	0	5
ENT - INPT	102	X_{15}	X_{16}	X_{17}	X_{18}	X_{19}	X_{20}	X_{21}	4	0	4
ENT - OPT	102	X_{22}	X_{23}	X_{24}	X_{25}	X_{26}	X_{27}	X_{28}	6	0	4
AICD & PACERS - INPT	119	X_{29}	X_{30}	X_{31}	X_{32}	X_{33}	X_{34}	X_{35}	29	4	10
PACERS - OPT	79	X_{36}	X_{37}	X_{38}	X_{39}	X_{40}	X_{41}	X_{42}	3	0	4
EYE - OPT	67	X_{43}	X_{44}	X_{45}	X_{46}	X_{47}	X_{48}	X_{49}	12	0	10
GENERAL/GI - INPT	140	X_{50}	X_{51}	X_{52}	X_{53}	X_{54}	X_{55}	X_{56}	20	2	5
GENERAL/GI - OPT	115	X_{57}	X_{58}	X_{59}	X_{60}	X_{61}	X_{62}	X_{63}	37	5	12
GU - INPT	135	X_{64}	X_{65}	X_{66}	X_{67}	X_{68}	X_{69}	X_{70}	13	2	5
GU - OPT	135	X_{71}	X_{72}	X_{73}	X_{74}	X_{75}	X_{76}	X_{77}	5	0	4
NEUROSURGERY - INPT	240	X_{78}	X_{79}	X_{80}	X_{81}	X_{82}	X_{83}	X_{84}	1	0	1
NEUROSURGERY - OPT	240	X_{85}	X_{86}	X_{87}	X_{88}	X_{89}	X_{90}	X_{91}	1	0	1
ORTHOPEDICS - INPT	110	X_{92}	X_{93}	X_{94}	X_{95}	X_{96}	X_{97}	X_{98}	20	0	5

(Continued)

TABLE 3.1 (CONTINUED)
Optimum Scheduling Derivation Worksheet for Monday–Friday Schedule

Hospital: Sample Hospital | **Schedule Option: MONDAY-FRIDAY** | **Prepared by: Murray V. Calichman**

SURGICAL CATEGORIES	AVE. CASE TIME (INCL. TRNRND)	OR TIME AVAILABILITY @ 87.5%							AVE. WKLY VOL.	MIN. CASES PER DAY M - Su	MAX. CASES PER DAY M - Sun
		M	T	W	TH	F	S	SU			
ORTHOPEDICS - OPT	110	X_{99}	X_{100}	X_{101}	X_{102}	X_{103}	X_{104}	X_{105}	13	0	5
PLASTIC - INPT	155	X_{106}	X_{107}	X_{108}	X_{109}	X_{110}	X_{111}	X_{112}	3	0	3
PLASTIC - OPT	149	X_{113}	X_{114}	X_{115}	X_{116}	X_{117}	X_{118}	X_{119}	2	0	3
ENT - OPT	30	X_{120}	X_{121}	X_{122}	X_{123}	X_{124}	X_{125}	X_{126}	35	0	12
THORACIC - INPT	165	X_{127}	X_{128}	X_{129}	X_{130}	X_{131}	X_{132}	X_{133}	9	0	4
THORACIC - OPT	100	X_{134}	X_{135}	X_{136}	X_{137}	X_{138}	X_{139}	X_{140}	0	0	0
VASCULAR - INPT	176	X_{141}	X_{142}	X_{143}	X_{144}	X_{145}	X_{146}	X_{147}	37	0	12
VASCULAR - OPT	127	X_{148}	X_{149}	X_{150}	X_{151}	X_{152}	X_{153}	X_{154}	2	0	3
EPS	75	X_{155}	X_{156}	X_{157}	X_{158}	X_{159}	X_{160}	X_{161}	17	2	6
ANGIO & CATHS - INPT	75	X_{162}	X_{163}	X_{164}	X_{165}	X_{166}	X_{167}	X_{168}	110	0	45
ANGIO & CATHS - OPT	75	X_{169}	X_{170}	X_{171}	X_{172}	X_{173}	X_{174}	X_{175}	65	0	25
TOTALS									476	17	196
TOTAL SURG MINS AVL.		9,923	9,923	9,923	9,923	9,923					
TOTAL CATH MINS AVL.		4,200	4,200	4,200	4,200	4,200	900	0			

TABLE 3.2
Percent of Patients Requiring Beds Each Day, by Grouped Days, During LOS (Maximized at 21 Days) for Monday–Friday Schedule

	-1,+6,+13	Op,+7,+14	+1,+8,+15	+2,+9,+16	+3,+10,+17	+4,+11,+18	+5,+12,+19	Max'd @ 21 Days
OHS - INPT	1.421	1.498	1.352	1.274	1.218	1.153	0.956	8.87
CARDIOLOGY - INPT	0.873	1.048	0.571	0.413	0.254	0.159	0.111	3.43
ENT - INPT	0.3529	1.0588	0.2353	0.2353	0.1765	0.1765	0.1176	2.35
ENT - OPT								
AICD & PACERS - INPT	0.872	1.038	0.524	0.261	0.141	0.090	0.074	3.00
PACERS - OPT								
EYE - OPT								
GENERAL/GI - INPT	0.8076	1.1520	0.8045	0.6114	0.4857	0.3753	0.2914	4.53
GENERAL/GI - OPT								
GU - INPT	0.327	1.077	0.692	0.433	0.327	0.221	0.173	3.25
GU - OPT								
NEUROSURGERY - INPT	0.8636	1.1364	0.8636	0.5909	0.5000	0.4545	0.4091	4.82
NEUROSURGERY - OPT								
ORTHOPEDICS - INPT	0.5606	1.1515	1.0455	0.9545	0.5303	0.4091	0.3030	4.95
ORTHOPEDICS - OPT								
PLASTIC - INPT	0.750	1.350	0.900	0.700	0.450	0.450	0.450	5.05
PLASTIC - OPT								
PMT - OPT								
THORACIC - INPT	1.127	1.444	1.143	1.063	0.937	0.873	0.730	7.32
THORACIC - OPT								
VASCULAR – INPT	0.682	1.159	0.773	0.611	0.466	0.347	0.284	4.32
VASCULAR - OPT								
EPS	0.903	1.049	0.562	0.357	0.205	0.162	0.103	3.34
ANGIO & CATHS – INPT	0.511	1.034	0.317	0.225	0.144	0.100	0.067	2.40
ANGIO & CATHS – OPT								

be limited to Monday–Friday, the unknown cases to be scheduled each Saturday and Sunday will be set equal to zero in the linear programming application.

The data is broken down by the following:

1. Surgical Category: The normal surgical categories that are referred to in the Operating Room (OR) and labs, including both inpatient and outpatient.
2. Average Case Time (including Turnaround Time): This is the average number of minutes the patient is in the OR plus the average time to clean the room and set it up for the next case. Thus the algebraic equation to comply with the OR time constraint for Mondays, for example, in the linear programming model would be

$$372X_1 + 0X_8 + 102X_{15} + \cdots + 127X_{148} \leq 9923$$

3. Average Weekly Volume: The number of cases to be performed each week to maintain volume at existing levels or algebraically in the linear programming model; for example, $X_1 + X_2 + X_3 + X_4 + X_5 + X_6 + X_7 = 30$ as the weekly volume equation for open-heart surgery (OHS).
4. Minimum Cases per Day, Monday– Friday: The minimum number of OHS cases required to be scheduled each day, Monday–Friday, or $X_1{:}X_5 \geq 2$ in the linear programming model. The Saturday and Sunday cases had already been set to zero (0).
5. Maximum Cases per Day, Monday–Friday: The maximum number of OHS cases not to be exceeded each day, Monday–Friday, or $X_1{:}X_5 \leq 8$ in the linear programming model.
6. Also listed are the number of staffed minutes available each day in the OR and in the Cath lab.

B. Table 3.2 – Percent of Patients Requiring Beds Each Day during Patient Stay: These are the statistics that lock patient bed use to the OR schedule. The distribution of lengths of stay (LOS) for each surgical category has to be developed from historical records. It is recommended that data from six months of operations be retrieved and sorted by surgical category and patient type for use in this table. The table consists of eight columns.
   1. Surgical Category: The first column is a duplication of the same column in Table 3.1.
   2. −1,+6,+13: These are the percentage of patients operated upon that require beds the day before surgery, six days after surgery (excluding the Op Day), and 13 days after surgery, all of which fall on the same day of the week. For example, if a patient is scheduled for a procedure on a Monday, Sunday becomes the day prior to surgery, as well as six days after the surgical day (for those operated on the previous week) and 13 days after the surgical day (for those operated on two weeks previously).

3. Op,+7,+14: These are the percentage of patients requiring beds the day of surgery, seven days after surgery, and 14 days after surgery, all on the same day of the week. In the previous example, where Monday is the day of the operation, these are the beds needed on Mondays.
4. +1,+8,+15: These are the percentage of patients requiring beds the day after surgery, eight days after surgery, and 15 days after surgery, all on the same day of the week. In the previous example, where Monday is the day of the operation, these are the beds needed on Tuesdays.
5. +2,+9,+16: These are the percentage of patients requiring beds two days after surgery, nine days after surgery, and 16 days after surgery, all on the same day of the week. In the previous example, where Monday is the day of the operation, these are the beds needed on Wednesdays.
6. +3,+10,+17: These are the percentage of patients requiring beds three days after surgery, ten days after surgery, and 17 days after surgery, all on the same day of the week. In the previous example, where Monday is the day of the operation, these are the beds needed on Thursdays.
7. +4,+11,+18: These are the percentage of patients requiring beds four days after surgery, 11 days after surgery, and 18 days after surgery, all on the same day of the week. In the previous example, where Monday is the day of the operation, these are the beds needed on Fridays.
8. +5,+12,+19: These are the percentage of patients requiring beds five days after surgery, 12 days after surgery, and 19 days after surgery, all on the same day of the week. In the previous example, where Monday is the day of the operation, these are the beds needed on Saturdays.

In the linear programming model, each day of the week would require beds using all of the aforementioned statistics, as they represent the beds needed on the Op Day for new patients, as well as the beds needed for all patients operated on previously on that day of the week in the past two weeks. In addition, there is a need to account for the patients operated on all of the other days in the week, this week, and the previous weeks, all conforming to the 21-day range, from −1 to +19, including Op Day.

C. Thus very long algebraic equations for each day of the week are entered into the linear programming model. These equations represent all surgical patients expected to be in the hospital on each day. The sum of all of these patients each day must be less than or equal to the bed constraint, which completes the seven algebraic equations (i.e., ≤ 177). Note that the 177 beds is, in itself, a derived number. Start the linear programming application with a number of beds in excess of what is anticipated, say 200 beds, for example. Then for each further iteration, reduce the number of beds until the solution becomes unfeasible. In this example, 176 beds would not be a feasible solution, so the minimum number of beds to satisfy all of the constraints in this problem is 177. Table 3.3 – Daily OR Time Availability: This table presents the OR time constraint, Monday–Friday. There are 189 clock hours staffed in the hospital's 17 ORs.

TABLE 3.3
Daily or Time Available by Day of Week (Hours) for Monday–Friday Schedule

	M	T	W	TH	F	S	SU
MAIN 1 (7:00–19:30)	12.5	12.5	12.5	12.5	12.5		
MAIN 2 (7:00–19:30)	12.5	12.5	12.5	12.5	12.5		
MAIN 3 (7:00–23:00)	16.0	16.0	16.0	16.0	16.0		
MAIN 4 (7:00–17:00)	10.0	10.0	10.0	10.0	10.0		
MAIN 5 (7:00–17:00)	10.0	10.0	10.0	10.0	10.0		
MAIN 6 (7:00–17:00)	10.0	10.0	10.0	10.0	10.0		
MAIN 7 (7:00–17:00)	10.0	10.0	10.0	10.0	10.0		
MAIN 8 (7:00–15:00)	8.0	8.0	8.0	8.0	8.0		
MAIN 9 (7:00–15:00)	8.0	8.0	8.0	8.0	8.0		
MAIN 10 (7:00–15:00)	8.0	8.0	8.0	8.0	8.0		
MAIN 11 (7:00–15:00)	8.0	8.0	8.0	8.0	8.0		
MAIN 12 (7:00–15:00)	8.0	8.0	8.0	8.0	8.0		
HC 1 (7:00–15:00)	8.0	8.0	8.0	8.0	8.0		
HC 2 (7:00–19:50)	12.5	12.5	12.5	12.5	12.5		
HC 3 (7:00–21:00)	14.0	14.0	14.0	14.0	14.0		
HC 1 (7:00–23:00)	16.0	16.0	16.0	16.0	16.0		
HC 1 (7:00–00:30)	17.5	17.5	17.5	17.5	17.5		
Total Hours Staffed	189.0	189.0	189.0	189.0	189.0		
Total Hours Staffed @ 87.5% Util	165.4	165.4	165.4	165.4	165.4		
Total OR Minutes @ 87.5%	**9,923**	**9,923**	**9,923**	**9,923**	**9,923**	**0.0**	**0.0**

Assuming a utilization of no greater than 87.5% of available time yields 165.4 available hours or 9,923 available minutes each day. Thus 9,923 becomes the time constraint each day, a number not to be exceeded in the resulting schedule.

D. Table 3.4 – Optimum OR Schedule: This table represents the resulting optimum patient schedule for a Monday–Friday operation. It was derived using Premium Solver linear programming software.

The objective of the schedule was to maintain the total number of cases each week at 476 given the appropriate daily and weekly volume constraints by surgical category, the OR daily time constraint of 9,923 minutes, and the bed constraint of 177 beds each day. A review of the last two columns indicates that the daily volumes comply with the minimum/maximum constraints for each surgical category. The other validations require the use of additional tables: Table 3.5 for weekly volumes, Table 3.6 for OR time, and Table 3.7 for bed use.

E. Table 3.5 – Total Weekly Volume versus Volume Constraints: As can be seen, the total weekly volume is equal to the weekly target (constraints), for each surgical category. Note that the constraint column, target range, is divided into two, with the first being ≥ the required weekly average (minimum numbers) and the second column being ≤ whatever increased volumes are planned (maximum numbers). In this example, the column was left blank, as the hospital opted to be volume neutral. Its goal was to get its bed use aligned correctly to eliminate patient overcrowding.
F. Table 3.6 – Daily OR Time Used: The table depicts the anticipated use of the OR (and other areas) Monday–Friday, compared to the time made available each day in the service areas. It is quite obvious that the results not only comply with the weekly volume constraints but also indicate the possibility of closing ORs on various days, especially on Tuesdays, when possibly half the ORs could be closed. Smaller savings could be achieved in the EPS and cath labs, as well.
G. Table 3.7 – Beds Required by Day of Week: This table illustrates the bed use with the derived optimum Monday–Friday patient schedule. As previously indicated, 177 beds is the minimum number of beds per day that provide a feasible solution to the problem. Note that 177 beds were required each night, Mondays–Fridays. Saturdays require 138 beds and Sundays 146 beds. The hospital will have an excess bed capacity of 39 on Saturday and 31 on Sunday.

The next step is to extend the OR schedule to a sixth day each week, in this case adding Saturdays to the schedule to see if the hospital could become even more efficient.

TABLE 3.4
Optimum or Schedule Solution for Monday–Friday Schedule

	M	T	W	TH	F	S	SU	TOTAL 476	MIN/DAY = Total Cases	MAX/DAY
OHS – INPT	8	3	8	4	7	0	0	30	2	8
CARDIOLOGY – INPT	2	0	0	0	0	0	0	2	0	5
ENT – INPT	0	4	0	0	0	0	0	4	0	4
ENT – OPT	2	0	0	4	0	0	0	6	0	4
AICD & PACERS – INPT	10	4	4	4	7	0	0	29	4	10
PACERS – OPT	3	0	0	0	0	0	0	3	0	4
EYE – OPT	0	0	0	10	2	0	0	12	0	10
GENERAL/GI – INPT	4	2	5	4	5	0	0	20	2	5
GENERAL/GI – OPT	10	5	5	12	5	0	0	37	5	12
GU – INPT	2	2	2	2	5	0	0	13	2	5
GU – OPT	1	0	0	4	0	0	0	5	0	4
NEUROSURGERY – INPT	0	0	0	0	1	0	0	1	0	1
NEUROSURGERY – OPT	0	0	0	1	0	0	0	1	0	1
ORTHOPEDICS – INPT	3	5	2	5	5	0	0	20	0	5
ORTHOPEDICS – OPT	4	0	5	4	0	0	0	13	0	5
PLASTIC – INPT	0	0	0	0	3	0	0	3	0	3
PLASTIC – OPT	0	0	0	2	0	0	0	2	0	3
PMT – OPT	0	0	11	12	12	0	0	35	0	12
THORACIC – INPT	0	1	4	0	4	0	0	9	0	4
THORACIC – OPT	0	0	0	0	0	0	0	0	0	0
VASCULAR – INPT	2	0	11	12	12	0	0	37	0	12
VASCULAR – OPT	0	1	0	1	0	0	0	2	0	3
EPS	6	2	2	2	5	0	0	17	2	6
ANGIO & CATHS – INPT	39	23	24	2	10	12	0	110	0	45
ANGIO & CATHS – OPT	0	25	25	15	0	0	0	65	0	25
TOTAL NO. OF CASES	96	77	108	100	83	12	0	476		

TABLE 3.5
Daily Operating Room Time Used by Day of Week for Monday–Friday Schedule

	M	T	W	TH	F	S	SU
OHS - INPT	2,976	1,116	2,976	1,488	2,604	0	0
CARDIOLOGY – INPT	0	0	0	0	0	0	0
ENT – INPT	0	408	0	0	0	0	0
ENT – OPT	204	0	0	408	0	0	0
AICD & PACERS – INPT	1,190	476	476	476	833	0	0
PACERS – OPT	237	0	0	0	0	0	0
EYE – OPT	0	0	0	670	134	0	0
GENERAL/GI – INPT	560	280	700	560	700	0	0
GENERAL/GI – OPT	1,150	575	575	1,380	575	0	0
GU – INPT	270	270	270	270	675	0	0
GU – OPT	135	0	0	540	0	0	0
NEUROSURGERY – INPT	0	0	0	0	240	0	0
NEUROSURGERY – OPT	0	0	0	240	0	0	0
ORTHOPEDICS – INPT	330	550	220	550	550	0	0
ORTHOPEDICS – OPT	440	0	550	440	0	0	0
PLASTIC – INPT	0	0	0	0	465	0	0
PLASTIC – OPT	0	0	0	298	0	0	0
PMT – OPT	0	0	330	360	360	0	0
THORACIC – INPT	0	165	660	0	660	0	0
THORACIC – OPT	0	0	0	0	0	0	0
VASCULAR – INPT	352	0	1,936	2,112	2,112	0	0
VASCULAR – OPT	0	127	0	127	0	0	0
Total OR Minutes Used	**7,844**	**3,967**	**8,693**	**9,919**	**9,908**	**0**	**0**
OR Minutes Available	**9,923**	**9,923**	**9,923**	**9,923**	**9,923**	**0**	**0**
EPS Minutes Used	**450**	**150**	**150**	**150**	**375**	**0**	**0**
ANGIO & CATHS – INPT	2,925	1,725	1,800	150	750	900	0
ANGIO & CATHS – OPT	0	1,875	1,875	1,125	0	0	0
Total Cath Minutes Used	**2,925**	**3,600**	**3,675**	**1,275**	**750**	**900**	**0**
Cath Minutes Available	**4,200**	**4,200**	**4,200**	**4,200**	**4,200**	**900**	

TABLE 3.6
Scheduled Weekly Volume (By Day) Versus Volume Constraints for Monday–Friday Schedule

Volume	M	T	W	TH	F	S	SU	Total	Target	Pt Days/ Each	Pt Days Total
OHS – INPT	8	3	8	4	7	0	0	30	30	8.87	266.2
CARDIOLOGY – INPT	2	0	0	0	0	0	0	2	2	3.43	6.9
ENT – INPT	0	4	0	0	0	0	0	4	4	2.35	9.4
ENT – OPT	2	0	0	4	0	0	0	6	6	0.00	0.0
AICD & PACERS – INPT	10	4	4	4	7	0	0	29	29	3.00	87.0
PACERS – OPT	3	0	0	0	0	0	0	3	3	0.00	0.0
EYE – OPT	0	0	0	10	2	0	0	12	12	0.00	0.0
GENERAL/GI – INPT	4	2	5	4	5	0	0	20	20	4.53	90.6
GENERAL/GI – OPT	10	5	5	12	5	0	0	37	37	0.00	0.0
GU – INPT	2	2	2	2	5	0	0	13	13	3.25	42.3
GU – OPT	1	0	0	4	0	0	0	5	5	0.00	0.0
NEUROSURGERY – INPT	0	0	0	0	1	0	0	1	1	4.82	4.8
NEUROSURGERY – OPT	0	0	0	1	0	0	0	1	1	0.00	0.0
ORTHOPEDICS – INPT	3	5	2	5	5	0	0	20	20	4.95	99.1
ORTHOPEDICS – OPT	4	0	5	4	0	0	0	13	13	0.00	0.0
PLASTIC – INPT	0	0	0	0	3	0	0	3	3	5.05	15.2
PLASTIC – OPT	0	0	0	2	0	0	0	2	2	0.00	0.0
MT – OPT	0	0	11	12	12	0	0	35	35	0.00	0.0
THORACIC – INPT	0	1	4	0	4	0	0	9	9	7.32	65.9
THORACIC – OPT	0	0	0	0	0	0	0	0	0	0.00	0.0
VASCULAR – INPT	2	0	11	12	12	0	0	37	37	4.32	159.9
VASCULAR – OPT	0	1	0	1	0	0	0	2	2	0.00	0.0
EPS	6	2	2	2	5	0	0	17	17	3.34	56.8
ANGIO & CATHS – INPT	39	23	24	2	10	12	0	110	110	2.40	263.7
ANGIO & CATHS – OPT	0	25	25	15	0	0	0	65	65	0.00	0.0
TOTAL CASES/WEEK	96	77	108	100	83	12	0	476	476	3.958	1168

TABLE 3.7
Beds Required by Day of Week for Monday–Friday Schedule

By Category	M	T	W	TH	F	S	SU	TOTAL	DAILY BED CONSTRAINT
OHS – INPT	37.0	38.6	38.6	40.3	39.0	35.4	37.2	266.2	
CARDIOLOGY – INPT	2.1	1.1	0.8	0.5	0.3	0.2	1.7	6.9	
ENT – INPT	1.4	4.2	0.9	0.9	0.7	0.7	0.5	9.4	
AICD & PACERS – INPT	15.5	13.8	12.9	14.8	11.9	6.4	11.8	87.0	
GENERAL/GI – INPT	11.6	12.6	14.5	15.8	14.5	10.8	10.7	90.6	
GU – INPT	5.2	5.6	5.9	6.7	8.7	5.8	4.3	42.3	
NEUROSURGERY – INPT	0.5	0.5	0.4	0.9	1.1	0.9	0.6	4.8	
ORTHOPEDICS – INPT	11.6	13.6	14.7	17.0	16.8	14.0	11.4	99.1	
PLASTIC – INPT	1.4	1.4	1.4	2.3	4.1	2.7	2.1	15.2	
THORACIC – INPT	7.8	9.4	9.8	10.1	11.0	9.2	8.5	65.9	
VASCULAR – INPT	15.2	16.6	25.6	31.5	30.6	22.3	18.1	159.9	
EPS	9.7	8.3	7.7	9.7	8.5	4.9	8.1	56.8	
ANGIO & CATHS – INPT	58.0	51.3	43.8	26.4	29.7	24.4	30.2	263.7	
TOTAL BEDS REQ'D.	**177.0**	**177.0**	**177.0**	**177.0**	**176.8**	**137.6**	**145.2**	**1167.6**	≤ **177**
TOTAL ROUNDED UP	**177**	**177**	**177**	**177**	**177**	**138**	**146**	**166.8**	**3.958** = ave los

SECTION 2: A REPEATING MONDAY–SATURDAY WEEKLY OR SCHEDULE

As will be shown, adding a sixth day to the optimum five-day-per-week schedule makes the hospital even more efficient in that it enables the hospital to further minimize its use of beds while performing the same 476 cases each week.

The following tables (Tables 3.8–3.10) illustrate the additional statistics required to derive an optimum weekly patient schedule Monday–Saturday.

A. Table 3.8 – Derivation Worksheet (Problem Setup): This is the problem setup for a Monday–Saturday optimum patient schedule. The table is almost exactly the same as for the five-day-per-week schedule. The difference is that it provides staffing for an additional 9,923 minutes in the OR , an additional 3,300 minutes in the cath lab, and an additional 150 minutes in the EPS lab, on Saturdays. The time available in the OR and cath and EPS labs on Sundays remains at zero in the linear programming application. Inasmuch as there was much excess time available in the OR with a five-day-per-week schedule, there is even more excess time with a six-day-per-week schedule. Therefore, many ORs could be closed during the week with a commensurate savings in staff. It is quite obvious that the time constraint in the OR could be reduced considerably.
B. Table 3.9 – Percent of Patients Requiring Beds Each Day during Patient Stay: This table is exactly the same as Table 3.3. It doesn't change because it is based on the distribution of patient LOS and not on the day of the week.
C. Table 3.10 – Daily OR Time Availability: This table presents the OR time constraint Monday–Sunday. The table was updated by adding 9,923 to the OR 3,300 minutes to the cath lab (not shown) and 150 minutes to the EPS lab (not shown) on Saturdays.
D. Table 3.11 – Optimum OR Schedule (Monday–Saturday): This table represents the resulting optimum patient schedule for a Monday–Saturday operation. It was derived by including the Saturday variables in the linear programming model and by reducing the new bed constraint. It was determined that the bed constraint for Monday–Sunday could be reduced to 170 and still result in a feasible schedule.

As before, a review of the last two columns indicates that the daily volumes comply with the daily minimum/maximum constraints.

And, as before, the validations for the volume, bed, and time constraints are available in Table 3.12 for OR time, Table 3.13 for weekly volumes, and Table 3.14 for bed use.

E. Table 3.12 – Daily OR Time Used: The table depicts the anticipated use of the OR (and other areas) Monday–Saturday, compared to the time made available each day with the optimum Monday–Saturday schedule.
F. Table 3.13 – Total Weekly Volume versus Volume Constraints: The total weekly volume is equal to the adjacent column totals, the weekly constraints, as required.

TABLE 3.8
Optimum Scheduling Derivation Worksheet for Monday–Saturday Schedule

Hospital: Sample Hospital | **Schedule Option: MONDAY-SATURDAY** | **Prepared by: Murray V. Calichman**

SURGICAL CATEGORIES	AVE. CASE TIME (INCL. TRNRND)	OR TIME AVAILABILITY @ 87.5%							AVE. WKLY VOL.	MIN. CASES PER DAY M - Su	MAX. CASES PER DAY M - Su
		M	T	W	TH	F	S	SU			
OHS – INPT	372	X_1	X_2	X_3	X_4	X_5	X_6	X_7	30	2	8
CARDIOLOGY – INPT	0	X_8	X_9	X_{10}	X_{11}	X_{12}	X_{13}	X_{14}	2	0	5
ENT – INPT	102	X_{15}	X_{16}	X_{17}	X_{18}	X_{19}	X_{20}	X_{21}	4	0	4
ENT – OPT	102	X_{22}	X_{23}	X_{24}	X_{25}	X_{26}	X_{27}	X_{28}	6	0	4
AICD & PACERS – INPT	119	X_{29}	X_{30}	X_{31}	X_{32}	X_{33}	X_{34}	X_{35}	29	4	10
PACERS – OPT	79	X_{36}	X_{37}	X_{38}	X_{39}	X_{40}	X_{41}	X_{42}	3	0	4
EYE – OPT	67	X_{43}	X_{44}	X_{45}	X_{46}	X_{47}	X_{48}	X_{49}	12	0	10
GENERAL/GI – INPT	140	X_{50}	X_{51}	X_{52}	X_{53}	X_{54}	X_{55}	X_{56}	20	2	5
GENERAL/GI – OPT	115	X_{57}	X_{58}	X_{59}	X_{60}	X_{61}	X_{62}	X_{63}	37	5	12
GU – INPT	135	X_{64}	X_{65}	X_{66}	X_{67}	X_{68}	X_{69}	X_{70}	13	2	5
GU – OPT	135	X_{71}	X_{72}	X_{73}	X_{74}	X_{75}	X_{76}	X_{77}	5	0	4
NEUROSURGERY – INPT	240	X_{78}	X_{79}	X_{80}	X_{81}	X_{82}	X_{83}	X_{84}	1	0	1
NEUROSURGERY – OPT	240	X_{85}	X_{86}	X_{87}	X_{88}	X_{89}	X_{90}	X_{91}	1	0	1
ORTHOPEDICS – INPT	110	X_{92}	X_{93}	X_{94}	X_{95}	X_{96}	X_{97}	X_{98}	20	0	5
ORTHOPEDICS – OPT	110	X_{99}	X_{100}	X_{101}	X_{102}	X_{103}	X_{104}	X_{105}	13	0	5

(Continued)

TABLE 3.8 (Continued)

Hospital: Sample Hospital | **Schedule Option: MONDAY-SATURDAY** | **Prepared by: Murray V. Calichman**

SURGICAL CATEGORIES	AVE. CASE TIME (INCL. TRNRND)	OR TIME AVAILABILITY @ 87.5%							AVE. WKLY VOL.	MIN. CASES PER DAY M - Su	MAX. CASES PER DAY M - Su
		M	T	W	TH	F	S	SU			
PLASTIC – INPT	155	X_{106}	X_{107}	X_{108}	X_{109}	X_{110}	X_{111}	X_{112}	3	0	3
PLASTIC – OPT	149	X_{113}	X_{114}	X_{115}	X_{116}	X_{117}	X_{118}	X_{119}	2	0	3
PMT – OPT	30	X_{120}	X_{121}	X_{122}	X_{123}	X_{124}	X_{125}	X_{126}	35	0	12
THORACIC – INPT	165	X_{127}	X_{128}	X_{129}	X_{130}	X_{131}	X_{132}	X_{133}	9	0	4
THORACIC – OPT	100	X_{134}	X_{135}	X_{136}	X_{137}	X_{138}	X_{139}	X_{140}	0	0	0
VASCULAR – INPT	176	X_{141}	X_{142}	X_{143}	X_{144}	X_{145}	X_{146}	X_{147}	37	0	12
VASCULAR – OPT	127	X_{148}	X_{149}	X_{150}	X_{151}	X_{152}	X_{153}	X_{154}	2	0	3
EPS	75	X_{155}	X_{156}	X_{157}	X_{158}	X_{159}	X_{160}	X_{161}	17	2	6
ANGIO & CATHS – INPT	75	X_{162}	X_{163}	X_{164}	X_{165}	X_{166}	X_{167}	X_{168}	110	0	45
ANGIO & CATHS – OPT	75	X_{169}	X_{170}	X_{171}	X_{172}	X_{173}	X_{174}	X_{175}	65	0	25
TOTALS									476	17	196
TOTAL SURG MINS AVL.		9,923	9,923	9,923	9,923	9,923	9,923	0			
TOTAL CATH MINS AVL.		4,200	4,200	4,200	4,200	4,200	4,200	0			

TABLE 3.9
Percent of Patients Requiring Beds Each Day, by Grouped Days, During LOS (Maximized at 21 Days) for Monday–Saturday Schedule

	-1,+6,+13	Op,+7,+14	+1,+8,+15	+2,+9,+16	+3,+10,+17	+4,+11,+18	+5,+12,+19	(max'd @ 21 days)
OHS – INPT	1.421	1.498	1.352	1.274	1.218	1.153	0.956	8.87
CARDIOLOGY – INPT	0.873	1.048	0.571	0.413	0.254	0.159	0.111	3.43
ENT – INPT	0.3529	1.0588	0.2353	0.2353	0.1765	0.1765	0.1176	2.35
ENT – OPT								
AICD & PACERS – INPT	0.872	1.038	0.524	0.261	0.141	0.090	0.074	3.00
PACERS – OPT								
EYE – OPT								
GENERAL/GI – INPT	0.8076	1.1520	0.8045	0.6114	0.4857	0.3753	0.2914	4.53
GENERAL/GI – OPT								
GU – INPT	0.327	1.077	0.692	0.433	0.327	0.221	0.173	3.25
GU – OPT								
NEUROSURGERY – INPT	0.8636	1.1364	0.8636	0.5909	0.5000	0.4545	0.4091	4.82
NEUROSURGERY – OPT								
ORTHOPEDICS – INPT	0.5606	1.1515	1.0455	0.9545	0.5303	0.4091	0.3030	4.95
ORTHOPEDICS – OPT								
PLASTIC – INPT	0.750	1.350	0.900	0.700	0.450	0.450	0.450	5.05
PLASTIC – OPT								
PMT – OPT								
THORACIC – INPT	1.127	1.444	1.143	1.063	0.937	0.873	0.730	7.32
THORACIC – OPT								
VASCULAR – INPT	0.682	1.159	0.773	0.611	0.466	0.347	0.284	4.32
VASCULAR – OPT								
EPS	0.903	1.049	0.562	0.357	0.205	0.162	0.103	3.34
ANGIO & CATHS – INPT	0.511	1.034	0.317	0.225	0.144	0.100	0.067	2.40
ANGIO & CATHS – OPT								

TABLE 3.10
Daily or Time Available (Hours) for Monday–Saturday Schedule

	M	T	W	TH	F	S	SU
MAIN 1 (7:00–19:30)	12.5	12.5	12.5	12.5	12.5	12.5	
MAIN 2 (7:00–19:30)	12.5	12.5	12.5	12.5	12.5	12.5	
MAIN 3 (7:00–23:00)	16.0	16.0	16.0	16.0	16.0	15.0	
MAIN 4 (7:00–17:00)	10.0	10.0	10.0	10.0	10.0	10.0	
MAIN 5 (7:00–17:00)	10.0	10.0	10.0	10.0	10.0	10.0	
MAIN 6 (7:00–17:00)	10.0	10.0	10.0	10.0	10.0	10.0	
MAIN 7 (7:00–17:00)	10.0	10.0	10.0	10.0	10.0	10.0	
MAIN 8 (7:00–15:00)	8.0	8.0	8.0	8.0	8.0	8.0	
MAIN 9 (7:00–15:00)	8.0	8.0	8.0	8.0	8.0	8.0	
MAIN 10 (7:00–15:00)	8.0	8.0	8.0	8.0	8.0	8.0	
MAIN 11 (7:00–15:00)	8.0	8.0	8.0	8.0	8.0	8.0	
MAIN 12 (7:00–15:00)	8.0	8.0	8.0	8.0	8.0	8.0	
EC 1 (7:00–15:00)	8.0	8.0	8.0	8.0	8.0	8.0	
EC 2 (7:00–19:50)	12.5	12.5	12.5	12.5	12.5	12.5	
EC 3 (7:00–21:00)	14.0	14.0	14.0	14.0	14.0	14.0	
HC 1 (7:00–23:00)	16.0	16.0	16.0	16.0	16.0	16.0	
HC 1 (7:00–00:30)	17.5	17.5	17.5	17.5	17.5	17.5	
Total Hours Staffed	189.0	189.0	189.0	189.0	189.0	189.0	
Total Hours Staffed @ 87.5% Util	165.4	165.4	165.4	165.4	165.4	165.4	
Total OR Minutes @ 87.5%	**9,923**	**9,923**	**9,923**	**9,923**	**9,923**	**9,923**	

TABLE 3.11
Optimum or Schedule Solution for Monday–Saturday Schedule

	M	T	W	TH	F	S	SU	TOTAL 476	MIN/DAY = Total Cs	MAX/DAY
CHS – INPT	7	3	7	3	8	2	0	30	2	8
CARDIOLOGY – INPT	2	0	0	0	0	0	0	2	0	5
ENT – INPT	0	4	0	0	0	0	0	4	0	4
ENT – OPT	0	0	0	0	4	2	0	6	0	4
AICD & PACERS – INPT	9	4	4	4	4	4	0	29	4	10
PACERS – OPT	0	0	0	0	3	0	0	3	0	4
EYE – OPT	0	0	0	2	0	10	0	12	0	10
GENERAL/GI – INPT	5	2	4	2	2	5	0	20	2	5
GENERAL/GI – OPT	5	5	5	5	12	5	0	37	5	12
GU – INPT	2	2	2	2	2	3	0	13	2	5
GU – OPT	0	0	0	0	4	1	0	5	0	4
NEUROSURGERY – INPT	1	0	0	0	0	0	0	1	0	1
NEUROSURGERY – OPT	0	0	0	0	1	0	0	1	0	1
ORTHOPEDICS – INPT	4	5	0	1	5	5	0	20	0	5
ORTHOPEDICS – OPT	0	0	0	3	5	5	0	13	0	5
PLASTIC – INPT	0	0	2	0	1	0	0	3	0	3
PLASTIC – OPT	0	0	0	0	2	0	0	2	0	3
PMT – OPT	0	0	0	11	12	12	0	35	0	12
THORACIC – INPT	4	0	4	0	0	1	0	9	0	4
THORACIC – OPT	0	0	0	0	0	0	0	0	0	0
VASCULAR – INPT	12	7	11	7	0	0	0	37	0	12
VASCULAR – OPT	0	0	0	0	2	0	0	2	0	3
EPS	5	3	3	2	2	2	0	17	2	6
ANGIO & CATHS – INPT	13	12	7	24	9	45	0	110	0	45
ANGIO & CATHS – OPT	15	0	0	25	25	0	0	65	0	25
TOTAL NO. OF CASES	84	47	49	91	103	102	0	476		

TABLE 3.12
Daily or Time Used for Monday–Saturday Schedule

	M	T	W	TH	F	S	SU
HS – INPT	2,604	1,116	2,604	1,116	2,976	744	0
CARDIOLOGY – INPT	0	0	0	0	0	0	0
ENT – INPT	0	408	0	0	0	0	0
ENT – OPT	0	0	0	0	408	204	0
AICD & PACERS – INPT	1,071	476	476	476	476	476	0
PACERS – OPT	0	0	0	0	237	0	0
EYE – OPT	0	0	0	134	0	670	0
GENERAL/GI – INPT	700	280	560	280	280	700	0
GENERAL/GI – OPT	575	575	575	575	1,380	575	0
GU – INPT	270	270	270	270	270	405	0
GU – OPT	0	0	0	0	540	135	0
NEUROSURGERY – INPT	240	0	0	0	0	0	0
NEUROSURGERY – OPT	0	0	0	0	240	0	0
ORTHOPEDICS – INPT	440	550	0	110	550	550	0
ORTHOPEDICS – OPT	0	0	0	330	550	550	0
PLASTIC – INPT	0	0	310	0	155	0	0
PLASTIC – OPT	0	0	0	0	298	0	0
PMT – OPT	0	0	0	330	360	360	0
THORACIC – INPT	660	0	660	0	0	165	0
THORACIC – OPT	0	0	0	0	0	0	0
VASCULAR – INPT	2,112	1,232	1,936	1,232	0	0	0
VASCULAR – OPT	0	0	0	0	254	0	0
Total OR Minutes Used	8,672	4,907	7,391	4,853	8,974	5,534	0
OR Minutes Available	9,923	9,923	9,923	9,923	9,923	9,923	0
EPS Minutes Used	375	225	225	150	150	150	0
ANGIO & CATHS – INPT	975	900	525	1,800	675	3,375	0
ANGIO & CATHS – OPT	1,125	0	0	1,875	1,875	0	0
Total Cath Minutes Used	2,100	900	525	3,675	2,550	3,375	0
Cath Minutes Available	4,200	4,200	4,200	4,200	4,200	4,200	0

G. Table 3.14 – Beds Required by Day of Week: This table illustrates the bed use with the derived optimum Monday–Saturday patient schedule. As mentioned previously, 170 beds was the minimum number of beds per day that provides a feasible solution to the problem, a reduction of seven beds each day versus the Monday–Friday schedule, or 49 beds each week. All 170 beds are used Monday–Saturday, with a reduction in the need for only 148 beds on Sunday. It should be mentioned that if the hospital were to agree to reduce the maximum number of cases each day for a number of surgical categories, it is more than likely that the hospital could reduce its bed need a bit further and achieve 100% utilization of beds seven days each week.

SECTION 3: THE ACTUAL WEEKLY OR PERFORMANCE

Although every week's actual OR performance differed from the previous week's performance, an average caseload was derived from past statistics. The average number of cases performed each day was loaded into the "derived optimum schedule" template to derive comparable statistics for bed use and OR time used.

The tables presented in this section will provide statistics similar to the other sections (i.e., the resulting schedule, the daily OR time used, the daily/weekly volume, and, most importantly, the daily/weekly need for beds). The OR time and bed use were calculated numbers using the same formulas that were used to derive time and bed use for the optimum solutions. The average number of cases performed each week, by surgical category and patient type (476), was calculated from this exercise (and kept constant for the two previous scheduling scenarios).

A. Table 3.15 – Actual OR Use: The table provides the actual OR use over a 12-week period.
B. Table 3.16 – Daily OR Time Used: This table depicts the actual calculated use of the OR (and other areas) each day, compared to the time made available each day. It is interesting to note the more balanced use of OR time Monday–Friday than in either of the two derived optimum schedules. However, that balance benefits the hospital very little. It is the need for beds that has to be balanced to eliminate problems with patient flow.
C. Table 3.17 – Total Weekly Volume versus Volume Constraints: And again, the total weekly volume is equal to the adjacent column totals, the weekly constraints, or weekly volume target.
D. Table 3.18 – Beds Required by Day of Week: This table illustrates the calculated bed use each day with the average existing daily caseload in the OR and other procedural areas. As previously determined, 177 beds is the maximum number of beds required for a Monday–Friday schedule (see Table 3.7). Therefore, there presently exists more inpatients than beds on Mondays (2), Tuesdays (7), Wednesdays (15), and Thursdays (12).
E. Since these are the average number of cases performed each day, many weeks the number of patients in-house, Tuesday–Thursday, are even greater than indicated. And this is why this hospital is beset by patient flow problems. It (unknowingly) schedules patients for the right operations but on the wrong days, leading to an uneven need for beds during the week.

TABLE 3.13

Scheduled Weekly Volume versus Volume Constraints for Monday–Saturday Schedule

Volume	M	T	W	TH	F	S	SU	WEEKLY TOTALS	Wkly Vol. Constraints
									<=
CHS – INPT	7	3	7	3	8	2	0	30	30
CARDIOLOGY – INPT	2	0	0	0	0	0	0	2	2
ENT – INPT	0	4	0	0	0	0	0	4	4
ENT – OPT	0	0	0	0	4	2	0	6	6
AICD & PACERS – INPT	9	4	4	4	4	4	0	29	29
PACERS – OPT	0	0	0	0	3	0	0	3	3
EYE – OPT	0	0	0	2	0	10	0	12	12
GENERAL/GI – INPT	5	2	4	2	2	5	0	20	20
GENERAL/GI – OPT	5	5	5	5	12	5	0	37	37
GU – INPT	2	2	2	2	2	3	0	13	13
GU – OPT	0	0	0	0	4	1	0	5	5
NEUROSURGERY – INPT	1	0	0	0	0	0	0	1	1
NEUROSURGERY – OPT	0	0	0	0	1	0	0	1	1
ORTHOPEDICS – INPT	4	5	0	1	5	5	0	20	20
ORTHOPEDICS – OPT	0	0	0	3	5	5	0	13	13
PLASTIC – INPT	0	0	2	0	1	0	0	3	3
PLASTIC – OPT	0	0	0	0	2	0	0	2	2
PMT – OPT	0	0	0	11	12	12	0	35	35
THORACIC – INPT	4	0	4	0	0	1	0	9	9
THORACIC – OPT	0	0	0	0	0	0	0	0	0
VASCULAR – INPT	12	7	11	7	0	0	0	37	37
VASCULAR – OPT	0	0	0	0	2	0	0	2	2
EPS	5	3	3	2	2	2	0	17	17
ANGIO & CATHS – INPT	13	12	7	24	9	45	0	110	110
ANGIO & CATHS – OPT	15	0	0	25	25	0	0	65	65
Total Inpt Cases	64	42	44	45	33	67	0	295	295
Total Opt Cases	20	5	5	46	70	35	0	181	181
TOTAL CASES/WEEK	84	47	49	91	103	102	0	476	476

TABLE 3.14
Beds Required by Day of Week for Monday–Saturday Schedule

By Category	M	T	W	TH	F	S	SU	TOTAL
OHS - INPT	37.2	38.4	37.7	39.6	39.5	36.3	37.4	
CARDIOLOGY – INPT	2.1	1.1	0.8	0.5	0.3	0.2	1.7	
ENT - INPT	1.4	4.2	0.9	0.9	0.7	0.7	0.5	
AICD & PACERS – INPT	15.1	13.6	12.7	12.3	12.2	8.9	12.2	
GENERAL/GI – INPT	13.3	13.3	13.3	12.2	13.2	12.7	12.3	
GU - INPT	5.5	6.0	6.1	6.2	6.5	6.9	5.0	
NEUROSURGERY – INPT	1.1	0.9	0.6	0.5	0.5	0.4	0.9	
ORTHOPEDICS – INPT	15.2	14.9	13.2	12.4	13.9	15.2	14.3	
PLASTIC - INPT	1.4	2.0	3.2	2.6	2.8	1.8	1.6	
THORACIC – INPT	9.8	10.0	10.9	9.0	8.9	8.1	9.1	
VASCULAR – INPT	24.2	26.9	30.3	26.5	19.5	15.2	17.2	
EPS	9.7	9.6	9.0	7.9	7.5	5.6	7.6	
ANGIO & CATHS – INPT	33.9	29.1	31.3	39.2	44.5	57.8	27.9	
TOTAL BEDS REQ'D.	170.0	170.0	170.0	169.9	170.0	169.9	147.8	1167.6
	170	170	170	170	170	170	148	166.8
								3.985 = ave los

TABLE 3.15
Actual Average Weekly Cases Performed

EXISTING AVERAGE	M	T	W	TH	F	S	SU	TOTAL 476 = Total Cs
CHS – INPT	6.1	5.6	6.2	5.6	4.7	0.5	0.3	30.0
CARDIOLOGY – INPT	0.2	0.3	0.8	0.4	0.2	0.1	0.0	2.0
ENT – INPT	0.9	0.9	0.7	0.7	0.8	0.0	0.0	4.0
ENT – OPT	1.2	1.0	0.7	1.3	1.8	0.0	0.0	6.0
AICD & PACERS – INPT	6.3	8.1	5.5	5.6	3.5	0.0	0.0	29.0
PACERS – OPT	0.5	0.9	0.6	1.0	0.0	0.0	0.0	3.0
EYE – OPT	2.1	1.4	4.2	4.3	0.0	0.0	0.0	12.0
GENERAL/GI – INPT	8.4	1.5	3.2	1.9	5.0	0.0	0.0	20.0
GENERAL/GI – OPT	6.4	5.6	10.5	4.5	9.3	0.7	0.0	37.0
GU – INPT	1.8	2.4	4.8	2.2	1.8	0.0	0.0	13.0
GU – OPT	1.2	0.0	0.0	2.2	1.6	0.0	0.0	5.0
NEUROSURGERY – INPT	0.8	0.0	0.2	0.0	0.0	0.0	0.0	1.0
NEUROSURGERY – OPT	0.3	0.0	0.0	0.7	0.0	0.0	0.0	1.0
ORTHOPEDICS – INPT	6.6	5.1	1.2	5.7	1.4	0.0	0.0	20.0
ORTHOPEDICS – OPT	2.1	1.6	3.5	1.6	3.7	0.3	0.2	13.0
PLASTIC – INPT	0.4	0.7	1.1	0.0	0.8	0.0	0.0	3.0
PLASTIC – OPT	0.2	0.4	0.5	0.6	0.2	0.1	0.0	2.0

PMT – OPT	5.1	8.5	9.1	5.3	7.0	0.0	0.0	35.0
THORACIC – INPT	1.3	1.1	1.9	2.8	1.4	0.4	0.1	9.0
THORACIC – OPT	0.0	0.0	0.0	0.0	0.0	0.0	0.0	0.0
VASCULAR – INPT	8.3	5.8	9.9	5.2	5.9	1.9	0.0	37.0
VASCULAR – OPT	0.0	0.0	1.0	0.0	1.0	0.0	0.0	2.0
EPS	5.6	1.7	6.1	2.3	1.3	0.0	0.0	17.0
ANGIO & CATHS – INPT	28.8	20.2	15.5	21.7	14.7	9.1	0.0	110.0
ANGIO & CATHS – OPT	15.3	7.5	12.3	14.7	15.2	0.0	0.0	65.0
TOTAL	109.9	81.3	99.5	90.3	81.3	13.1	0.6	476.0

TABLE 3.16

Daily OR Time Used for Actual Cases Performed

Daily OR Time Used	M	T	W	TH	F	S	SU
OHS – INPT	2,269	2,455	2,306	2,083	1,748	186	112
CARDIOLOGY – INPT	0	0	0	0	0	0	0
ENT – INPT	92	92	71	71	82	0	0
ENT – OPT	122	102	71	133	184	0	0
AICD & PACERS – INPT	750	964	655	666	417	0	0
PACERS – OPT	40	71	47	79	0	0	0
EYE – OPT	141	94	281	288	0	0	0
GENERAL/GI – INPT	1,176	210	448	266	700	0	0
GENERAL/GI – OPT	736	644	1,208	518	1,070	81	0
GU – INPT	243	324	648	297	243	0	0
GU – OPT	162	0	0	297	216	0	0
NEUROSURGERY – INPT	192	0	48	0	0	0	0
NEUROSURGERY – OPT	72	0	0	168	0	0	0
ORTHOPEDICS – INPT	726	561	132	627	154	0	0
ORTHOPEDICS – OPT	231	176	385	176	407	33	22
PLASTIC – INPT	62	109	171	0	124	0	0
PLASTIC – OPT	30	60	75	89	30	15	0
PMT – OPT	153	255	273	159	210	0	0
THORACIC – INPT	215	182	314	462	231	66	17

THORACIC – OPT	0	0	0	0	0	0	0
VASCULAR – INPT	1,461	1,021	1,742	915	1,038	334	0
VASCULAR – OPT	0	0	127	0	127	0	0
Total OR Minutes Used	8,871	7,318	9,002	7,295	6,980	715	150
OR Minutes Available	9,923	9,923	9,923	9,923	9,923	0	0
EPS Minutes Used	420	128	458	173	98	0	0
ANGIO & CATHS – INPT	2,160	1,515	1,163	1,628	1,103	683	0
ANGIO & CATHS – OPT	1,148	563	923	1,103	1,140	0	0
Total Cath Minutes Used	3,308	2,078	2,085	2,730	2,243	683	0
Cath Minutes Available	4,200	4,200	4,200	4,200	4,200	900	

TABLE 3.17
Weekly Volume versus Volume Constraints for Actual Cases Performed

Volume	M	T	W	TH	F	S	SU	TOTAL	Target <=
[illegible]HS – INPT	6.1	6.6	6.2	5.6	4.7	0.5	0.3	30.0	30.0
CARDIOLOGY – INPT	0.2	0.3	0.8	0.4	0.2	0.1	0.0	2.0	2.0
ENT – INPT	0.9	0.9	0.7	0.7	0.8	0.0	0.0	4.0	4.0
ENT – OPT	1.2	1.0	0.7	1.3	1.8	0.0	0.0	6.0	6.0
AICD & PACERS – INPT	6.3	8.1	5.5	5.6	3.5	0.0	0.0	29.0	29.0
PACERS – OPT	0.5	0.9	0.6	1.0	0.0	0.0	0.0	3.0	3.0
EYE – OPT	2.1	1.4	4.2	4.3	0.0	0.0	0.0	12.0	12.0
GENERAL/GI – INPT	8.4	1.5	3.2	1.9	5.0	0.0	0.0	20.0	20.0
GENERAL/GI – OPT	6.4	5.6	10.5	4.5	9.3	0.7	0.0	37.0	37.0
GU – INPT	1.8	2.4	4.8	2.2	1.8	0.0	0.0	13.0	13.0
GU – OPT	1.2	0.0	0.0	2.2	1.6	0.0	0.0	5.0	5.0
NEUROSURGERY – INPT	0.8	0.0	0.2	0.0	0.0	0.0	0.0	1.0	1.0
NEUROSURGERY – OPT	0.3	0.0	0.0	0.7	0.0	0.0	0.0	1.0	1.0
ORTHOPEDICS – INPT	6.6	5.1	1.2	5.7	1.4	0.0	0.0	20.0	20.0
ORTHOPEDICS – OPT	2.1	1.6	3.5	1.6	3.7	0.3	0.2	13.0	13.0
PLASTIC – INPT	0.4	0.7	1.1	0.0	0.8	0.0	0.0	3.0	3.0
PLASTIC – OPT	0.2	0.4	0.5	0.6	0.2	0.1	0.0	2.0	2.0
PEIT – OPT	5.1	8.5	9.1	5.3	7.0	0.0	0.0	35.0	35.0

THORACIC – INPT	1.3	1.1	1.9	2.8	1.4	0.4	0.1	9.0	9.0
THORACIC – OPT	0.0	0.0	0.0	0.0	0.0	0.0	0.0	0.0	0.0
VASCULAR – INPT	8.3	5.8	9.9	5.2	5.9	1.9	0.0	37.0	37.0
VASCULAR – OPT	0.0	0.0	1.0	0.0	1.0	0.0	0.0	2.0	2.0
EPS	5.6	1.7	6.1	2.3	1.3	0.0	0.0	17.0	17.0
ANGIO & CATHS – INPT	28.8	20.2	15.5	21.7	14.7	9.1	0.0	110.0	110.0
ANGIO & CATHS – OPT	15.3	7.5	12.3	14.7	15.2	0.0	0.0	65.0	65.0
TOTAL CASES/WEEK	109.9	81.3	99.5	90.3	81.3	13.1	0.6	476.0	476.0

TABLE 3.18
Beds Required by Day of Week for Actual Cases Performed

By Category	M	T	W	TH	F	S	SU	TOTAL
OHS – INPT	37.7	38.7	39.4	40.1	38.6	35.7	36.1	266.2
CARDIOLOGY – INPT	0.7	1.2	1.5	1.2	1.0	0.7	0.6	6.9
ENT – INPT	1.6	1.6	1.5	1.6	1.5	0.7	0.9	9.4
AICD & PACERS – INPT	15.0	17.2	16.7	14.8	9.7	5.3	8.3	87.0
GENERAL/GI – INPT	15.0	13.5	13.0	13.8	13.1	9.7	12.4	90.6
GU – INPT	4.6	6.2	8.6	7.9	6.7	4.6	3.6	42.3
NEUROSURGERY – INPT	1.0	0.9	0.7	0.6	0.5	0.4	0.8	4.8
ORTHOPEDICS – INPT	13.9	15.7	16.6	17.0	14.1	11.6	10.1	99.1
PLASTIC – INPT	1.9	2.5	2.8	2.3	2.3	1.7	1.7	15.2
THORACIC – INPT	8.8	9.0	10.0	10.6	9.9	9.0	8.6	65.9
VASCULAR – INPT	22.1	24.3	26.9	25.6	23.8	18.9	18.2	159.9
EPS	8.7	10.9	11.6	8.8	6.1	3.7	7.2	56.8
ANGIO & CATHS – INPT	47.5	42.2	41.9	44.2	36.0	25.1	26.9	263.7
TOTAL BEDS REQ'D.	178.4	183.9	191.2	188.3	163.3	127.1	135.3	1167.6
	179	184	192	189	164	128	136	166.8
								3.985 = alos

SECTION 4: THE ACTUAL AVERAGE WEEKLY OPERATING ROOM PERFORMANCE VERSUS OPTIMUM SCHEDULES

This section compares the actual use of the OR and other procedural areas to their anticipated use with the Monday–Friday and Monday–Saturday optimum elective patient schedules. The more pertinent statistics are presented in a comparative table. Bed use is presented in a graphical format, as well, to emphasize the difference in beds required each day by the three scheduling scenarios presented.

A. Table 3.19 – Comparative Data: This table presents statistical results that are affected by the introduction of scientific scheduling.
   a. Number of Cases: Although the weekly number of cases is the same in all three scheduling scenarios, simply using time in the OR as the only operational constraint, as most hospitals presently do, results in a mismatch between the allocation of a fixed number of beds per day (the supply) and the actual daily census (the demand for beds). The existing system requires a maximum of 192 beds. The Monday–Friday optimum schedule reduces the need for beds to 177, which is further reduced to 170 with the Monday–Saturday optimum schedule.
   b. Total Beds Used: This set of data clearly shows the primary benefit of scheduling scientifically, as the Monday–Friday schedule provides a balanced need for beds five days a week. On the other hand, the Monday–Saturday schedule indicates a balanced need for patient beds six days each week. Compare that with the existing need for many more beds Tuesdays–Thursdays and leaving many more beds vacant Friday, Saturday, and Sunday nights. Not only does this existing misalignment of beds wreak havoc with patient flow; it also does the same with staffing.
   c. Maximum Beds Required: Again, compare the maximum beds per week that are used now for the elective procedural patients of 192 versus the need for a maximum of 177 beds with a five-day-per-week optimum schedule and the need for a maximum of 170 beds with a six-day-per-week optimum schedule.
   d. OR Minutes Used: These statistics indicate the potential for reducing the number of staffed ORs during the week no matter the scheduling scenario, a fact that most ORs take advantage of. The OR time constraint should have been set at a much lower value.
   e. Catheterization Lab Minutes Used: See comments for ORs.
   f. EPS Lab Minutes Used: See comment for ORs.

B. Figure 3.1: As the following graph clearly indicates, there is significant opportunity to reduce the allocation of beds for elective surgical patients by using historical data and scientific means to perform the scheduling.
   g. The optimum six-day-per-week schedule requires an allocation of 170 beds per week, sufficient for 1,190 (170 × 7) patient days. Rounding up the patient days that would be used at 1,168 results in a weekly bed utilization rate of 98.2%.

TABLE 3.19
Comparative Scheduling Data for the Three Scheduling Options with 486 Cases Per Week

THREE SCHEDULING PLANS FOR PERFORMING 486 CASES PER WEEK IN THE OPERATING ROOM, CATHETER AND EPS LABS

		M	T	W	TH	F	SA	SU	TOTALS
NO. OF CASES	ACTUAL USE	109.9	81.3	99.5	90.3	81.3	13.1	0.6	476
	OPTIMUM M–F SCHED	96	77	108	100	83	12	-	476
	OPTIMUM M–Sa SCHED	84	47	49	91	103	102	-	476
TOTAL BEDS USED	ACTUAL USE	178.4	183.9	191.2	188.3	163.3	127.1	135.3	1,167.5
	OPTIMUM M–F SCHED	177.0	177.0	177.0	177.0	176.8	137.6	145.2	1,167.6
	OPTIMUM M–Sa SCHED	170.0	170.0	170.0	169.9	170.0	169.9	147.8	1,167.6
MAX. BEDS REQ'D.	ACTUAL USE	192	192	192	192	192	192	192	1,344
	OPTIMUM M–F SCHED	177	177	177	177	177	177	177	1,239
	OPTIMUM M–Sa SCHED	170	170	170	170	170	170	170	1,190
OR MINS USED	ACTUAL USE	8,871	7,318	9,002	7,295	6,980	715	150	40,331
	OPTIMUM M–F SCHED	7,844	3,967	8,693	9,919	9,908	-	-	40,331
	OPTIMUM M–Sa SCHED	8,672	4,907	7,391	4,853	8,974	5,534	-	40,331
CATH MINS USED	ACTUAL USE	3,308	2,078	2,085	2,730	2,243	683	-	13,127
	OPTIMUM M–F SCHED	4,200	3,600	1,875	-	2,550	900	-	13,125
	OPTIMUM M–Sa SCHED	2,100	900	525	3,675	2,550	3,375	-	13,125
EPS MINS USED	ACTUAL USE	420	128	458	173	98	-	-	1,275
	OPTIMUM M–F SCHED	450	150	150	150	375	-	-	1,275
	OPTIMUM M–Sa SCHED	375	225	225	150	150	150	-	1,275

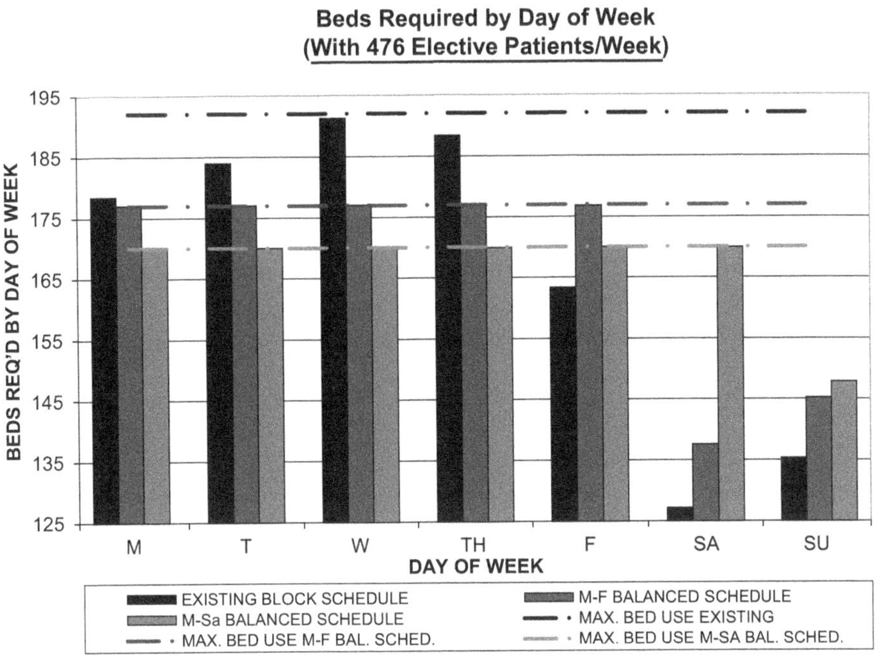

FIGURE 3.1 Beds Required by Day of Week for the Three Scheduling Options.

h. The optimum five-day-per-week schedule requires an allocation of 177 beds per week, sufficient for 1,239 (177 × 7) patient days. Rounding up the patient days that would be used at 1,168 results in a weekly bed utilization rate of 94.3%.

i. The existing schedule requires an allocation of 192 beds per week, sufficient for 1,344 (192 × 7) patient days. Rounding up the patient days that would be used at 1,168 results in a weekly bed utilization rate of 86.9%.

SUMMARY

To avoid the many problems associated with patient overcrowding in hospitals, it is imperative to recognize the role that the surgical elective schedule plays in their creation. The elective schedule should comply with the three primary operational constraints each day. It should not exceed the time available in the OR suite. It should not schedule more or fewer cases each week than the existing average weekly volumes. And it should not result in more (or less) patients in-house than the number of beds available for those patients on any given day.

As the results indicate, the existing block schedule results in an average of 191.2 patients on Wednesdays and 188.3 patients on Thursdays. These are the days the hospital is likely to be overwhelmed with patients and problems associated with the overcrowding. The five-day-per-week (Monday–Friday) schedule provides the required balance with its 177 beds, as there will be a bed for each expected patient on each of those five days and more than enough beds on weekends, resulting in an increase in weekly bed utilization from 86.9% to 94.3%. The six-day-per-week schedule (Monday–Saturday) provides an even more effective schedule in that it reduces the number of elective surgical patients in the hospital each of those six days to 170 and has more than a sufficient number of beds for patients on Sundays, resulting in an increase in weekly bed utilization to 98.2%.

To return to unimpeded patient flow to, through, and out of the OR, it is incumbent upon hospitals to derive patient schedules on the basis of data and mathematics rather than on the unsuitable basis of assigning block time by seniority or some other inappropriate means.

Before ending this chapter, it is worth noting the following:

1. Many years ago, hospitals had minimum patient flow issues. Back then, patients arrived at the hospital one day prior to surgery for testing. If a bed was not available for the arriving patient, the case was canceled, to the consternation of the surgeon, the patient, the patient's family, the OR, and the hospital's finance department. Hospitals, back then, simply didn't allow elective patients to be admitted if beds were not available, and as a result, they controlled patient flow. However, that practice changed when the government mandated that hospitals would no longer be reimbursed on a *per diem* basis. The change to a fixed reimbursement by case (by diagnosis-related group) resulted in patients being scheduled for tests early in the morning on their day of surgery (to save one day on each patient's eventual LOS) and by discontinuing the practice of canceling surgery because of the unavailability of beds.
2. Hopefully, the problems hospitals are now having with scheduling elective surgical patients given the necessary priority of COVID-19 patients will simply be a painful memory by the time this book is published. However, if there becomes a mandated need for hospitals to reduce the number of beds allocated to elective surgery, the optimum schedules discussed in this chapter will enable hospitals to comply with the mandate in a very effective fashion.

4 Reducing the Deterioration Rate of Inventory through Preservation Technology Investment under Fuzzy and Cloud Fuzzy Environment

Nita H. Shah and Milan B. Patel
Gujarat University
Ahmedabad, Gujarat, India

CONTENTS

4.1 INTRODUCTION

Inventory management is considered to be one of the most interesting and significant areas of operations research. Management of an inventory system becomes even more important when the inventory suffers deterioration. Deterioration is a process in which the items/raw material/goods become worse in the condition, which results in decreasing the usefulness of the original product. Acknowledging the fact that deterioration can have an adverse effect on the total profits of the firm, one cannot overlook or neglect the effect of deterioration while formulating the inventory model.

The first economic ordered quantity (EOQ) model developed by Harris (1913) is missing the effect of deterioration. Whitin (1957) made the first attempt to formulate an inventory model by introducing the impact of deterioration. Ghare and Schrader (1963) introduced a mathematical model for exponentially decaying inventories by assuming the constant deterioration rate. Nahmias (1982) conducted a brief review on perishable inventory having both deterministic and stochastic demand. A comprehensive survey on continuously deteriorating inventory has been done by Raafat (1991), Shah and Shah (2000) and Goyal and Giri (2001).

How to control or reduce the rate of deterioration or decrease the effect of deterioration has been the most fundamental and crucial question for the decision makers dealing with deteriorating items. To handle the effects of deterioration, industries apply various techniques, such as preservations or by providing suitable environmental conditions to the product. The aforementioned papers and reviews examined the deterioration rate as a constant or a random variable. However, recognizing the fact that deterioration can be controlled by applying preservation technology (PT), one must consolidate preservation technology investment while formulating the inventory model. Lee (2005) developed a model to decide on supporting investment strategies regarding inventory and preventive maintenance in an imperfect production system. Dye and Hsieh (2012) formulated a model with time varying rates of deterioration and partial backlogging with PT investment and replenishment schedule as decision variables. Tayal et al. (2014) examined two echelon supply chain models for deteriorating items under the effect of PT investment. Shah et al. (2016) investigated an inventory model for deteriorating items with PT with a trapezoidal demand rate. Mashud et al. (2020) investigated problems related to non-instantaneous deteriorating items under the effect of trade credit, PT and advertisement policy. Recently, Chaudhari et al. (2020) formulated the inventory model for deteriorating items and PT with an advance payment scheme under quadratic demand.

It is notable that there is always some ambiguity in human analysis and decision making. Moreover, reality cannot be calculated or assumed with precision except up to some partial extent. Therefore, to get a closer approach to reality, one must incorporate fuzzy quantities while formulating the inventory model. By introducing the concept of membership function, Zadeh (1965) invented fuzzy set theory. Bellman and Zadeh (1970) worked on decision making in a fuzzy environment with the help of dynamic programming. Yao et al. (2000) formulated an inventory model without back order by considering demand and order quantities as triangular fuzzy numbers (TFN). Shekarian et al. (2017) conducted a systematic and comprehensive review in the field of fuzzy inventory management. Raula et al. (2018) formulated an inventory model with stock-dependent demand by considering demand, holding cost and deterioration rates as pentagonal fuzzy numbers. They used the signed distance method for defuzzification.

In any enterprise or industry, it is observed that as the manager gains experience, uncertainty gets reduce with time. To justify this phenomenon, a new type of fuzzy number – namely, the cloud fuzzy number – is being practised in this chapter. By considering the learning effect of the practitioner, Maity et al. (2019) developed an EOQ model under a lock fuzzy environment.

De and Mahata (2017) introduced the cloudy fuzzy number and used it to solve the back order EOQ model. Again De and Mahata (2019) have used it to formulate the EOQ model for imperfect quality items. Karmakar et al. (2018) developed an EOQ model under a cloudy fuzzy demand rate. Maiti (2019) worked on an economic production lot size model with a cloudy fuzzy demand rate in which the production rate is dependent on demand.

Since the estimation of demand is the most challenging task which depends on many uncertain events, the researchers have taken demand as TFN to formulate the fuzzy model and a cloud triangular fuzzy number (CTFN) to formulate the cloud fuzzy model. In this chapter, Yager's (1981) method for ranking fuzzy numbers is used for defuzzification. A numerical example, along with a sensitivity analysis, is carried out under different environments. To justify the newly introduced concept (i.e. cloud fuzzy environment), the calculation of the degree of fuzziness and the cloud index is also presented. In addition, the optimal results under various environments are compared.

4.2 NOTATIONS AND ASSUMPTIONS

The proposed problem is developed with the following notations and assumptions.

4.2.1 Notations

(i) h: holding cost of inventory per unit (\$)
(ii) s: sales price per unit (\$)
(iii) s_b: back-ordered sales price (\$)
(iv) c: purchase cost per unit (\$)
(v) A: ordering cost per unit (\$)
(vi) r: lost sales penalty cost per unit (\$)
(vii) R: rate of demand per year
(viii) R_f: fuzzy rate of demand per year
(ix) R_c: cloud fuzzy rate of demand per year
(x) T: total cycle time (year)
(xi) t_1: critical time/shortage time at which inventory levels become zero (year); $0 < t_1 \leq T$
(xii) P: PT investment cost (\$)
(xiii) θ: rate of deterioration; $0 \leq \theta < 1$
(xiv) $\theta_1(P)$: reduced rate of deterioration because of PT; $0 \leq \theta_1(P) < 1$

(xv) $\sigma(v)$: probability that a customer will wait for 'v' time to purchase the item
(xvi) $I(t)$: inventory level during time $[0, t]$
(xvii) Q_1: quantity without back order
(xviii) Q_2: back-ordered quantity
(xix) Pr: total profit per unit time ($)

4.2.2 Assumptions

i) Rate of demand is constant for the crisp and fuzzy model
ii) During a given cycle time, no replacement or repairing of deteriorating items occur
iii) Cycle time is greater than critical time
iv) Probability of customer's back ordered is $\sigma(v) = 1 - \frac{v}{T}$, where v is the waiting time for customer to get the order.
v) Sales price is assumed to be constant and back-ordered sales prices are linearly dependent on sales prices such that $s_b = \mu s$, $\mu \in (0, 1)$
vi) Holding cost, purchase cost and ordering cost are assumed to be constant during the planning horizon

4.3 PRELIMINARY CONCEPTS

4.4 FUZZY SET

A fuzzy set $\bar{X}$ is a set of well-defined ordered pairs over a universal set U defined as

$$\bar{X} = \left\{\left(x, \phi(x)\right) / x \in U\right\},$$

where $\phi(x)$ is a function ϕ: $U \to [0, 1]$ $\forall x \in U$ known as the membership function.

4.5 TFN

A fuzzy number expressed as $\bar{R} = (R_1, R_2, R_3)$ is known as TFN if the value of its membership function is strictly increasing in $[R_1, R_2]$ and strictly decreasing in $[R_2, R_3]$ and equal to 1 on R_2. The membership function of TFN is given by

$$\phi(\bar{R}) = \begin{cases} \frac{x - R_1}{R_2 - R_1}, R_1 \le x \le R_2 \\ \frac{x - R_3}{R_2 - R_3}, R_2 \le x \le R_3 \\ 0, \ otherwise \end{cases} \tag{4.1}$$

4.6 α– CUT

The α– cut is a crisp set consisting of those elements of the universal set U whose membership value is greater than or equal to α.

$$i.e.\ \alpha - cut = \{x \in U / \phi(x) \geq \alpha\}, \alpha \in [0,1] \tag{4.2}$$

4.7 LEFT AND RIGHT α– CUT OF TFN

Left α– cut is the smallest value of α– cut expressed as

$$L_\alpha = R_1 - \alpha(R_1 - R_2) \tag{4.3}$$

Right α– cut is the highest value of α– cut expressed as

$$R_\alpha = R_3 - \alpha(R_3 - R_2) \tag{4.4}$$

Left α– cut and right α– cut are shown graphically in Figure 4.1.

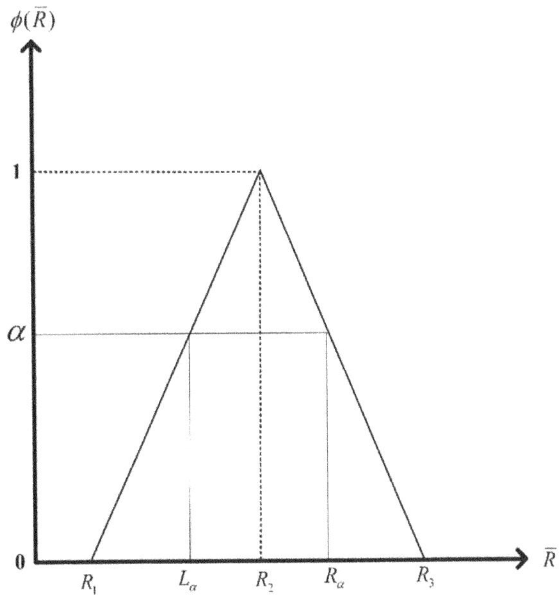

FIGURE 4.1 Left and Right α– Cut of TFN.

4.8 CTFN

A TFN is known as CTFN if the set converges to a crisp number as time tends towards infinity.

$$\tilde{R} = \left(R_2\left(1 - \frac{\beta}{1+t}\right), R_2, R_2\left(1 + \frac{\gamma}{1+t}\right) \right) \tag{4.5}$$

where $\beta, \gamma \in (0,1)$ and $t > 0$. From equation (4.5), it can be seen that as $t \to \infty, \tilde{R} \to \{R_2\}$.

The membership function of CTFN can be defined as follow:

$$\phi(\tilde{R}, t) = \begin{cases} \dfrac{x - R_2\left(1 - \frac{\beta}{1+t}\right)}{\frac{\beta R_2}{1+t}}, R_2\left(1 - \frac{\beta}{1+t}\right) \le x \le R_2 \\ \dfrac{R_2\left(1 + \frac{\gamma}{1+t}\right) - x}{\frac{\gamma R_2}{1+t}}, R_2 \le x \le R_2\left(1 + \frac{\gamma}{1+t}\right) \\ 0 \quad , otherwise \end{cases} \tag{4.6}$$

It can be graphically represented as show in Figure 4.2.

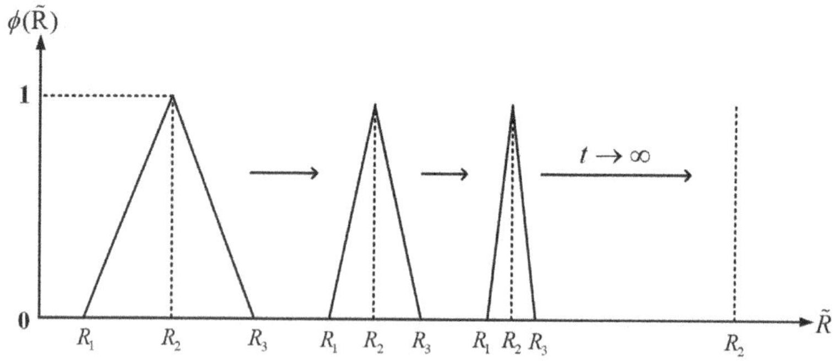

FIGURE 4.2 Membership Function of CTFN.

4.9 LEFT AND RIGHT α– CUT OF CTFN

As per the definition of left and right α– cut of TFN (see equations (4.2) and (4.3)), left and right α– cut of CTFN can be expressed as

$$L_{\alpha,t} = R_2\left(1-\frac{\beta}{1+t}\right)+\frac{\alpha\beta}{1+t}R_2 \quad \& \quad R_{\alpha,t} = R_2\left(1+\frac{\gamma}{1+t}\right)-\frac{\alpha\gamma}{1+t}R_2 \tag{4.7}$$

respectively.

4.10 YAGER'S RANKING INDEX METHOD (1981)

According to Yager's Ranking Index Method, defuzzification for a TFN can be given by

$$YRI\left(\bar{R}\right)=\frac{1}{2}\int_0^1\left(L_\alpha + R_\alpha\right)d\alpha \tag{4.8}$$

where L_α and R_α are left and right α– cut of TFN, respectively.

Using equations (4.3) and (4.4), equation (4.8) reduces to

$$YRI\left(\bar{R}\right)=\frac{1}{4}\left(R_1+2R_2+R_3\right) \tag{4.9}$$

4.11 EXTENSION OF YAGER'S RANKING INDEX METHOD

By applying the extension of Yager's Ranking Index Method given by De and Mahata (2017), defuzzification of CTFN can be given by

$$YRI\left(\tilde{R}\right)=\frac{1}{2T}\int_{\alpha=0}^{\alpha=1}\int_{t=0}^{t=T}\left(L_{\alpha,t}+R_{\alpha,t}\right)d\alpha\, dt \tag{4.10}$$

Substituting the value of the left and right α– cut of CTFN from equations (4.7) and (4.10) reduces to

$$YRI\left(\tilde{R}\right)=R_2\left(1-\frac{(\beta-\gamma)}{4}\frac{\log(1+T)}{T}\right) \tag{4.11}$$

4.12 FORMULATION OF THE CRISP MATHEMATICAL MODEL

To formulate the crisp model, the inventory with deteriorating items with a constant deterioration rate θ is considered. After investing \$$P$ as the PT investment, the reduced rate of deterioration is taken as

$$\theta_1(P) = \theta\left(1 - e^{-\lambda P}\right), \lambda \geq 0 \tag{4.12}$$

which is a function of P bounded above by the initial deterioration rate θ.

Assume that initially we have ordered Q quantity out of which Q_2 is back ordered. The inventory level decreases with time and becomes zero at the critical time (t_1).

The rate of change in the inventory level can be explained by

$$\frac{dI}{dt} = -R - \left[\theta - \theta_1(P)\right] I(t) \tag{4.13}$$

with the initial condition $I(t_1) = 0$, the solution of equation (4.13) is given by

$$I(t) = \frac{R}{\left[\theta\left(1 - e^{-\lambda P}\right) - \theta\right]}\left(1 - e^{(t - t_1)\left[\theta\left(1 - e^{-\lambda P}\right) - \theta\right]}\right) \tag{4.14}$$

The total profit function per cycle consists of the following costs:

1) Inventory Holding Cost: $IHC = h\int_0^{t_1} I(t)\,dt$

$$= \frac{hR}{\left[\theta\left(1 - e^{-\lambda P}\right) - \theta\right]}\left(t_1 - \frac{1}{\theta\left(1 - e^{-\lambda P}\right) - \theta} + \frac{e^{(-t_1)\left[\theta\left(1 - e^{-\lambda P}\right) - \theta\right]}}{\theta\left(1 - e^{-\lambda P}\right) - \theta}\right)$$

2) Lost Sale Cost: $LSC = R\left(\frac{T}{2} + \frac{t_1^2}{2T} - t_1\right) r$

3) Sales Revenue: $SR = sQ_1 + \mu s Q_2 = sRt_1 + \frac{R}{2T}\left(T^2 - t_1^2\right)\mu s$

4) Purchase Cost: $PC = \left\{\frac{R}{\left[\theta\left(1 - e^{-\lambda P}\right) - \theta\right]}\left(1 - e^{(-t_1)\left[\theta\left(1 - e^{-\lambda P}\right) - \theta\right]}\right) + \frac{R}{2T}\left(T^2 - t_1^2\right)\right\} c$

5) Ordering Cost: $OC = A$

6) PT Investment: $PTI = P$

By using the previous expressions, total profit per unit time is given by

$$\Pr(T,t_1,P) = \frac{1}{T}\left[SR-(IHC+PC+LSC+OC+PTI)\right]$$

$$= \frac{1}{T}\left\{\begin{array}{l} \dfrac{R(T^2-t_1^2)\mu s}{2T}+Rt_1 s-c\left(\dfrac{R\left(1-e^{\left(\theta-\theta\left(1-e^{-\lambda P}\right)\right)t_1}\right)}{\theta\left(1-e^{-\lambda P}\right)-\theta}+\dfrac{R(T^2-t_1^2)}{2T}\right) \\ -rR\left(\dfrac{T}{2}+\dfrac{t_1^2}{2T}-t_1\right)-\dfrac{hR\left(t_1-\dfrac{1}{\theta\left(1-e^{-\lambda P}\right)-\theta}+\dfrac{e^{\left(\theta-\theta\left(1-e^{-\lambda P}\right)\right)t_1}}{\theta\left(1-e^{-\lambda P}\right)-\theta}\right)}{\theta\left(1-e^{-\lambda P}\right)-\theta}-A-P \end{array}\right\} \tag{4.15}$$

Total profit per unit time is a continuous function of three variables – namely, cycle time (T), critical time (t_1) and PT investment (P).

4.13 SOLUTION PROCEDURE FOR CRISP MODEL

To maximize total profit, differentiate the total profit function given in equation (4.15) partially with respect to T, t_1 and P and equate them to zero.

$$\text{i.e.}\quad \frac{\partial \Pr(T,t_1,P)}{\partial T}=0,\quad \frac{\partial \Pr(T,t_1,P)}{\partial t_1}=0,\quad \frac{\partial \Pr(T,t_1,P)}{\partial P}=0 \tag{4.16}$$

With assigned values of inventory parameters, equation (4.16) gives optimal solutions of cycle time (T), critical time (t_1) and PT investment (P).

4.14 FORMULATION OF FUZZY MATHEMATICAL MODEL

In the real-life scenario, it is observed that the demand rate cannot be predicted precisely. Hence, to formulate the fuzzy model, the demand rate is considered to be a TFN.

Consider $R_f = (R_1, R_2, R_3)$ as a TFN for the fuzzy demand rate. Hence the total profit function reduces to

$$\Pr_f(T,t_1,P)=\frac{1}{T}\left\{\begin{array}{l} \dfrac{R_f(T^2-t_1^2)\mu s}{2T}+R_f t_1 s-c\left(\dfrac{R_f\left(1-e^{\left(\theta-\theta\left(1-e^{-\lambda P}\right)\right)t_1}\right)}{\theta\left(1-e^{-\lambda P}\right)-\theta}+\dfrac{R_f(T^2-t_1^2)}{2T}\right) \\ -rR_f\left(\dfrac{T}{2}+\dfrac{t_1^2}{2T}-t_1\right)-\dfrac{hR_f\left(t_1-\dfrac{1}{\theta\left(1-e^{-\lambda P}\right)-\theta}+\dfrac{e^{\left(\theta-\theta\left(1-e^{-\lambda P}\right)\right)t_1}}{\theta\left(1-e^{-\lambda P}\right)-\theta}\right)}{\theta\left(1-e^{-\lambda P}\right)-\theta} \\ -A-P \end{array}\right. \tag{4.17}$$

The membership function for total fuzzy profit can be written as

$$\phi_1(\mathrm{Pr}) = \begin{cases} \dfrac{\mathrm{Pr}-\mathrm{Pr}_1}{\mathrm{Pr}_2-\mathrm{Pr}_1}, \mathrm{Pr}_1 \le \mathrm{Pr} \le \mathrm{Pr}_2 \\ \dfrac{\mathrm{Pr}-\mathrm{Pr}_3}{\mathrm{Pr}_2-\mathrm{Pr}_3}, \mathrm{Pr}_2 \le \mathrm{Pr} \le \mathrm{Pr}_3 \\ 0, \quad \textit{otherwise} \end{cases} \tag{4.18}$$

where Pr_i, $i = 1, 2, 3$ can be obtained by replacing R_f with R_i, $i = 1, 2, 3$ in equation (4.17). By using the defuzzification method explained in Section 4.10, the defuzzified value of fuzzy objective function is obtained and expressed as

$$I(\mathrm{Pr}_f) = \left(\begin{array}{l} \dfrac{1}{4T}(R_1+2R_2+R_3)\left\{ \begin{array}{l} \dfrac{(T^2-t_1^2)\mu s}{2T} + t_1 s - c\left(\dfrac{\left(1-e^{\left(\theta-\theta\left(1-e^{-\lambda P}\right)\right)t_1}\right)}{\theta\left(1-e^{-\lambda P}\right)-\theta} + \dfrac{(T^2-t_1^2)}{2T} \right) \\ -r\left(\dfrac{T}{2}+\dfrac{t_1^2}{2T}-t_1\right) - \dfrac{h\left(t_1 - \dfrac{1}{\theta\left(1-e^{-\lambda P}\right)-\theta} + \dfrac{e^{\left(\theta-\theta\left(1-e^{-\lambda P}\right)\right)t_1}}{\theta\left(1-e^{-\lambda P}\right)-\theta}\right)}{\theta\left(1-e^{-\lambda P}\right)-\theta} \end{array} \right\} \\ -\dfrac{1}{T}(A+P) \end{array} \right) \tag{4.19}$$

4.15 FORMULATION OF CLOUD FUZZY MATHEMATICAL MODEL

To formulate the cloud fuzzy model, consider the demand rate as a CTFN.

$$R_c = \left(R_2\left(1-\frac{\beta}{1+t}\right), R_2, R_2\left(1+\frac{\gamma}{1+t}\right) \right),$$

where $\beta, \gamma \in (0,1)$, $t > 0$

The total cloud fuzzy profit function can be expressed as

$$\mathrm{Pr}_c(T, t_1, P) = \frac{1}{T}\left\{ \begin{array}{l} \dfrac{R_c(T^2-t_1^2)\mu s}{2T} + R_c t_1 s - c\left(\dfrac{R_c\left(1-e^{\left(\theta-\theta\left(1-e^{-\lambda P}\right)\right)t_1}\right)}{\theta\left(1-e^{-\lambda P}\right)-\theta} + \dfrac{R_c(T^2-t_1^2)}{2T} \right) \\ -rR_c\left(\dfrac{T}{2}+\dfrac{t_1^2}{2T}-t_1\right) - \dfrac{hR_c\left(t_1 - \dfrac{1}{\theta\left(1-e^{-\lambda P}\right)-\theta} + \dfrac{e^{\left(\theta-\theta\left(1-e^{-\lambda P}\right)\right)t_1}}{\theta\left(1-e^{-\lambda P}\right)-\theta}\right)}{\theta\left(1-e^{-\lambda P}\right)-\theta} \\ -A-P \end{array} \right\} \tag{4.20}$$

The membership function for the total cloud fuzzy profit can be given by

$$\phi_2(\mathrm{Pr},t)=\begin{cases}\dfrac{\mathrm{Pr}-\mathrm{Pr}_1}{\mathrm{Pr}_2-\mathrm{Pr}_1}, \mathrm{Pr}_1\le \mathrm{Pr}\le \mathrm{Pr}_2\\ \dfrac{\mathrm{Pr}-\mathrm{Pr}_3}{\mathrm{Pr}_2-\mathrm{Pr}_3}, \mathrm{Pr}_2\le \mathrm{Pr}\le \mathrm{Pr}_3\\ 0, \quad otherwise\end{cases}, \tag{4.21}$$

where Pr_1 can be obtained by replacing R_c with $R_2\left(1-\frac{\beta}{1+t}\right)$, Pr_2 by replacing R_c with R_2 and Pr_3 by replacing R_c with $R_2\left(1+\frac{\gamma}{1+t}\right)$ in equation (4.20). By using the defuzzification method explained in Section 4.11, the defuzzified value of the total cloud fuzzy profit function can be derived.

4.16 NUMERICAL EXAMPLE AND PROOF OF CONCAVITY

For the crisp mode:

$$R=100, \mu=0.08, \theta=0.02, h=0.05, \lambda=0.01, s=30, c=10, A=120, r=35$$

For the fuzzy model:

$R_1=60, R_2=100, R_3=150$ and keep other parameters, as in the crisp model.

For the cloud fuzzy model:

$\beta=0.12,\ \gamma=0.15$ and keep other parameters, as in the crisp and fuzzy model.

By using methods explained in Section 4.13, optimal values of cycle time (T), critical time (t_1) and PT investment (P) are obtained under different environments and shown in Table 4.1. Values of decision variables under different environments over several

TABLE 4.1
Comparison of Optimal Solutions under Different Environments

Environment	PT Investment (P) (in $)	Cycle Time (T) (in year)	Critical Time (t_1) (in year)	Quantity without Back Order (Q_1)	Back Ordered Quantity (Q_2)	Total Profit (Pr) (in $)
Crisp	234.24	12.95	10.05	1,005.19	258.09	2,105.86
Fuzzy	234.19	12.79	9.92	1,016.80	261.16	2,159.20
Cloud Fuzzy	200.85	10.95	8.49	851.20	217.87	8,442.10

TABLE 4.2
Value of Decision Variables for Different Environments over Several Cycle Time

Environment	Crisp			Fuzzy			Cloud Fuzzy		
T (year)	P ($)	t_1 (year)	Pr ($)	P ($)	t_1 (year)	Pr ($)	P ($)	t_1 (year)	Pr ($)
10.91	199.95	8.46	2,105.5434	202.40	8.47	2,158.9179	200.18	8.46	8,442.1034
10.92	200.13	8.47	2,105.5466	202.59	8.47	2,158.9210	200.36	8.47	8,442.1036
10.93	200.32	8.48	2,105.5499	202.77	8.48	2,158.9241	200.55	8.48	8,442.1036
10.94	200.50	8.49	2,105.5532	202.95	8.49	2,158.9272	200.73	8.49	8,442.1037
10.95	200.68	8.50	2,105.5564	203.13	8.50	2,158.9302	200.91	8.50	*8,442.1039*
10.96	200.86	8.50	2,105.5596	203.32	8.50	2,158.9333	201.09	8.50	8,442.1038
10.97	201.04	8.51	2,105.5628	203.50	8.51	2,158.9363	201.27	8.51	8,442.1037
10.98	201.22	8.52	2,105.5660	203.68	8.52	2,158.9393	201.46	8.52	8,442.1034
10.99	201.41	8.53	2,105.5692	203.86	8.53	2,158.9423	201.64	8.53	8,442.1034
11.00	201.59	8.53	2,105.5723	204.04	8.53	2,158.9452	201.82	8.53	8,442.1030

N.B.: Italic font represents optimal solution in cloud fuzzy environment.

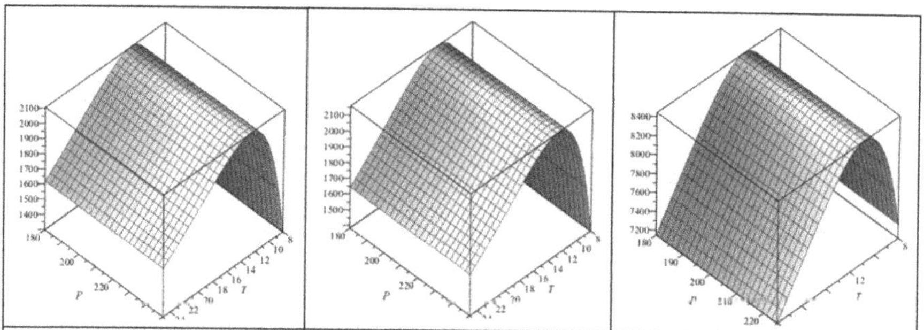

Crisp Environment | Fuzzy Environment | Cloud Fuzzy Environment

FIGURE 4.3 Proof of Concavity of Profit Function under Different Environments.

cycle time are shown in Table 4.2. Figure 4.3 show the concavity of objective functions under different environments.

The degree of fuzziness under a fuzzy environment can be calculated using the formula $D_f = \dfrac{U-L}{2M}$, where, U and L are upper bound and lower bound of TFN, respectively, and $M = 3(median) - 2(mean)$ is the mode of the TFN. The degree of fuzziness under the cloud fuzzy environment known as the cloud index can be obtained using the formula $I_c = \frac{\log(1+T)}{T}$. We have a fuzzy demand rate $R_f = (60, 100, 150)$. Therefore, upper bound is $U = 150$ and lower bound is $L = 60$. The median of R_f is 100, and the mean is 103.33; therefore, mode $M = 93.34$ and degree of fuzziness $D_f = 0.48$. Since the optimal cycle time under the cloud fuzzy environment is $T = 10.95$ and cloud index $I_c = 0.10$, which proves that ambiguity associated with the cloud fuzzy environment is less than in the fuzzy environment.

4.17 SENSITIVITY ANALYSIS

To identify the most sensitive parameters under different environments, sensitivity analysis is carried out and shown graphically in Figure 4.4, Figure 4.5 and Figure 4.6.

Since the uncertainty prevails in a smaller range, sensitivity analysis is being carried out in the range of –20% to 20%. It is observed that sales prices and demand rates are the most sensitive parameters in all environments. Purchase cost is a less sensitive parameter as compare to demand rate and sales price. With the incremental increase in purchase cost, the final profit decreases.

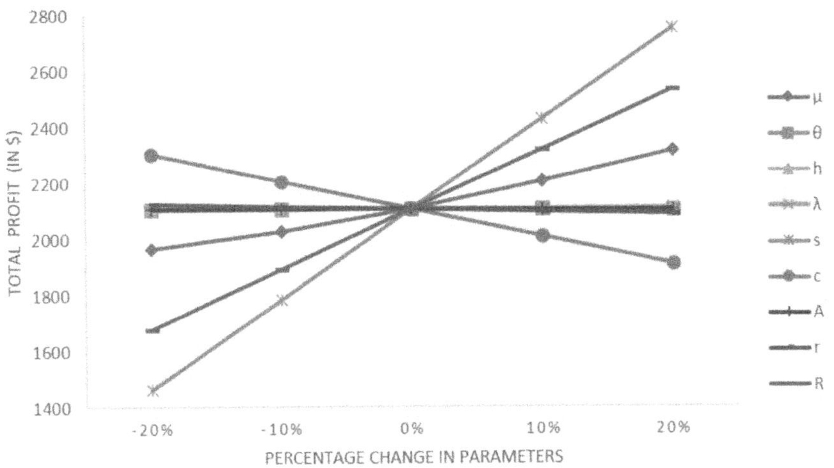

FIGURE 4.4 Sensitivity Analysis under Crisp Environment.

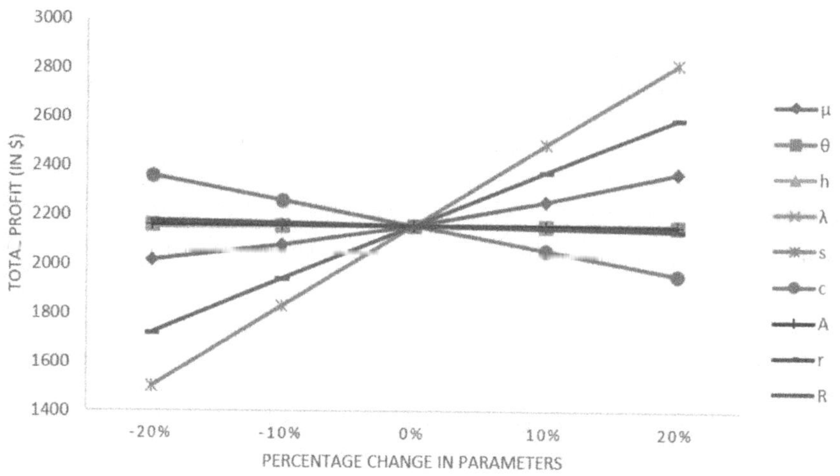

FIGURE 4.5 Sensitivity Analysis under Fuzzy Environment.

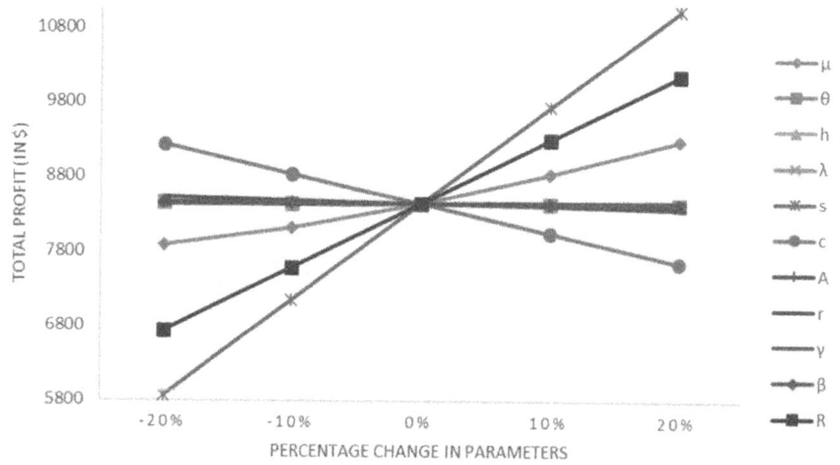

FIGURE 4.6 Sensitivity Analysis under Cloud Fuzzy Environment.

4.18 CONCLUSION

In this chapter, we have studied an inventory model of deteriorating items with the effect of PT investment under crisp, fuzzy and cloud fuzzy environments. Cloud fuzzy is a recently introduced concept in fuzzy literature. This study aims to find the most appropriate environment to resolve the proposed problems. Fuzzy model assumes the inventory parameters to be fuzzy forever (i.e. over the course of time, fuzziness remains constant), which is an unrealistic assumption. The cloud fuzzy environment allows uncertainty to vary with time. This study reveals the superiority of the cloud fuzzy model as compared to the fuzzy and crisp models. Besides, this analysis shows that less uncertainty does not imply more profit. A numerical example was also presented to validate the results. For this chapter, demand is considered to be fuzzy and cloud fuzzy. This model can also be extended by considering some other parameters as fuzzy and cloud fuzzy.

ACKNOWLEDGMENTS

The authors thank DST-FIST file # MSI-097 for the technical support to the department.

REFERENCES

Bellman, R. E., & Zadeh, L. A. (1970). Decision-making in a fuzzy environment. *Management Science*, 17(4): B-141.

Chaudhari, U., Shah, N. H., & Jani, M. Y. (2020). Inventory modelling of deteriorating item and preservation technology with advance payment scheme under quadratic demand. *Optimization and Inventory Management*, 69–79.

De, S. K., & Mahata, G. C. (2017). Decision of a fuzzy inventory with fuzzy backorder model under cloudy fuzzy demand rate. *International Journal of Applied and Computational Mathematics*, 3(3): 2593–2609.

De, S. K., & Mahata, G. C. (2019). A cloudy fuzzy economic order quantity model for imperfect-quality items with allowable proportionate discounts. *Journal of Industrial Engineering International*, 15: 571–583.

Dye, C. Y., & Hsieh, T. P. (2012). An optimal replenishment policy for deteriorating items with effective investment in preservation technology. *European Journal of Operational Research*, 218(1): 106–112.

Ghare, P. M., & Schrader, G. F. (1963). An inventory model for exponentially deteriorating items. *Journal of Industrial Engineering*, 14(2): 238–243.

Goyal, S. K., & Giri, B. C. (2001). Recent trends in modeling of deteriorating inventory. *European Journal of Operational Research*, 134(1): 1–16.

Harris, F. W. (1913). How many parts to make at once. *Factory, the Magazine of Management*, 10(2): 135–136.

Karmakar, S., De, S. K., & Goswami, A. (2018). A study of an EOQ model under cloudy fuzzy demand rate. *Communications in Computer and Information Science*, 834: 149–163.

Lee, H. H. (2005). A cost/benefit model for investments in inventory and preventive maintenance in an imperfect production system. *Computers & Industrial Engineering*, 48(1): 55–68.

Maiti, A. K. (2019). Fuzzy economic production lot-size model under imperfect production process with cloudy fuzzy demand rate. *International Journal of Advanced Scientific Research and Management*, 4(3): 111–126.

Maity, S., De, S. K., & Mondal, S. P. (2019). A study of an EOQ model under lock fuzzy environment. *Mathematics*, 7(1): 75.

Mashud, A. H. M., Hasan, M. R., Wee, H. M., & Daryanto, Y. (2020). Non-instantaneous deteriorating inventory model under the joined effect of trade-credit, preservation technology and advertisement policy. *Kybernetes*, 49(6): 1645–1674.

Nahmias, S. (1982). Perishable inventory theory: A review. *Operations Research*, 30(4): 680–708.

Raafat, F. (1991). Survey of literature on continuously deteriorating inventory models. *Journal of the Operational Research Society*, 42(1): 27–37.

Raula, P., Indrajitsingha, S. K., Samanta, P. N., Raju, L. K., & Misra, U. K. (2018). Inventory model of deteriorating items for supermarket: A fuzzy approach. *Asian Journal of Mathematics and Computational Research*, 23(4): 231–239.

Shah, N. H., & Shah, Y. K. (2000). Literature survey on inventory models for deteriorating items. *Economic Annals*, 44(1): 221–237.

Shah, N. H., Shah, D. B., & Patel, D. G. (2016). Optimal preservation investment, pricing and ordering policies for deteriorating inventory with trapezoidal demand. *International Journal of Operational Research*, 26(3): 367–381.

Shekarian, E., Kazemi, N., Abdul-Rashid, S. H., & Olugu, E. U. (2017). Fuzzy inventory models: A comprehensive review. *Applied Soft Computing*, 55. 588–621.

Tayal, S., Singh, S. R., Sharma, R., & Chauhan, A. (2014). Two echelon supply chain model for deteriorating items with effective investment in preservation technology. *International Journal of Mathematics in Operational Research*, 6(1): 84–105.

Whitin, T. M. (1957). *Theory of inventory management.* Princeton, NJ: Princeton University Press.

Yager, R. R. (1981). A procedure for ordering fuzzy subsets of the unit interval. *Information Sciences*, 24(2): 143–161.

Yao, J. S., Chang, S. C., & Su, J. S. (2000). Fuzzy inventory without back order for fuzzy order quantity and fuzzy total demand quantity. *Computers & Operations Research*, 27(10): 935–962.

Zadeh, L. A. (1965). Fuzzy sets. *Information and Control*, 8(3): 338–353.

5 Image Formation Using Deep Convolutional Generative Adversarial Networks

Astha Bathla and Aakanshi Gupta
Guru Gobind Singh Indraprastha University
New Delhi, India

CONTENTS

5.1 INTRODUCTION

Unsupervised representation learning is an erudite concept, especially in the realm of research analysing images. In the area of research, it has been conventional to learn reusable feature representations from a large dataset of images. One can employ a multitudinous aggregate of unspecified videos and images to study magnificent intermediary representations. As a consequence, the concerned can perform particular image-related tasks such as image editing or image construction. Some of these tasks embrace the generation and restoration of images. Image generation is examining a hierarchy of representations and their applications, in particular, to employ them to construct new images from a set of already existing ones. Likewise, image restoration [1], also known as image inpainting, is a repairing technique which practices filling in deteriorated parts to structure a complete image.

Regeneration of damaged regions is a content-aware computation. The term content-aware signifies filling the targeted portions using their neighbouring pixel information. In the realm of image processing, image inpainting [2] is an operative research content. In this chapter, the focus is on achieving the previously explained image concomitant intents using deep convolutional generative adversarial networks (DCGAN) [3].

DCGAN is a new framework for evaluating generative models under an adversarial process in which one concurrently trains two models. The DCGAN-based algorithm acquires the CelebA dataset as input which consists of a collection of roughly 5,22,000 images of human faces from the web. It operates 80 per cent of this dataset for training and the remainder for testing and validation. Commencing the inpainting process, one uses an optimal input noise z to generate the best-fitting image for the image completion purpose. The applied approach uses the corresponding area to compliment missing segments of the target image. Specific methods exercise only texture replication technology, thus promoting the unfeasibility of satisfactory results, as there is a loss of information. For instance, after effacing the nose in an exclusive face image, the consequence of filling it with neighbouring deducted information is a reconstructed nose filled with human face skin. Therefore, both global and local image aspects are the essence of image restoration. From a global perspective, the repaired results compliment the human visual perception. While on the contrary, a local perspective help sharpens the texture details.

In the most recent years, scientists have implemented image restoration via deep learning to obtain breakthrough results, all of which benefit from big data training. In this chapter, the experiment results are sample images procured via learning features from a set of unlimited distribution. The generator model spawns new images. The discriminator model checks the proximity among generated and real images. Output also plots the losses in the two models.

5.2 THE APPROACH

5.2.1 Image Generation

This chapter exercises deep learning methods for image generation and restoration. DCGAN, an augmentation of the generative adversarial network, trains deep learning models to encapsulate features from training distribution to form raw data from an identical distribution. It avails itself of convolutional-transpose and convolutional layers in the generator and discriminator, respectively. The generator incorporates batch-norm layers, convolutional-transpose layers and ReLU activation layers, as shown in Figure 5.1. It maps the latent space vector (z) to data space. Seeing the data as an aggregate of images in this demonstration, converting latent space vector to data space indicates originating an RGB(red, green, blue) image ($3 \times 64 \times 64$) with a size similar to the training data in the final analysis. The discriminator, a binary classification network, inputs an image and produces a scalar probability as output that categorises the input image as fake or real. The discriminator takes a $3 \times 64 \times 64$ input image, operates it through a sequence of stridden convolution layers, batch-norm

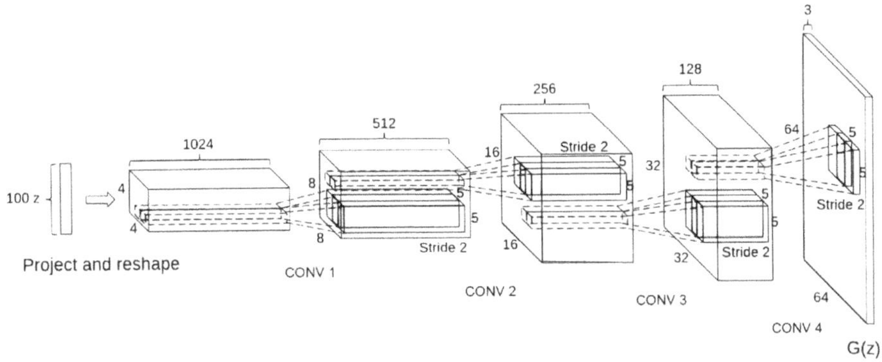

FIGURE 5.1 Generator and its Layers.

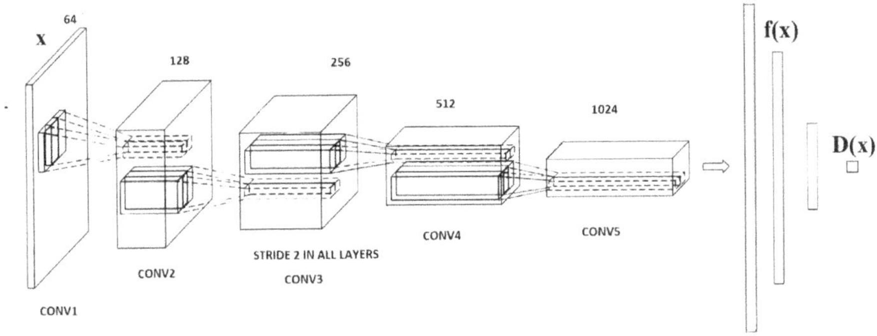

FIGURE 5.2 Discriminator and its Layers.

layers and LeakyReLU activation layer and uses a sigmoid activation function to induce eventual probability. Figures 5.2 and 5.3 depicts the structure of the discriminator model.

5.2.2 Optimisers and Loss Functions

As known, the discriminator model in DCGANs outputs a scalar probability that apprises the sample as real or fake. The real image comes by label 1 and the fake one by 0. The training procedure for the discriminator aims at maximising the probability of assigning accurate labels to the samples, and for the generator, it focuses on minimising the probability that the discriminator will proclaim generated images as fake. Consequently, it optimises the generator and discriminator using the following min-max loss function:

$$_{\text{min}-\text{max}} V(D,G) = E_y \sim P_{\text{data}}(y)\left[\log D(y)\right] + E_z \sim P_z(z)\left[\log\left(1-D\left(G(z)\right)\right)\right], \quad (5.1)$$

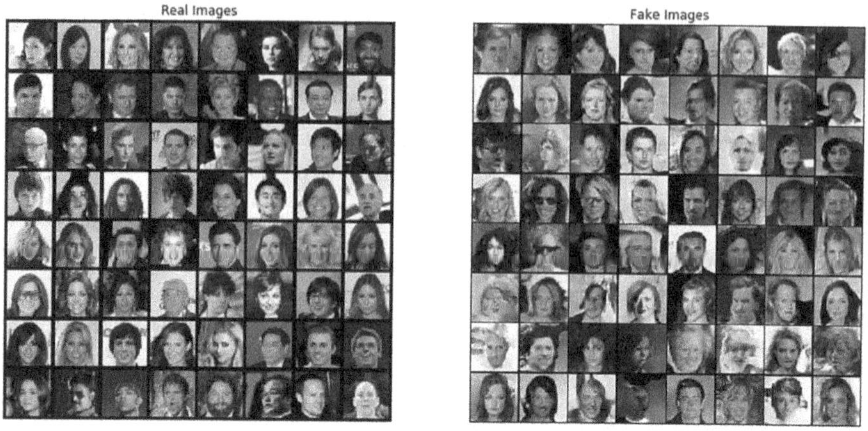

FIGURE 5.3 Real Images versus Fake Images.

where y is the real image, z is the input latent vector to the generator, P_{data} is the training distribution, Pz is the generator distribution, D(y) is the probability of created image and D(G(z)) is the probability of the real image.

5.2.3 Image Completion

The applied image completion procedure uses the simultaneously trained models, generator and discriminator. The desired image it uses for pursuing image completion requires an input (z). The DCGAN framework[4] patches the selected image with a square binary mask affixed in its centre. Next, it multiplies elements of the image with the elements of the mask. This element-wise product gives the original part of the image (M⊙y). Supposedly, some z^ (using the loss functions defined next) creates some image G(z^) and gives an appropriate construction of missing or targeted portions. Accordingly, the addition of original pixels to the reconstructed pixels G(z^)⊙(1−M) gives form to the reconstructed image.

$$x_{\text{reconstructed}} = y \odot M + G\left(z^{\wedge}\right) \odot \left(1 - M\right), \tag{5.2}$$

where M is the mask for target segments, y is the input image and G(z^) is the created sample from z^.

5.2.3.1 Contextual Loss

The DCGAN outlook for image restoration encircles the agenda of completing a missing segment of an image. Unharmed portions of the damaged image procure information about the target section. The surrounding pixel information succours in fixing the impaired. One calculates contextual loss for the stated merit. Despite being beneficial, one drawback of this loss is that it treats every pixel equally. This property is a hindrance to the completion process in many instances. By way of illustration, if someone masks the centre of a face image and uses the background information for image reconstruction, he or she will obtain off-target results. Thus to get satisfactory results, supplementary concentration should be on the portions close to the hole. The significance of an uncorrupted pixel is positively in parallel with the number of corrupted pixels surrounding it. The pixel distant from the masked segment is of utterly minimum importance in the restoration process.

Assuring the equivalence of known pixels in the input image to the pixels in the generator-created image maintains a context similar to the input image. It is vital to penalise G(z) for not constituting an identical image from the known pixels. An element-wise subtraction of pixels in y from G(z) and a record of their difference implements this as

$$L_{contextual}(z|y,M) = \|G(z) \odot M_1 - y \odot M_1\|^1, \quad (5.3)$$

where $\|x\|1 = \sum i|xi|$, y is the input image, M is the mask of target sections, ⊙ signifies the element-wise product, G(z) indicates the generated sample by z and M_1 is the square mask with a size equivalent to the sample.

5.2.3.2 Perceptual Loss

The functional role of the perceptual loss is to proclaim the realism of the spawned image. According to the followed outlook, the task of the discriminator model is to assign a probability to the generated sample. Labels for real and fake are 1 and 0, respectively. Higher probability depicts the correspondence of the sample to the real image. Therefore, perceptual loss aims to embolden the reconstructed images to be like real images. The definition of perceptual loss is as follows:

$$L_{perceptual}(z) = \log(1 - D(G(z))), \quad (5.4)$$

where G(z) indicates the generated sample by z and D(G(z)) is the output of the discriminator.

5.2.3.3 Inpainting

The best epoch output fills the damaged segment of the target image. The most identical epoch(z^) is selected using the losses calculated.

$$L(z) \equiv L_{contextual}(z) + \lambda L_{perceptual}(z) \quad (5.5)$$

$$z\hat{} \equiv \arg\ \min z L(z), \quad (5.6)$$

where λ is a hyper-parameter that controls how to import the contextual loss relative to the perceptual loss (λ = 0.1 by default).

Then as before, the reconstructed image compliments the absent values of y with G(z^):

$$x_{reconstructed} = (1 - M) \odot G(z\hat{}) + M \odot y \quad (5.7)$$

5.2.4 Implementation Details

Randomly initialise all model weights from a normal distribution with stdev = 0.02 and mean = 0. The generator's input is a latent space vector, and the output is an RGB image of size (3 × 64 × 64). A tanh function processes this output to put it back into the input range of [−1,1]. The discriminator's input is an image of size (3 × 64 × 64), and production is a probability of the sample being fake or real (0 = fake, 1 = real). Log in both optimisers with values of Beta 1 = 0.5 and learning rate 0.0002.

5.3 EXPERIMENTAL RESULTS

DCGAN course of action uses the CelebaA dataset for fabricating results. The later compromises approximately 5,22,000 images, 20 per cent of which execute testing. Demonstration ceases with three outcomes. The first is an accumulation of pseudo-samples from a data distribution. The model acquires the skill of representations from unlimited amounts of images and uses these to contrive new samples. Generated images compliment the real images but are below par. Besides, experiments induce inconsistent results.

The experiment's second output, depicted in Figure 5.4, portrays image completion or inpainting. The algorithm completes a masked sample of data distribution. Besides, it scrutinises the filled-in corrupted image [5] with the original to observe

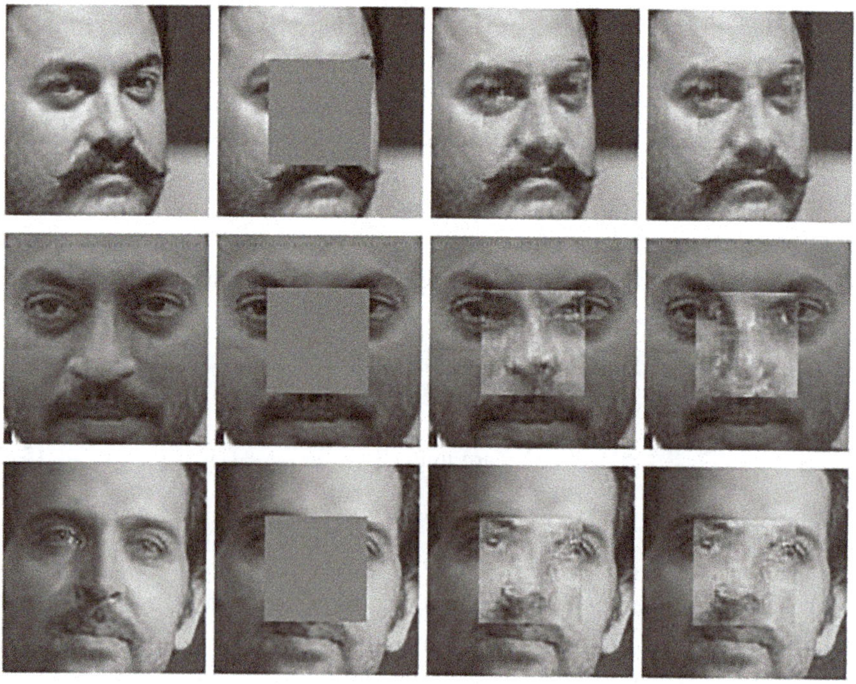

FIGURE 5.4 Process of Image Inpainting.

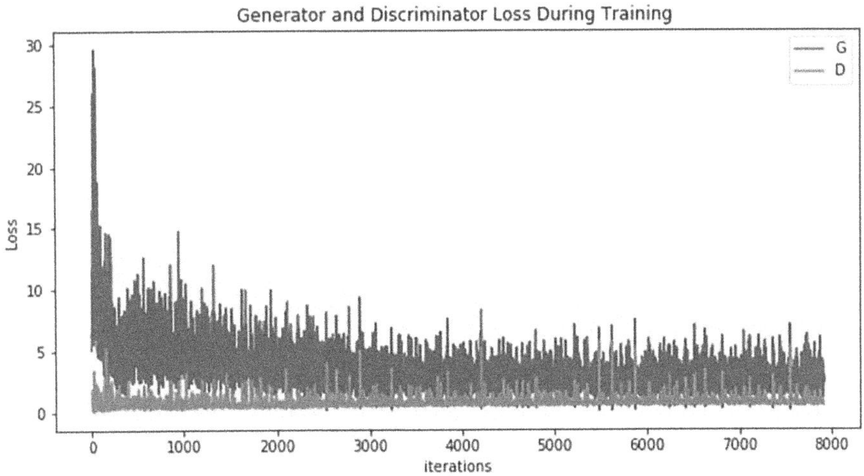

FIGURE 5.5 Generator and Discriminator Losses.

the equivalence of both images. The results exhibit standard reconstructed images. Studies show the prolonged training of models with an exorbitant number of images turns out superior quality outcomes.

The upshot of this examination is a plot between the losses of two models: generator and discriminator. In Figure 5.5, the blue region illustrates the generator loss, and the orange outlines the discriminator loss.

5.4 CONCLUSION

This chapter laid out a clear demonstration of image inpainting and generation techniques. The DCGAN framework trains two deep learning models. The generator originates samples from latent vectors, and the discriminator distinguishes the samples from training images. Image selection for the completion purpose commences the advance towards the inpainting technique. The chosen image is masked, and then the attacked region is filled in. Even though the DCGAN approach is efficacious, it still has a prolonged outcome. In addition, colour variation exists among pixels of targeted and surrounding regions.

REFERENCES

1. C. Guillemot and O. Le Meur, "Image Recording: An Overview and Recent Developments," *Signal Print Magazine IEEE*, vol. 31, no. 1, p. 127–144, 2014.
2. C. Barnes, E. Shechtman, A. Finkelstein, and D. M. B. Goldman, "Game Similarities: A Random Communication Algorithm for Graphical Modeling," *ACM Transmission in Graphics-TOG*, vol. 28, no. 3, pp. 24–33, 2009.
3. A. Radford, L. Metz, and S. Chintala, "An Undeniable Presentation Study on Deep Network Communication," in *Xiv preprint arXiv: 1511.06434*, pp. 2092–2096, 2015.

4. Goodfellow, J. Pouget-Abadie, M. Mirza, B. Xu, D. Warde-Farley, S. Ozair, A. Courville, and J. M. Bengio, *Information Processing Systems*, vol. 3, Montreal, QC, Canada, December 2014, pp. 2672–2680.
5. R. A. Yeh, C. Chen, T. Yian Lim, A. G. Schwing, M. Hasegawa-Johnson, and M. N. Do, "This semantic image has deep reproductive mechanisms." In *Proceedings – 30th IEEE Computer Conference Vision and Part Recognition, CVPR 2017*, Honolulu, HI, United States, November 2017, pp. 6882–6890.

6 Optimal Preservation Technology Investment and Price for the Deteriorating Inventory Model with Price-Sensitivity, Stock-Dependent Demand

Nita H. Shah
Gujarat University
Ahmedabad, Gujarat, India

Chetansinh R. Vaghela
Marwadi University
Rajkot, Gujarat, India

CONTENTS

6.1 INTRODUCTION

The classical economic order quantity is based on the assumption that items retain their utility throughout their lifetimes. However, decay, damage, and spoilage occur in foods, films, drugs, electronic goods, dairy products, etc. The deterioration may occur during storage and, therefore, the loss owing to deterioration cannot be neglected. Ghare and Schrader (1963) incorporated the effect of deterioration in inventory modeling. They computed economic order quantity with exponential deterioration. Covert and Philip (1973) extended the work of Ghare and Schrader (1963) with Weibull distribution and Gamma distribution. The review articles by Nahmias (1982), Raafat (1991), Shah and Shah (2000), Goyal and Giri (2001) and Bakker et al. (2012), Shah and Vaghela (2018b), Vaghela and Shah (2020) cited the available literature on deterioration and inventory modeling.

Setting optimal sale price and procurement quantity for deteriorating items was studied by Sana (2010), Ghosh et al. (2011), and Shah and Vaghela (2018a), including their cited references and many more. Sarkar and Sarkar (2013) formulated an economic manufacturing quantity model with probabilistic deterioration during a production system. These citations assumed the demand rate to be constant and deterministic. However, in the market, it is observed to be dependent on the selling price, time, stock displayed, advertisement, etc. Burwell et al. (1997), Wee (1997), Mondal et al. (2003), and You (2005) studied the effect of various parameters on the demand.

Levin et al. (1972) quoted that the number of items displayed increases the demand. The economic theory suggests that demand and price are inversely related. Thus when the demand rate is stock-dependent, the retailer should choose profitable price and replenishment strategies. Datta and Pal (2001) analyzed a finite horizon inventory model with the price and stock-dependent demand rate. The variations were studied by Teng and Chang (2005), Hou (2006), Hou and Lin (2006), Yang et al. (2010), Dye and Hsieh (2011), Pal et al. (2013), etc.

The management realized the need for significant investment to control the rate of deterioration of items. They started using improved and enhanced storage technology to increase the utility of the product. This helped to reduce unnecessary waste, minimize economic losses, and enhance business competitiveness. In recent years, many researchers started studying inventory models to control the deterioration rate.

Hsu et al. (2010) formulated an inventory model with a constant demand rate. They incorporated preservation technology investment to reduce the deterioration rate of items. Dye and Hsieh (2012) established optimal replenishment and investment in preservation when deterioration is time dependent. Hsieh and Dye (2013) applied particle swarm optimization to compute optimal production and preservation technology investment for minimizing the total cost. Zhang et al. (2014) developed a pricing policy for deteriorating items with preservation technology investment when shortages are not allowed. Mishra (2015) extended Zhang et al.'s (2014) model to allow shortages.

In this chapter, a more general inventory model is studied with (i) price and stock-dependent demand rate, (ii) shortages are not allowed, (iii) deterioration rate is constant, and (iv) preservation technology investment is used. The profit is maximized with respect to decision variables. The managerial issues are worked out with the help of sensitivity.

The chapter is organized as follows: In Section 6.2, basic notations and assumptions of the proposed problem are given. A mathematical model is formulated in Section 6.3. In Section 6.4, a numerical example is illustrated to support the proposed model and sensitivity analysis is carried out, followed by the conclusion in Section 6.5.

6.2 ASSUMPTIONS AND NOTATIONS

6.2.1 Notations

A: Ordering cost per order
C: Purchase cost per item (constant)
h: Holding cost per item per unit time
$I(t)$: Inventory level at any time t
n: Number of shipments (decision variable)
P: Selling price per item (decision variable.)
Q: Ordering quantity in each shipment
$R(P, I(t))$: Stock-dependent and price dependent demand rate
T: Replenishment time
u: Cost of preservation technology per unit time (decision variable)
$f(u)$: Proportion of reduced deterioration rate $0 \le f(u) \le 1$
θ: Rate of deterioration $0 < \theta < 1$
$\pi(n, P, u)$: Total profit during replenishment time

6.2.2 Assumptions

- The inventory system has a single item.
- The items in inventory deteriorate at a constant rate. There is no repair or replacement of deteriorated items during the considered replenishment time.
- The replenishment rate is infinite.
- Lead time is zero or negligible, and shortages are not allowed.
- Demand rate $R(P, I(t))$ is a function of selling price P and inventory level $I(t)$ (say)

$$R\left(P,I\left(t\right)\right)=\alpha-\beta P+\gamma I\left(t\right), \tag{6.1}$$

- where $\alpha > 0$ is the scale demand, and β is the parameter sensitive to price. γ, $0 \le \gamma \le 1$ is the parameter of the stock-dependent consumption rate.
- The function $f\left(u\right)=1-\dfrac{1}{1+\lambda u}$ is the continuous, increasing, and concave function of u for $\lambda > 0$. Set $f(0) = 0$.

6.3 MATHEMATICAL MODEL

In this study, the inventory is consumed because of stock- and price-dependent demand and deterioration at a constant rate θ. The number of shipments ordered during $[0, T]$ are n. Each shipment contains the ordering quantity Q. By investing amount u in preservation technology, the deterioration rate can be decreased for items in the warehouse. Under the assumption that $f(u)$ is the reduction in the deterioration rate owing to investment u, then the governing differential equation for inventory level $I(t)$ at any time t, where $0 \le t \le \frac{T}{n}$ is given by

$$\frac{dI(t)}{dt} = -R(P, I(t)) - \theta I(t)\mu, 0 \le t \le \frac{T}{n}, \tag{6.2}$$

where $\mu = f(u) = 1 - \frac{1}{1+\lambda u}$ with $\lambda > 0$ with boundary condition $I\left(\frac{T}{n}\right) = 0$ and initial condition $I(0) = Q$.

The solution of the differential equation (6.2) is given by

$$I(t) = \frac{(\alpha - \beta P)}{(\theta\mu + \gamma)}\left[e^{(\theta\mu+\gamma)\left(\frac{T}{n}-t\right)} - 1\right], 0 \le t \le \frac{T}{n}, \tag{6.3}$$

and the initial inventory level is given by

$$Q = I(0) = \frac{(\alpha - \beta P)}{(\theta\mu + \gamma)}\left[e^{(\theta\mu+\gamma)\left(\frac{T}{n}\right)} - 1\right]. \tag{6.4}$$

The following costs are incurred by a decision maker in n shipments.

(a) Net sales revenue is

$$SR = n \cdot P \cdot Q = \frac{nP(\alpha - \beta P)}{(\theta\mu + \gamma)}\left[e^{(\theta\mu+\gamma)\left(\frac{T}{n}\right)} - 1\right]. \tag{6.5}$$

(b) Purchase cost:

$$PC = n \cdot C \cdot Q = \frac{nC(\alpha - \beta P)}{(\theta\mu + \gamma)}\left[e^{(\theta\mu+\gamma)\left(\frac{T}{n}\right)} - 1\right]. \tag{6.6}$$

(c) Inventory holding cost:

$$IHC = n \cdot h \cdot \left(\int_0^{T/n} I(t)\,dt\right) = -\frac{h(-\alpha + \beta P)}{(\theta\mu + \gamma)^2}\left[ne^{(\theta\mu+\gamma)\left(\frac{T}{n}\right)} - (\theta\mu + \gamma)T - n\right]. \tag{6.7}$$

(d) Ordering cost:

$$OC = n \cdot A. \tag{6.8}$$

(e) Preservation technology investment:

$$PTI = u \cdot T. \tag{6.9}$$

Therefore, the total profit is given by

$$\pi(n, P, u) = SR - (PC + IHC + OC + PTI). \tag{6.10}$$

Total profit is the function of discrete variable n and two continuous variables P and u.

For fixed n to maximize total profit with respect to selling price P and preservation technology cost u, the necessary conditions are

$$\frac{\partial \pi(P, u)}{\partial P} = 0 \text{ and } \frac{\partial \pi(P, u)}{\partial u} = 0. \tag{6.11}$$

The algorithm to get optimum values of the decision variables is as follows:

Step 1: Assign values to the inventory parameters.
Step 2: Set $n = 1$.
Step 3: Find critical point (P^*, u^*) by solving $\frac{\partial \pi(P, u)}{\partial P} = 0$ and $\frac{\partial \pi(P, u)}{\partial u} = 0$ simultaneously using mathematical software Maple 13.
Step 4: For maximum profit $\frac{\partial^2 \pi(P, u)}{\partial P^2} \cdot \frac{\partial^2 \pi(P, u)}{\partial u^2} - \frac{\partial^2 \pi(P, u)}{\partial P \partial u} \cdot \frac{\partial^2 \pi(P, u)}{\partial u \partial P} > 0$ and $\frac{\partial^2 \pi(P, u)}{\partial P^2} < 0$, $\frac{\partial^2 \pi(P, u)}{\partial u^2} < 0$ should be satisfied.
Step 5: Increment n by $n + 1$.
Step 6: Continue steps 3 to 5 until $\pi(n-2, P^*, u^*) \leq \pi(n-1, P^*, u^*)$ and $\pi(n-1, P^*, u^*) \geq \pi(n, P^*, u^*)$ is satisfied.
Call this value n^*.
Step 7: Calculate profit $\pi(n^*, P^*, u^*)$ for n^* number of shipments from equation (6.10). Next, we validate the model with a numerical example for inventory parameters.

6.4 NUMERICAL EXAMPLE AND SENSITIVITY ANALYSIS

Example 6.1: We consider an inventory system with following parameters.

$T = 2$ years, $A = \$250$ per order, $C = \$20$ per item, $h = \$15$ per unit, $\theta = 10\%$, $\lambda = 0.2$, $\alpha = 250$, $\beta = 6$, $\gamma = 0.02$.

TABLE 6.1
Optimum Solutions

n	P^*(in \$)	u^* (in \$)	$\pi(n^*, P^*, u^*)$(in \$)
1	38.16	4.83	48.53
2	34.54	5.37	421.41
3	33.31	4.27	547.17
$n^* = 4$	**32.69**	**3.37**	**572.99**
5	32.32	2.66	552.77

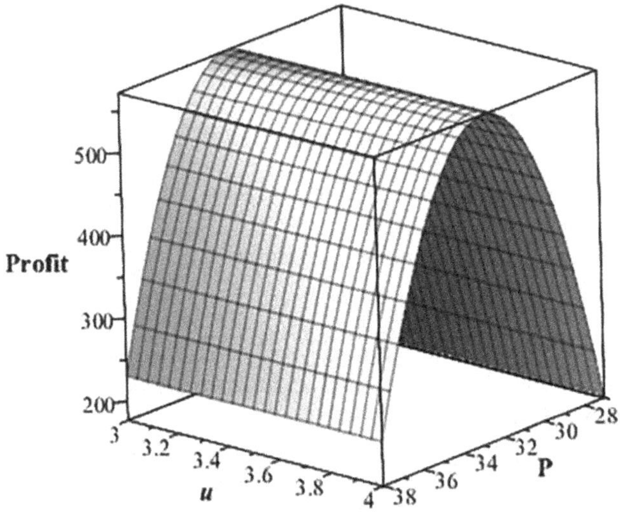

FIGURE 6.1 Concavity of Profit Function with Respect to P and u for $n = 4$.

Using the algorithm shown in Section 6.3, we calculate the optimal solution (n^*, P^*, u^*) and maximum profit for different values of n. From Table 6.1, it is observed that for four shipments, by setting the selling price at \$32.69 and investing \$3.37 in preservation technology, the decision maker attains the maximum profit of \$572.99. The concavity of the profit function is shown in Figure 6.1.

Now, for the data used in example 6.1, we observe the effects of different inventory parameters on the decision variables selling price (P), preservation technology investment (u), and total profit by varying the values as −20%, −10%, 10%, and 20%.

Now, the following observations can be made from the previous sensitivity analysis.

From Figure 6.2, we observe that

- Demand scale α has a large effect on selling price P;
- A noticcable decrease in selling price is caused by price-sensitive parameter β;

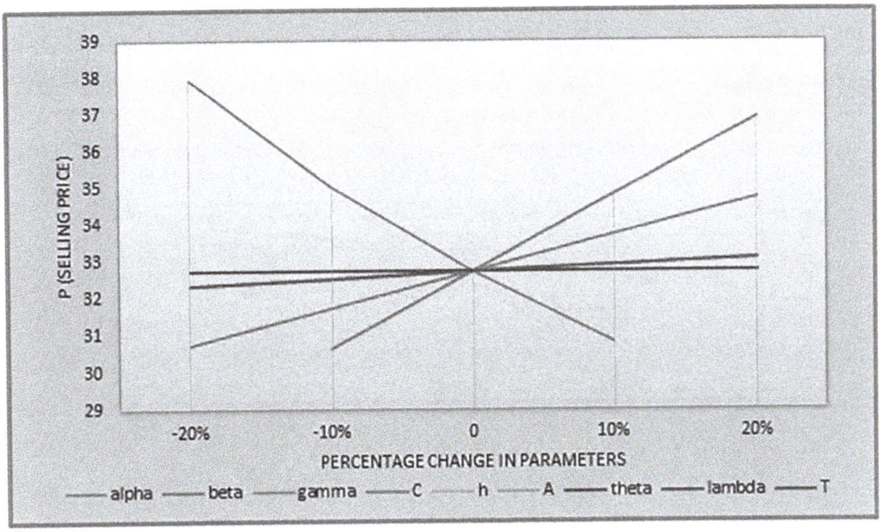

FIGURE 6.2 Effect on Selling Price (P) with Respect to Inventory Parameters.

- Stock-dependent parameter γ, deterioration rate θ, and λ cause minor decrement in selling price P;
- Positive effect on selling price is observed when the purchase cost C, holding cost h, and replenishment time T increases; and
- Ordering cost A has no effect on selling price P.

From Figure 6.3, the observations are

- Again, demand scale α has a large effect on preservation technology cost u;
- Noticeable decrease in preservation technology cost is observed when price-sensitive parameter β and purchase cost C are increased;
- A marginal increase is observed in preservation technology cost when stock-dependent parameter γ, deterioration rate θ, replenishment time T, and λ is increased;
- Minor negative effect on preservation technology cost is observed when holding cost h is increased; and
- Ordering cost A has no effect on preservation technology cost.

From Figure 6.4, we can observe that

- Demand scale α has large impact on profit function;
- Profit decreases as the price-sensitive parameter β increases;
- Stock-dependent parameter γ increases profit marginally;
- It is very obvious that when holding cost h, purchasing cost C, and ordering cost A increase, then total profit decreases; and
- Profit increases when replenishment time T is increased.

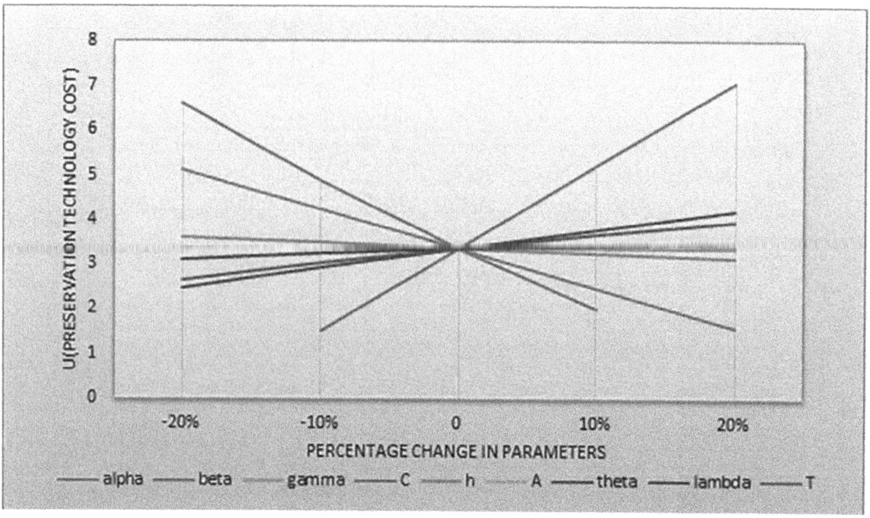

FIGURE 6.3 Effect on Preservation Technology cost (u) with Respect to Inventory Parameters.

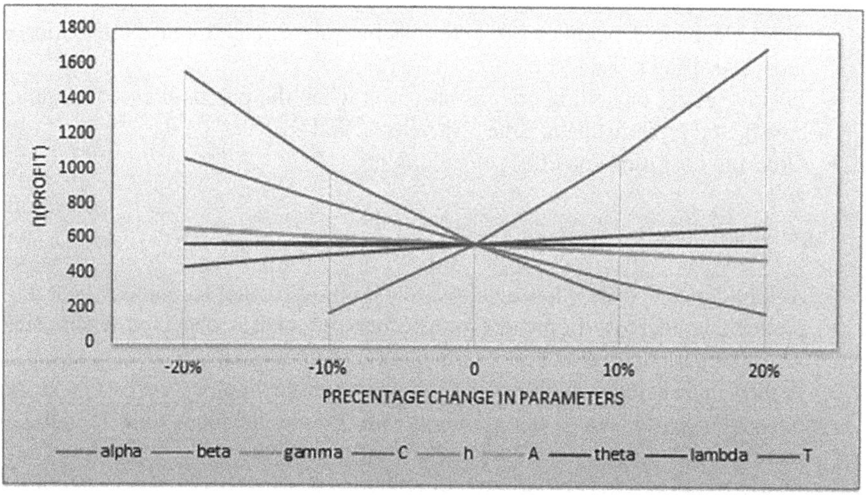

FIGURE 6.4 Effect on Total Profit with Respect to Inventory Parameters.

6.5 CONCLUSION

In this research, a deterministic inventory model is analyzed for price-sensitive, stock-dependent demand. The attempt is made to advice the player for investing in preservation to reduce the deterioration of items. The proposed model can be useful to the players dealing with food items, dairy products, etc., to determine optimal order transfers, investment required per unit, and selling price to maximize the profit. For future research, we can extend the model to probabilistic deterioration, fuzzy deterioration, etc.

ACKNOWLEDGMENTS

The authors thank DST-FIST 2014, File No: MSI-097 for technical support.

REFERENCES

Bakker, M., Riezebos, J., and Teunter, R. H. 2012. "Review of inventory systems with deterioration since 2001." *European Journal of Operational Research*, 221(2), 275–284.

Burwell, T. H. D., Dave Fitzpatrick, S. K. E., and Roy, M. R. 1997. "Economic lot size model for price-dependent demand under quantity and freight discounts," *International Journal of Production Economics*, 48(2), 141–155.

Covert, R. P., and Philip, G. C. 1973. "An EOQ model for items with Weibull distribution deterioration." *AIIE Transactions*, 5(4), 323–326.

Datta, T. K., and Pal, K. 2001. "An inventory system with stock dependent, price sensitive demand rate." *Production Planning & Control*, 12, 13–20.

Dye, C. Y., and Hsieh, T. P. 2011. "Deterministic ordering policy with price and stock dependent demand under fluctuating cost and limited capacity." *European Journal of Operational Research*, 218(1), 106–112.

Dye, C. Y., and Hsieh, T. P. 2012. "An optimal replenishment policy for deteriorating items with effective investment in preservation technology." *Expert Systems with Applications*, 38(12), 14976–14983.

Ghare, P. M., and Schrader, G. F. 1963. "A model for exponentially decaying inventory." *Journal of Industrial and Engineering Chemistry*, 14(2), 238–243.

Ghosh, S. K., Khanra, S., and Chaudhuri, K. S. 2011. "Optimal price and lot size determination for a perishable product under conditions of finite production, partial backordering and lost sale." *Applied Mathematics and Computation*, 217(13), 6047–6053.

Goyal, S. K., and Giri, B. C. 2001. "Recent trend in modelling of deteriorating inventory." *European Journal of Operational Research Society*, 134(1), 1–16.

Hou, K. L. 2006. "An inventory model for deteriorating items with stock-dependent consumption rate and shortages under inflation and time discounting." *European Journal of Operational Research*, 168(2), 463–474.

Hou, K. L., and Lin, L. C. 2006. "An EOQ model for deteriorating items with price and stock dependent selling rates under inflation and time value of money." *International Journal of Systems Science*, 37(15), 1131–1139.

Hsieh, T. P., and Dye, C. Y. 2013. "A production inventory model incorporating the effect of preservation technology investment when demand is fluctuating with time." *Journal of Computational and Applied Mathematics*, 239(1), 25–36.

Hsu, P. H. H., Wee, H. M., and Teng, H. M. 2010. "Preservation technology investment for deteriorating inventory." *International Journal of Production Economics*, 124(2), 388–394.

Levin, R. I., McLaughlin, C. P., Lamone, R. P., and Kottas, J. F. 1972. *Productions/operations management: Contemporary policy for managing operating systems*. New York: McGraw-Hill, p. 373.

Mishra, U. 2015. "An inventory model for deteriorating items under trapezoidal type demand and controllable deterioration rate." *Production Engineering – Research and Development*. doi:10.1007/s11740015-0625-8.

Mondal, B., Bhunia, A. K., and Maiti, M. 2003. "An inventory system of ameliorating items for price dependent demand rate." *Computers and Industrial Engineering*, 45(3), 443–456.

Nahmias, S. 1982. "Perishable inventory theory: A review," *Operations Research*, 30(4), 680–708.

Pal, S., Mahapatra, G. S., and Samanta, G. P. 2013. "An inventory model of price and stock dependent demand rate with deterioration under inflation and delay in payment." *International Journal of System Assurance and Engineering Management*, 5(4), 591–601.

Raafat, F. 1991. "Survey of literature on continuously deteriorating inventory models." *Journal of the Operational Research Society*, 42(1), 27–37.

Sana, S. S. 2010. "Optimal selling price and lot size with time varying deterioration and partial backlogging." *Applied Mathematics and Computation*, 217(1), 185–194.

Sarkar, M., and Sarkar, B. 2013. "An economic manufacturing quantity model with probabilistic deterioration in a production system." *Economic Modelling*, 31, 245–252.

Shah, N. H., and Shah, Y. K. 2000. "Literature survey on inventory models for deteriorating items." *Economic Annals*, 44(1), 221–237.

Shah, N. H., and Vaghela, C. R. 2018a. "Retailer's optimal pricing and replenishment policy for new product and optimal take-back quantity of used product." *Yugoslav Journal of Operations Research*, 28(3), 345–353.

Shah, N. H., and Vaghela, C. R. 2018b. "An EPQ model for deteriorating items with price dependent demand and two level trade credit financing." *Investigación Operacional*, 39(2), 170–180.

Teng, J. T., and Chang, C. T. 2005. "Economic production quantity models for deteriorating items with price and stock dependent demand." *Computers and Operations Research*, 32(2), 297–308.

Vaghela, C. R., and Shah, N. H. 2020. "Dynamic pricing, advertisement investment and replenishment model for deteriorating items." In N. H. Shah and M. Mittal (Eds.), *Optimization and Inventory Management* (pp. 81–92). Singapore: Springer.

Wee, H. M. 1997. "A replenishment policy for items with a price dependent demand and a varying rate of deterioration." *Production Planning & Control*, 8(5), 494–499.

Yang, H. L., Teng, J. T., and Chern, M. S. 2010. "An inventory model under inflation for deteriorating items with stock dependent consumption rate and partial backlogging shortages." *International Journal of Production Economics*, 123(1), 8–19.

You, S. P. 2005. "Inventory policy for products with price and time dependent demands." *Journal of the Operational Research Society*, 56(7), 870–873.

Zhang, J., Bai, Z., and Tang, W. 2014. "Optimal pricing policy for deteriorating items with preservation technology investment." *Journal of Industrial and Management Optimization*, 10(4), 1261–1277.

7 EOQ with Shortages and Learning Effect

Isha Sangal and Mahesh Kumar Jayaswal
Banasthali Vidyapith
Jaipur, Rajasthan, India

Rita Yadav
Banasthali Vidyapith
Jaipur, Rajasthan, India

Mandeep Mittal
AIAS, Amity University
Uttar Pradesh, Noida, India

Faizan Ahmad Khan
University of Tabuk
Tabuk, Saudhi Arabia, UAE

CONTENTS

7.1 INTRODUCTION

It is considered that all the items in an ordered lot are of good quality. Owing to this fact, this chapter deals with an EOQ inventory model that proposed a learning effect with partial backlogging for deteriorating items when the deteriorating rate is time-dependent. The aim of this chapter is to study the impact of learning on the total cost under the influence of shortages for deteriorating items. Further, the proposed model optimizes order quantity by minimizing the retailer's expected total cost per cycle. The sensitivity analysis has been presented as a consequence of numerical examples. Ford Harris (1915) developed the initial study of a stock scheme. This was formulated by minimizing the total of stock ordering and holding costs where the order rate was discerned and stable, shortages were not permitted, replenishment rate was endless, and lead time was insignificant. Covert and Philip (1973) improved a mathematical stock model for the decaying things under Weibull distribution. The learning curve is a mathematical tool developed by Wright (1936). In his first attempt, he calculated the mathematical formula which derives the relationship between learning changeable in quantitative form and derived the result in the proposition of the learning curve (LC). Baloff (1966) discussed the mathematical behavior of the learning theory. Jaber et al. (1996a) has worked on assuming the optimal lot sizing using the conditions of bounded learning cases and focused on inventory level and the minimization of the entire store costs that are keeping the LCs into reflection. Jaggi et al. (2013) discussed making inventory representation with trade financing periods of defective things under the suitable backlogging case. Sangal et al. (2016) discussed the working policy with the help of fuzzy concepts with demand, which is not fulfilled under learning effect. Agarwal et al. (2017) developed an inventory model in which deterioration rate follows the Weibull distribution and learning effect had incorporated on holding cost. The backlogging rate had been considered as an exponential function of time and minimized the total cost. Lashgari et al. (2016) explained an inventory model for decaying items with the help of back-ordering and financial considerations under a two-level credit financing policy. Pal et al. (2016) presented a two-echelon competitive integrated supply chain model under price and credit-period dependent demand. Navarro et al. (2019) proposed a supply chain model for multiple members and multiple products by considering demand as a function of the marketing effort. They optimize the lot size and marking effort demand to achieve the maximum profit by an analytical method. Mashud et al. (2019) presented an integrated price-sensitive inventory model with shortages under the two-level credit policies. An EOQ fuzzy mathematical production model has been developed by Moghdani et al. (2020) for the multiple items under multiple deliveries. A mathematical production model has been developed by Taleizadeh et al. (2019) for the multiple products with the help of some policies, such as reworking and screening. Jayaswal et al. (2019) discussed the learning phenomenon of seller ordering strategy for defective quality articles with permissible delay in payment.

In practical implications, the components of an inventory model are tentative and inaccurate, and the determination of maximum cycle length is difficult, as it is a non-stochastic in the distinguishable managerial process. Such types of problems have been tried to be solved by this model and the verification of the result has been illustrated by the appropriate examples. In this chapter, we optimize cycle length and corresponding total cost under the learning effect for deteriorating items with partial backlogging. In this chapter, the deterioration rate is also considered time dependent. Sensitivity analysis, as well as the conclusion, is presented in the last section. The contributions of this chapter are highlighted below:

- An inventory model has been developed based on the learning effect concept with the occurrence of shortages for deteriorating items.
- The deterioration rate is time dependent.

- The chapter attempts to develop the mathematical formula to find the optimal cycle length and corresponding cost for the buyer with learning effects on holding costs and ordering costs.
- From the results, the presence of the learning effect in the present model has an affirmative effect on the retailer ordering policy.
- The numerical example with sensitivity analysis is exhibited.

7.2 ASSUMPTIONS AND NOTATION

Assumptions and notations are provided next:

ASSUMPTIONS

- The rate of replenishment is infinite.
- The demand is not fulfilled, and shortages are not being met.
- The lead time is zero.
- The time horizon is finite.
- It can be assumed that the carrying cost is followed by the learning effect, as discussed by Jayaswal et al. (2019).
- It can be also be assumed that the ordering cost is followed by the learning effects discussed by Jayaswal et al. (2019).
- It can be considered that the rate of deterioration is time dependent (i.e., $r = \theta\ t,\ 0 < \theta < 1$).
- It is also assumed that the negative inventory backlogging rate is $B(t) = \frac{1}{1+\delta(T-t)}$; $\delta > 0$ denote the backlogging parameter and $t_1 \le t \le T$.

NOTATION

D: Demand (order require) rate, which is a function of time (in unit)
c: The purchase cost per unit
Q = IM + IB: The order quantity during cycle length$T$
IM: The maximum inventory level during $t \in [0, T]$
IB: The maximum back-ordered units during the stock-out period $t \in [0, T]$
C_k: Ordering cost \$/order
t_1: The point at which the inventory level reaches zero (decision variable)
t_2: The length of the period during which shortages are allowed (decision variable)
$T = t_1 + t_2$: Cycle length in year
C_1: The back-ordered cost \$/unit/unit time
C_2: The cost of lost sales per unit
C_h: Holding cost \$/unit items
r: Deterioration rate
δ: Backlogging parameter
α: Learning factor
$I_1(t)$: The level of positive inventory at time $t \in [0, t_1]$
$I_2(t)$: The level of positive inventory at time $t \in [t_1, t_1 + t_2]$
$\Psi(t_1, t_2)$: Retailer's whole cost per unit time

7.3 THE CONCEPTS WHICH ARE RELATED TO ASSUMPTION

In this part, the following concepts have been taken to relate the following assumptions:

- As per the assumption, holding cost follows the learning effect mathematically presented by (Jayaswal et al., 2019)

$$C_h(n) = C_{ho} + \frac{C_{h1}}{n^{\alpha}},$$

where n is number of shipments and α is a learning factor.

- As per assumption, ordering cost obeys the learning effect proposed by (Jayaswal et al., 2019)

$$C_k(n) = C_{ko} + \frac{C_{k1}}{n^{\alpha}},$$

where n is number of shipments.

7.4 OBJECTIVE OF THIS CHAPTER

The goal of this chapter is to determine the optimal cycle length and to analyze the effect of learning on total cost for the buyer considering the shortages for deteriorating items. The whole cost per unit time can be found with the following formula:

$$\text{Total cost per cycle} = \frac{\text{Total cost}}{\text{Time}}$$

7.5 MATHEMATICAL MODEL

The formulation of the present model has been discussed under credit policy in cash under the learning effect for decaying things with the linear demand rate. We are considering $I_1(t)$ to be the stock stage in the time $t \in [0, t_1]$, which is reducing because of demand and decaying the stock stage in the time $t \in [t_1, t_1 + t_2]$ under the condition with shortages. The current inventory level and stock level are governed by the linear differential equation with boundary conditions, which is provided in Figure 7.1

$$\frac{dI_1(t)}{dt} = -D - r(t)I_1(t),\ t \in [0, t_1] \tag{7.1}$$

with boundary condition. $I_1(t) = 0$, at $t = t_1$

By assumptions,

$$r(t) = \theta t,$$

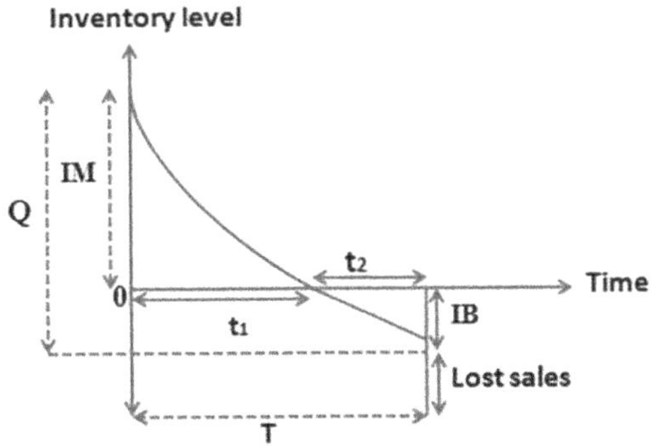

FIGURE 7.1 Inventory Process.

Subsituting the value of $r(t)$in equation (7.1), we get,

$$\frac{dI_1(t)}{dt} = -D - r\theta I_1(t),\ t \in [0, t_1] \tag{7.2}$$

with boundary condition. $I_1(t) = 0$, at $t = t_1$

Now, to get a closed form expression of $I_1(t)$, we use the exponential function expansion in terms of infinite series, and ignoring the second- and higher-order powers of θ. After simplification, the solution of the differential equation (7.2) is given below:

$$I_1(t) = D\left[\left(1 - \frac{\theta t^2}{2}\right)\left((t_1 - t) + \frac{\theta(t_1^3 - t)}{6}\right)\right], 0 \le t \le t_1 \tag{7.3}$$

At t_1, the inventory level is zero, and after t_1 shortages occur. The inventory level depends on the demand and fraction of demand is backlogged during the interval $[t_1, t_1 + t_1]$. The state of stock during $[t_1, t_1 + t_1]$ can be denoted by the differential equation

$$\frac{dI_2(t)}{dt} = -\frac{D}{1 + \delta(t_1 + t_2 - t)},\ t_1 \le t \le t_1 + t_2 \tag{7.4}$$

Further, the solution of the ordinary differential equation (7.) with boundary condition, $I_2(t) = 0$ at $t = t_1$ is given by

$$I_2(t) = \frac{a}{\delta}\left(\log\left(1 + \delta(t_1 + t_2 - t)\right) - \log(1 + \delta t_2)\right) \tag{7.5}$$

The given mathematical model is extended by assuming the previous consideration and notation with an inventory level which is delivered to a retailer with purchasing cost *PC*, fixed ordering cost *OC*, holding cost *IHC*, back-ordered cost *BC*, and cost of lost sales *LS*. The retailer's whole cost is shown in the following equation:

$\Psi(t_1, t_2)$=, now the buyer's total cost per unit time is

$$\Psi\left(t_1, t_2\right) = \frac{OC + BC + PC + LS + IHC}{t_1 + t_2} \tag{7.6}$$

Now the maximum inventory level is

$$IM = I_1\left(0\right) = D\left[\frac{6t_1 + \theta t_1^{\,3}}{6}\right], \tag{7.7}$$

and the maximum back-ordered units

$$IB = -I_2\left(t_1 + t_2\right) = \frac{D}{\delta}\log\left(1 + \delta t_2\right) \tag{7.8}$$

Here, the order quantity, Q, during the interval $[0, T]$ is given as follows:

$$Q = IM + IB = \left[\left(1 - \frac{\theta t^2}{2}\right)\left(\left(t_1 - t\right) + \frac{\theta\left(t_1^{\,3} - t\right)}{6}\right)\right] \tag{7.9}$$

The detailed calculations of all the components used in the equation (7.) are given next:

$$IHC = C_h\left[\int_0^{t_1} I_1\left(t\right)dt\right] = C_h D\left[\frac{t_1^{\,2}}{2} + \frac{\theta}{6}\left(t_1^{\,4} - \frac{t_1^{\,2}}{2}\right) - \frac{\theta t_1^{\,4}}{24}\right], \tag{7.10}$$

$$BC = C_1\left[\int_0^{t_1} -I_2\left(t\right)dt\right] = \frac{C_1 D}{\delta^2}\left[\delta t_2 - \log\left(1 + \delta t_2\right)\right], \tag{7.11}$$

$$LS = C_2 D\left[\int_{t_1}^{t_1+t_2}\left(1 - \frac{1}{1 + \delta\left(t_1 + t_2 - t\right)}\right)dt\right] = \frac{C_2 D}{\delta}\left[\delta t_2 - \log\left(1 + \delta t_2\right)\right], \tag{7.12}$$

$$PC = cQ = cD\left[t_1 + \frac{\theta t_1^3}{6} + \frac{1}{\delta}\log\left(1+\delta t_2\right)\right], \tag{7.13}$$

$$OC = C_k = C_0 + \frac{C_k}{n^\alpha} \tag{7.14}$$

Using the values of *OC*, *BC*, *PC*, *IHC* and *LS* from equations (7.) to (7.1) and substituting equation (7.), we get

$$\Psi\left(t_1, t_2\right) = \frac{1}{t_1+t_2}\left[\begin{array}{l} C_0 + \frac{C_k}{n^\alpha} + \left(C_0 + \frac{C_h}{n^\alpha}\right)D\left[\frac{t_1^2}{2} + \frac{\theta}{6}\left(t_1^4 - \frac{t_1^4}{2}\right) - \frac{\theta t_1^4}{24}\right] \\ + \frac{C_1 D}{\delta^2}\left[\delta t_2 - \log\left(1+\delta t_2\right)\right] \\ + \frac{C_2 D}{\delta}\left[\delta t_2 - \log\left(1+\delta t_2\right)\right] \\ + cD\left[t_1 + \frac{\theta t_1^3}{6} + \frac{1}{\delta}\log\left(1+\delta t_2\right)\right] \end{array}\right]. \tag{7.15}$$

7.6 SOLUTION METHOD

On optimizing the whole worth with respect to t_1 and t_2 for the maximum profit, if we take $\frac{\partial\Psi\left(t_1, t_2\right)}{dt_1} = 0$, and $\frac{\partial\Psi\left(t_1, t_2\right)}{dt_2} = 0$, find out the value of t_1 and t_2. If the value of $t_1 = t_{11}$ and $t_2 = t_{22}$ (suppose), after that, we calculate $\frac{\partial^2\Psi\left(t_1, t_2\right)}{dt_1^2}$ and $\frac{\partial^2\Psi\left(t_1, t_2\right)}{dt_2^2}$, we put the value of $t_1 = t_{22}$ and $t_2 = t_{22}$ in the second derivative, and we get $\frac{\partial^2\Psi\left(t_{11}, t_{22}\right)}{\partial t_1^2} > 0$ and $\frac{\partial^2\Psi\left(t_{11}, t_{22}\right)}{\partial t_2^2} > 0$ the $t_1 = t_{11}$ and $t_2 = t_{22}$ are the maximum value of t_1 and t_2 which are represented by t_1* and t_2*, respectively. Suppose the minimum cost corresponding this value is represented by $\Psi(t_1*, t_2*)$. The whole cost per cycle can be convex if it satisfies the essential and sufficient conditions $\left[\frac{\partial^2\Psi\left(t_1, t_2\right)}{\partial t_1 \partial t_2}\right]^2 - \left[\frac{\partial^2\Psi\left(t_1, t_2\right)}{\partial t_1^2}\right]\left[\frac{\partial^2\Psi\left(t_1, t_2\right)}{\partial t_2^2}\right] \geq 0$ and $\frac{\partial^2\Psi\left(t_1, t_2\right)}{\partial t_1^2} > 0$, $\frac{\partial^2\Psi\left(t_1, t_2\right)}{\partial t_2^2} > 0$, and the calculation has been shown in the Appendix[A].The numerical solution has been found with the help of Mathematica 0.9 software.

7.7 NUMERICAL EXAMPLE

The following example is for finding the minimum total cost for subsequent t_1, and t_2. The input parameters used are taken from (Tripathi and Mishra, 2010).

$$
\begin{aligned}
&D = 40,\ \delta = 4,\ \theta = 0.04,\ h_0 = \$5\,/\,uni\,/\,year,\\
&h_1 = \$2\,/\,unit\,/\,year,\ k_0 = \$200\,/\,unit,\\
&k_1 = \$50\,/\,unit,\ n = 5,\ \alpha = 0.2,\ c = \$10\,/\,unit,\\
&C_1 = \$15\,/\,units,\ C_2 = \$20\,/\,units,\ t_1^* = 1.1821 year,\\
&t_2^* = 0.3647 year,\ \Psi\left(t_1^*, t_2^*\right) = 722\$
\end{aligned}
$$

and other components can be calculated on the numerical data

$$\frac{\partial^2\Psi\left(1.1821, 0.3647\right)}{\partial t_1^{\,2}} = 10.89 > 0, \frac{\partial^2\Psi\left(1.1821, 0.3647\right)}{\partial t_2^{\,2}} = 11.19 > 0 \text{ and } \frac{\partial^2\Psi}{\partial t_1 \delta t_2} = 15.67 > 0,$$

which implies that $\left[\frac{\partial^2\Psi\left(t_1, t_2\right)}{\partial t_1 \partial t_2}\right]^2 - \left[\frac{\partial^2\Psi\left(t_1, t_2\right)}{\partial t_1^{\,2}}\right]\left[\frac{\partial^2\Psi\left(t_1, t_2\right)}{\partial t_2^{\,2}}\right] = 123.69 \geq 0$, which shows the condition of convexity of the total cost function.

7.8 SENSITIVITY ANALYSIS

In this section, to analyze the effect of model parameters, such as the number of shipments, deterioration rate, demand rate, and learning factor on the total cost, we perform a sensitivity analysis. The results are given in Figures 7.2–7.5.

Observations and managerial implications are as follows:

- From Figure 7.2, it can be observed that if the learning factor increases, then the minimum cost of the organization decreases.
- From Figure 7.3, it can be analyzed that if the number of shipments increases, the organization's cost decreases because of the learning effect on holding cost and ordering cost.
- From Figure 7.4, it can be indicated that if the deterioration rate increases, the organization's total cost increases.
- From Figure 7.5, it can be shown that if the demand rate increases, the organization's total cost increases.

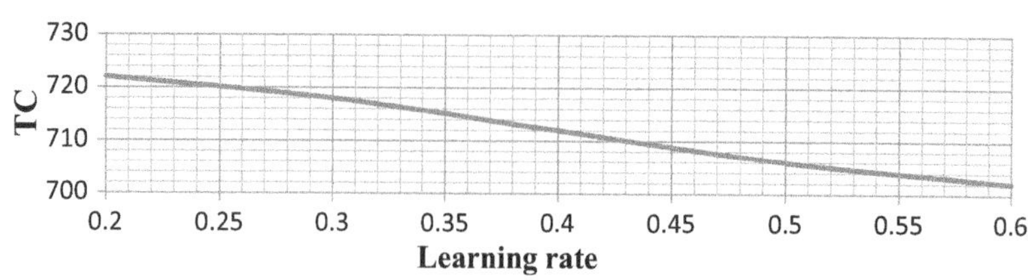

FIGURE 7.2 Effect of Learning Rate on Total Cost

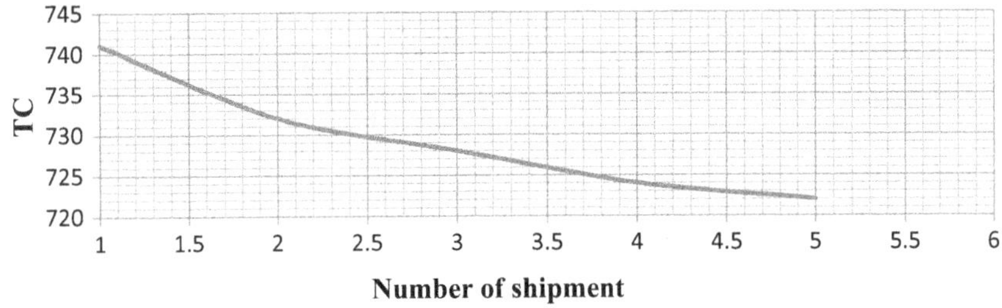

FIGURE 7.3 Effect of Shipment on Total Cost under Learning Effect.

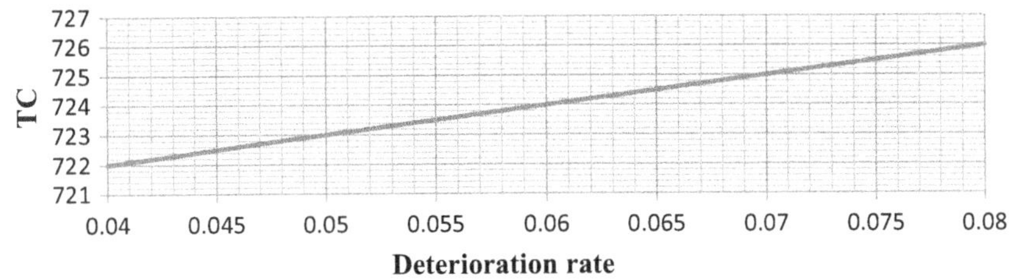

FIGURE 7.4 Effect of Deterioration Rate on Total Cost.

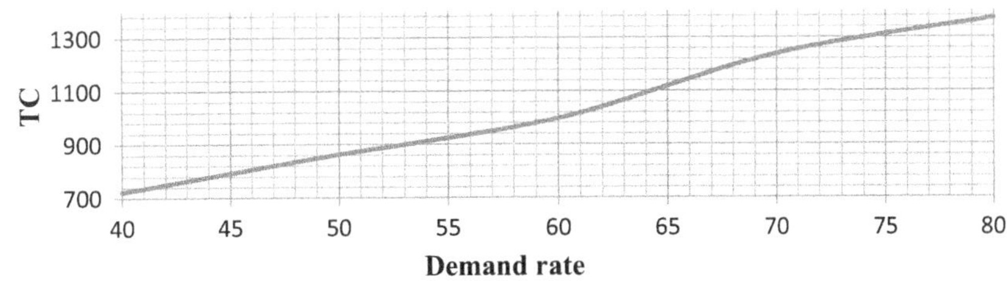

FIGURE 7.5 Effect of Demand Rate on Total Cost.

7.9 CONCLUSIONS

This chapter attempted to develop the mathematical formula to find the optimal cycle length and corresponding total cost under shortages for a buyer when there are learning effects on holding and ordering costs. Based on the observation of some results, we analyzed that the values of α, θ, D, and n act as good roles to minimize the total inventory cost. This chapter can be extended for more realistic situations, such as stock-dependent and stochastic demand with partial-trade credit.

REFERENCES

Agarwal, A., Sangal, I., & Singh, S. R. (2017). "Optimal policy for non-instantaneous decaying inventory model with learning effect and partial shortages." *International Journal of Computer Applications*, 161(10), 13–18.

Baloff, N. (1966). "Startups in machine-intensive production systems." *Journal of Industrial Engineering*, 17, 23–30.

Covert, R. P., & Philip, G. C. (1973). "An EOQ model for items with Weibull distribution deterioration." *AIIE transactions*, 5(4), 323–326.

Harris, F. W. (1915). "How many parts to make at once." *The Magazine of Management*, 10, 135–136.

Jaber, M. Y., & Bonney, M. (1996). "Optimal lot sizing under learning considerations: The bounded learning case." *Applied Mathematical Modelling*, 20(10), 750–755.

Jaggi, C. K., Goel, S. K., & Mittal, M. (2013). "Credit financing in economic ordering policies for defective items with allowable shortages." *Applied Mathematics and Computation*, 219(10), 5268–5282.

Lashgari, M., Taleizadeh, A. A., & Sana, S. S. (2016). "An inventory control problem for deteriorating items with back-ordering and financial considerations under two levels of trade credit linked to order quantity." *Journal of Industrial & Management Optimization*, 12(3), 1091.

Jayaswal, M., Sangal, I., Mittal, M., & Malik, S. (2019). "Effects of learning on retailer ordering policy for imperfect quality items with trade credit financing", *Uncertain Supply Chain Management*, 7, 49-62.

Mashud, A. H. M., Uddin, M. S., & Sana, S. S. (2019). "A two-level trade-credit approach to an integrated price-sensitive inventory model with shortages." *International Journal of Applied and Computational Mathematics*, 5(4), 121.

Moghdani, R., Sana, S. S., & Shahbandarzadeh, H. (2020). "Multi-item fuzzy economic production quantity model with multiple deliveries." *Soft Computing*, 24(14), 10363–10387.

Navarro, K. S., Chedid, J. A., Florez, W. F., Mateus, H. O., Cárdenas-Barrón, L. E., & Sana, S. S. (2019). "A collaborative EPQ inventory model for a three-echelon supply chain with multiple products considering the effect of marketing effort on demand."*Journal of Industrial & Management Optimization*, 272–283.

Pal, B., Sana, S. S., & Chaudhuri, K. (2016). "Two-echelon competitive integrated supply chain model with price and credit period dependent demand." *International Journal of Systems Science*, 47(5), 995–1007.

Taleizadeh, A. A., Yadegari, M., & Sana, S. S. (2019). "Production models of multiple products using a single machine under quality screening and reworking policies." *Journal of Modelling in Management*, 14(1), 232–259.

Tripathi, C. K., & Mishra, U. (2010). "An Inventory Model for Weibull Time-Dependence." *International Mathematical Forum*, 5(54), 2675–2687.

Wright, T. P. (1936). "Factors affecting the cost of airplanes." *Journal of the Aeronautical Sciences*.

APPENDIX A

$$\Psi\left(t_1, t_2\right)=\frac{1}{t_1+t_2}\begin{bmatrix} C_0+\dfrac{C_k}{n^\alpha}+\left(C_0+\dfrac{C_h}{n^\alpha}\right)D\left[\dfrac{t_1^{\,2}}{2}+\dfrac{\theta}{6}\left(t_1^{\,4}-\dfrac{t_1^{\,4}}{2}\right)-\dfrac{\theta t_1^{\,4}}{24}\right] \\ +\dfrac{C_1 D}{\delta^2}\left[\delta t_2-\log\left(1+\delta t_2\right)\right] \\ +\dfrac{C_2 D}{\delta}\left[\delta t_2-\log\left(1+\delta t_2\right)\right] \\ +cD\left[t_1+\dfrac{\theta t_1^{\,3}}{6}+\dfrac{1}{\delta}\log\left(1+\delta t_2\right)\right] \end{bmatrix} \tag{1}$$

$$\frac{\partial \Psi}{\partial t_1} = -\frac{1}{(t_1+t_2)^2}\left[\begin{array}{l} C_0 + \dfrac{C_k}{n^{\alpha}} \\ +\left(C_0+\dfrac{C_h}{n^{\alpha}}\right)D\left[\dfrac{t_1^2}{2}+\dfrac{\theta}{6}\left(t_1^4-\dfrac{t_1^4}{2}\right)-\dfrac{\theta t_1^4}{24}\right] \\ +\dfrac{C_1D}{\delta^2}\left[\delta t_2-\log(1+\delta t_2)\right] \\ +\dfrac{C_2D}{\delta}\left[\delta t_2-\log(1+\delta t_2)\right] \\ +cD\left[t_1+\dfrac{\theta t_1^3}{6}+\dfrac{1}{\delta}\log(1+\delta t_2)\right] \end{array}\right] +$$

$$\frac{1}{(t_1+t_2)}\left[\left(C_0+\frac{C_h}{n^{\alpha}}\right)D\left[\begin{array}{l}\dfrac{2t_1}{2} \\ +\dfrac{\theta}{6}\left(4t_1^3-\dfrac{4t_1^3}{2}\right) \\ -\dfrac{4\theta t_1^3}{24}\end{array}\right]+cD\left[1+\frac{3\theta t_1^2}{6}\right]\right]=0 \qquad (2)$$

$$\frac{\partial^2 \Psi}{\partial t_1^2} = \frac{2}{(t_1+t_2)^3}\left[\begin{array}{l} C_0 + \dfrac{C_k}{n^{\alpha}} \\ +\left(C_0+\dfrac{C_h}{n^{\alpha}}\right)D\left[\dfrac{t_1^2}{2}+\dfrac{\theta}{6}\left(t_1^4-\dfrac{t_1^4}{2}\right)-\dfrac{\theta t_1^4}{24}\right] \\ +\dfrac{C_1D}{\delta^2}\left[\delta t_2-\log(1+\delta t_2)\right] \\ +\dfrac{C_2D}{\delta}\left[\delta t_2-\log(1+\delta t_2)\right] \\ +cD\left[t_1+\dfrac{\theta t_1^3}{6}+\dfrac{1}{\delta}\log(1+\delta t_2)\right] \end{array}\right] - \frac{1}{(t_1+t_2)^2}$$

$$\left[\begin{array}{l}\left(C_0+\dfrac{C_h}{n^{\alpha}}\right)D\left[\begin{array}{l}\dfrac{2t_1}{2} \\ + \\ \dfrac{\theta}{6}\left(4t_1^3-\dfrac{4t_1^3}{2}\right) \\ - \\ \dfrac{4\theta t_1^3}{24}\end{array}\right] \\ +cD\left[1+\dfrac{3\theta t_1^2}{6}\right]\end{array}\right] + \frac{1}{(t_1+t_2)^2}\left[\begin{array}{l}\left(C_0+\dfrac{C_h}{n^{\alpha}}\right)D\left[\begin{array}{l}\dfrac{2t_1}{}+ \\ \dfrac{\theta}{6}\left(4t_1^3-\dfrac{2t_1^3}{1}\right) \\ -\dfrac{\theta t_1^3}{6}\end{array}\right] \\ +cD\left[1+\dfrac{\theta t_1^2}{3}\right]\end{array}\right.$$

$$+\frac{1}{(t_1+t_2)}\begin{bmatrix}\left(C_0+\frac{C_h}{n^{\alpha}}\right)D\begin{bmatrix}1+\frac{\theta}{6}\begin{pmatrix}12t_1^2\\-\frac{6t_1^2}{1}\end{pmatrix}\\-\frac{\theta t_1^2}{2}\end{bmatrix}\\+cD\left[\frac{\theta t_1}{1}\right]\end{bmatrix}>0 \tag{3}$$

$$\frac{\partial\Psi}{\partial t_2}=-\frac{1}{(t_1+t_2)^2}\begin{bmatrix}C_0+\frac{C_k}{n^{\alpha}}\\+\left(C_0+\frac{C_h}{n^{\alpha}}\right)D\left[\frac{t_1^2}{2}+\frac{\theta}{6}\left(t_1^4-\frac{t_1^4}{2}\right)-\frac{\theta t_1^4}{24}\right]\\+\\\frac{C_1D}{\delta^2}\left[\delta t_2-\log(1+\delta t_2)\right]+\\\frac{C_2D}{\delta}\left[\delta t_2-\log(1+\delta t_2)\right]+\\cD\left[t_1+\frac{\theta t_1^3}{6}+\frac{1}{\delta}\log(1+\delta t_2)\right]\end{bmatrix}+\frac{1}{t_1+t_2}\begin{bmatrix}\frac{C_1D}{\delta^2}\left[\delta-\frac{\delta}{1+\delta t_2}\right]+\\\frac{C_2D}{\delta}\left[\delta-\frac{\delta}{1+\delta t_2}\right]\\+cD\left[\frac{1}{(1+\delta t_2)}\right]\end{bmatrix}=0 \tag{4}$$

$$\frac{\partial^2\Psi}{\partial t_2^2}=\frac{2}{(t_1+t_2)^3}\begin{bmatrix}C_0+\frac{C_k}{n^{\alpha}}\\+\left(C_0+\frac{C_h}{n^{\alpha}}\right)D\begin{bmatrix}\frac{t_1^2}{2}+\frac{\theta}{6}\left(t_1^4-\frac{t_1^4}{2}\right)\\-\frac{\theta t_1^4}{24}\end{bmatrix}\\+\\\frac{C_1D}{\delta^2}\left[\delta t_2-\log(1+\delta t_2)\right]+\\\frac{C_2D}{\delta}\left[\delta t_2-\log(1+\delta t_2)\right]+\\cD\left[t_1+\frac{\theta t_1^3}{6}+\frac{1}{\delta}\log(1+\delta t_2)\right]\end{bmatrix}-\frac{1}{(t_1+t_2)^2}\begin{bmatrix}\frac{C_1D}{\delta^2}\left[\delta-\frac{\delta}{1+\delta t_2}\right]+\\\frac{C_2D}{\delta}\left[\delta-\frac{\delta}{1+\delta t_2}\right]\\+cD\left[\frac{1}{(1+\delta t_2)}\right]\end{bmatrix}$$

$$-\frac{1}{(t_1+t_2)^2}\begin{bmatrix}\frac{C_1D}{\delta^2}\left[\delta-\frac{\delta}{1+\delta t_2}\right]+\\ \frac{C_2D}{\delta}\left[\delta-\frac{\delta}{1+\delta t_2}\right]\\ +cD\left[\frac{1}{(1+\delta t_2)}\right]\end{bmatrix}+\frac{1}{t_1+t_2}\begin{bmatrix}\frac{C_1D}{\delta^2}\left[\frac{\delta^2}{(1+\delta t_2)^2}\right]+\\ \frac{C_2D}{\delta}\left[\frac{\delta^2}{(1+\delta t_2)}\right]\\ +cD\left[\frac{-\delta}{(1+\delta t_2)^2}\right]\end{bmatrix}>0 \tag{5}$$

$$\frac{\partial^2\Psi}{\partial t_1\delta t_2}=\frac{2}{(t_1+t_2)^3}\begin{bmatrix}C_0+\frac{C_k}{n^\alpha}+\\ \left(C_0+\frac{C_h}{n^\alpha}\right)D\left[\frac{t_1^2}{2}+\frac{\theta}{6}\left(t_1^4-\frac{t_1^4}{2}\right)-\frac{\theta t_1^4}{24}\right]\\ +\frac{C_1D}{\delta^2}\left[\delta t_2-\log(1+\delta t_2)\right]+\\ \frac{C_2D}{\delta}\left[\delta t_2-\log(1+\delta t_2)\right]\\ +cD\left[t_1+\frac{\theta t_1^3}{6}+\frac{1}{\delta}\log(1+\delta t_2)\right]\end{bmatrix}+\frac{1}{t_1+t_2}\begin{bmatrix}\frac{C_1D}{\delta^2}\left[\delta-\frac{\delta}{1+\delta t_2}\right]\\ +\\ \frac{C_2D}{\delta}\left[\delta-\frac{\delta}{1+\delta t_2}\right]\\ +cD\left[\frac{1}{1+\delta t_2}\right]\end{bmatrix}>0 \tag{6}$$

8 Optimal Production-Inventory Policies for Processed Fruit Juices Manufacturer and Multi-retailers with Trended Demand and Quality Degradation

Nita H. Shah
Gujarat University
Ahmedabad, Gujarat, India

Digeshkumar B. Shah
L. D. College of Engineering
Ahmedabad, Gujarat, India

Dushyantkumar G. Patel
Govt. Polytechnic
Ahmedabad, Gujarat, India

CONTENTS

8.1 INTRODUCTION

In this era of ready-made items and health consciousness, people are more tempted to intake ready-made fruit juices and enjoy instant energy. Non-availability of good fruits and seasonal fruits force customers to buy processed fruit juices, which they can have at any time without any effort. Fruits have a fixed life span, which motivates one to think about the preservation of fruits either in their original forms or in some other forms. According to a study (Gustavsson et al. 2011), fruits and vegetables are wasted from 37% to 55% because of a lack of coordination in the supply chain between different players. The mismatch of demand and supply can be addressed by harmonizing the inventory policies between different stakeholders.

To establish harmony in the supply chain between the different players is somewhat complicated, especially for food products, as they are deteriorating with respect to time and various storage processes. Ghare and Schrader (1963) introduced a mathematical model for deteriorating items. This model was followed by Park (1983), Raafat (1985), Yang and Wee (2003), Rau et al. (2004), Yu et al. (2012) and Taleizadeh et al. (2015) for making inventory policies for deteriorating items. Park (1983) established a model integrating inventory policy in which raw material decayed in storage before being in the process of production. Raafat (1985) extended Park's (1983) model to finished products, which are also subject to decay or deterioration. Further, Yang and Wee (2003) and Rau et al. (2004) advanced the model by considering a number of shipment policies for a multi-stage supply chain production cycle. Yu et al. (2012) and Taleizadeh et al. (2015) set up multi-stage models for raw materials and finished goods by incorporating different characteristics of deterioration.

All the cited articles adopted an instantaneous extinction approach to represent the characteristics of deterioration which are not suitable for food items. Wu et al. (2006) and Ouyang et al. (2006) studied non-instantaneous deteriorating items (i.e. inventory starts deteriorating after a certain period, which is more appropriate for food items such as fruits, vegetables and others). Chang et al. (2010) extended this study by considering the profit function to be optimized while Geetha and Uthayakumar (2010) allowed shortages with partial backlogs. Maihami and Nakhai Kamalabadi (2012), Shah et al. (2013), Soni (2013), Chung et al. (2014) and Wu et al. (2014) also incorporated the non-instantaneous depletion approach in their studies.

For food products, the quality or value of the products deteriorates over the planning horizon; on the other hand, the products still meet the minimum standard of consumption, as the quantity of the product is untouched. Blackburn and Scudder (2009) studied the value change in the supply chain for melon and sweet corn. They anticipated an approach of controlling the selection rate and the bulk of the delivery batch from the harvesting area to the controlled facility to optimize the loss of value in the supply chain. Rong et al. (2011) used a kinetic model to symbolize the loss of quality of peppers in a supply chain. One of the important parameters is temperature, which is used to maintain the quality of food as indicated in the production distribution policy.

Food products pass through different junctures across the supply chain where quality reduces with various behaviours of deterioration depending on their variety (Heldman 2011). For example, the quality of wheat flour as raw material in terms of gluten ingredients declines over its storage time (Mis 2003). Tsiros and Heilman (2005) determined that customers' readiness to pay for a product which is getting closer to its expiry date is reduced.

Mostly, all the articles considered the inventory model with a single vendor and single buyer. But in practice, the one vendor, one buyer situation is seen very rarely. Shah et al. (2011) and Fauza et al. (2016) derived the deteriorating inventory system with a single vendor and multiple buyers. In this chapter, we study the inventory system for a single vendor and multiple buyers under trending demand and deterioration.

8.2 ASSUMPTIONS AND NOTATIONS

We shall use the following notations and assumptions to develop the mathematical model of the problem under consideration.

8.2.1 Notations

N Number of buyers
$R_i(t)$ Annual demand rate of the ith buyer (units/unit time) $= a_i(1 + b_i t - c_i t^2)$, where $a_i > 0$ is scale demand and $0 \le b_i < 1$ denotes the linear rate of change of demand with respect to time, $0 \le c_i < 1$ denotes the quadratic rate of change of demand with respect to time
$\theta(t)$ Time varying deterioration rate at time t, where $0 \le \theta(t) \le 1$
m Maximum lifetime (in years) of the deteriorating item
T Vendor's cycle time (a decision variable)
n_i Number of deliveries per order cycle time T for the ith buyer (decision variable)
$I_v(t)$ Inventory level of vendors at any instant of time t, $0 \le t \le T$
$I_{bi}(t)$ Inventory level of ith buyer at any instant of time t, $0 \le t \le T$
I_{mv} The maximum inventory level of the vendor
I_{mbi} The maximum inventory level of ith buyer
C_v Vendor's purchase cost per unit
C_b Buyer's purchase cost per unit
A_v Vendor's ordering cost per order
A_b Buyer's ordering cost per order
I_v Vendor's inventory carrying charge fraction per unit per unit time
I_b Buyer's inventory carrying charge fraction per unit per unit time
TC_v Total cost of vendor per unit time
TC_b Total cost of all buyer's per unit time
TC Total cost for vendor-buyer inventory system when they take joint decisions
m Maximum lifetime (in years) of the deteriorating item

8.2.2 Assumptions

1. A supply chain system of a single vendor and multi-buyers is considered.
2. The inventory system under study deals with deteriorating items with expiry rates. The deterioration rate tends to be 1 when time tends to maximum lifetime m. Following Sarkar (2012) and Chen and Teng (2014), as well as Wang et al. (2014), the functional form for the deterioration rate is $\theta(t) = \dfrac{1}{1+m-t}$; $0 \le t \le T$. There is no repair or replacement of deteriorated items during the cycle time.
3. The planning horizon is infinite.
4. Shortages are not allowed. Lead time is zero or negligible.

8.3 MATHEMATICAL MODEL

In this section, we develop the mathematical model of an integrated inventory system for a single vendor and *N* buyers. In the supply chain system, different demands from all *N* buyers are observed by the vendor during cycle time *T*, which is satisfied from the inventory. The rate of change of the vendor's inventory level at any instant of time *t* is governed by the differential equation

$$\frac{dI_v(t)}{dt} = -\sum_{i=1}^{N} a_i\left(1+b_i t - c_i t^2\right) - \theta(t) I_v(t), 0 \le t \le T.$$

In Figure 8.1, the inventory level for the ith buyer at any instant of time t is governed by the differential equation

$$\frac{dI_{b_i}(t)}{dt} = -a_i\left(1+b_i t - c_i t^2\right) - \theta(t) I_{b_i}(t), 0 \le t \le \frac{T}{n_i} \text{ where } i = 1,2,\ldots..N.$$

Using various boundary conditions $I_v(T) = 0, I_{b_i}(T_{b_i}) = 0$ and $T_{b_i} = \frac{T}{n_i}$, the solutions of the previous differential equations are as follows:

$$I_v(T) = (1+m-t)\sum_{i=1}^{N} a_i \begin{pmatrix} \frac{c_i}{2}\left(T^2 - t^2\right) + \left(-b_i + c_i(1+m)\right) \\ (T-t) + R_i \, ln\left(\frac{1+m-t}{1+m-T}\right) \end{pmatrix}; 0 \le t \le T$$

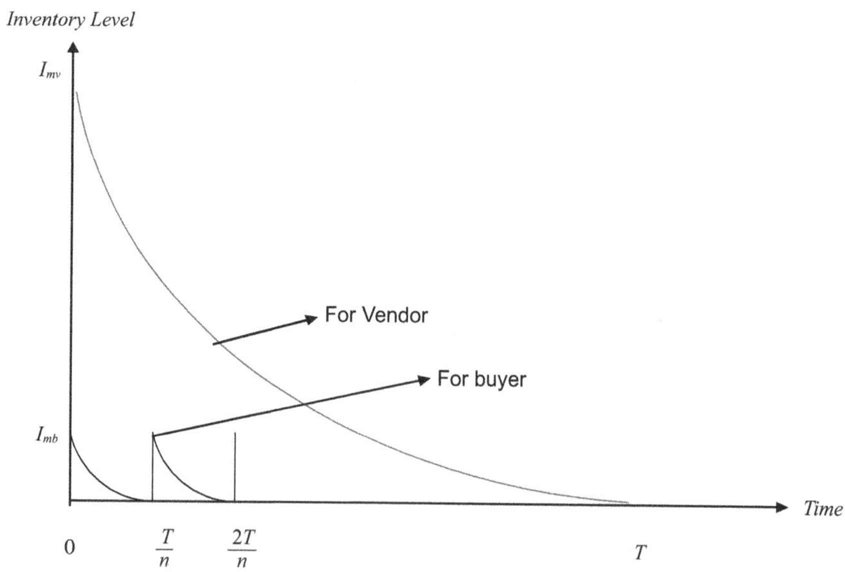

FIGURE 8.1 Vendor-Buyer Inventory Status.

$$I_{b_i}(t) = a_i(1+m-t)\left(\frac{c_i}{2}\left(\frac{T^2}{n_i^2}-t^2\right)+\left(-b_i+c_i(m+1)\right)\left(\frac{T}{n_i}-t\right)+R_i\, ln\left(\frac{1+m-t}{1+m-\frac{T}{n_i}}\right)\right);$$

$$0 \le t \le \frac{T}{n_i} \quad \text{where } i = 1,2,\ldots..N.$$

Using $I_v(0) = I_{mv}$ and $I_{bi}(0) = I_{mb_i}$, the purchase quantities for the vendor and the ith buyer are

$$I_{mv}(T) = (1+m)\sum_{i=1}^{N} a_i\left(\begin{array}{l}\frac{c_i}{2}T^2+\left(-b_i+c_i(1+m)\right)T+\left(1-(m+1)\left(-b_i+c_i(1+m)\right)\right)\\ ln\left(\frac{1+m}{1+m-T}\right)\end{array}\right)$$

$$I_{mb_i}(T) = (1+m)a_i\left(\begin{array}{l}\frac{c_i}{2ni^2}T^2+\frac{\left(-b_i+c_i(1+m)\right)T}{n_i}+\left(1-(m+1)\left(-b_i+c_i(1+m)\right)\right)\\ ln\left(\frac{1+m}{1+m-\frac{T}{n_i}}\right)\end{array}\right)$$

where $i = 1, 2, \ldots..N$.

During the cycle time [0, T], the ith buyer's inventory level is $\int_0^{\frac{T}{n_i}} I_{b_i}(t)dt$. Hence the average inventory level for all N buyers per time is $\frac{1}{T}\sum_{i=1}^{N} n_i \int_0^{\frac{T}{n_i}} I_{b_i}(t)dt$. Therefore, the inventory holding cost for all N buyers is

$$IHC_b = C_b I_b \frac{1}{T}\sum_{i=1}^{N} n_i \int_0^{\frac{T}{n_i}} I_{b_i}(t)dt.$$

The ordering cost for the ith buyer is $n_i A_b$. Hence the ordering cost for all N buyers per time unit is

$$OC_b = \frac{1}{T}\sum_{i=1}^{N} n_i A_b.$$

During cycle time T, the number of units deteriorated for the ith buyer is $n_i\left(I_{mb_i}-\frac{R\left(\frac{T}{n_i}\right)T}{n_i}\right)$.

Therefore, deterioration cost for all N buyers is

$$DC_b=\frac{C_b}{T}\sum_{i=1}^{N}n_i\left(I_{mb_i}-\frac{R\left(\frac{T}{n_i}\right)T}{n_i}\right).$$

Hence the buyer's total cost TC_b per time unit is

$$TC_b=IHC_b+OC_b+DC_b.$$

During cycle time T, the vendor's average inventory level per time unit is $\frac{1}{T}\int_0^T I_v(t)dt$.

The vendor's inventory in the joint two-echelon inventory model is the difference between the vendor-buyers combined inventory and all buyers' inventory. Therefore, the vendor's holding cost per time unit is

$$IHC_v=C_vI_v\left[\frac{1}{T}\int_0^T I_v(t)dt-\frac{1}{T}\sum_{i=1}^{N}n_i\int_0^{\frac{T}{n_i}}I_{b_i}(t)dt\right]$$

The vendor's ordering cost per time unit is

$$OC_v=\frac{A_v}{T}$$

The unit deteriorated at the vendor's inventory system is $\left(I_{mv}-\sum_{i=1}^{N}n_iI_{mb_i}\right)$. Hence cost owing to the deterioration of units is

$$DC_v=\frac{C_v}{T}\left(I_{mv}-\sum_{i=1}^{N}n_iI_{mb_i}\right).$$

Hence the vendor's total cost TC_v per time unit is

$$TC_v = IHC_v + OC_v + DC_v$$

The joint total cost TC is the sum of TC_b and TC_v.

$$TC = TC_b + TC_v \tag{8.1}$$

Here joint total cost TC is the function of discrete variable n and continuous variable T, where $i = 1, 2, \ldots N$.

8.4 COMPUTATIONAL PROCEDURE

Here the objective is to determine the value of n_i, which minimizes the joint total cost TC, where $i = 1, 2, \ldots, N$. Since the number of delivery n_i per order cycle T is a discrete variable, the following steps can be carried out to determine the value of n_i.

(i) For different values of n_i, differentiate the total cost function TC from (8.1) with respect to decision variable T and set it equal to zero (i.e. $\frac{\partial TC}{\partial T} = 0$). For each n_i, denote order cycle T by notation $T(n_i)$, where $i = 1, 2, \ldots, N$.

(ii) Find the optimal solution of n_i and T such that the following condition must satisfy

$$TC\left(n_i^* - 1, T\left(n_i^* - 1\right)\right) \geq TC\left(n_i^*, T\left(n_i^*\right)\right) \leq TC\left(n_i^* + 1, T\left(n_i^* + 1\right)\right).$$

Here $(n_1^*, n_2^*, n_3^*, \ldots\ldots, n_N^*)$ and $T(n_1^*, n_2^*, n_3^*, \ldots\ldots, n_N^*)$ produce the optimal solution.

8.5 NUMERICAL EXAMPLE AND SENSITIVITY ANALYSIS

To illustrate the proposed model in the simplest manner, the supply chain system with a single vendor and two buyers is assumed. For numerical analysis, the following parameter values are considered in proper units.

N = No. of buyers = 2
a_1 = Annual rate of constant demand of the first buyer = 8,000
a_2 = Annual rate of constant demand of the second buyer = 9,000
b_1 = Linear rate of change of demand of the first buyer = 0.05
b_2 = Linear rate of change of demand of the second buyer = 0.05
c_1 = Quadratic rate of change of demand of the first buyer = 0.1
c_2 = Quadratic rate of change of demand of the second buyer = 0.1
$c_v = 10$, $c_b = 13$, $I_v = 0.15$, $I_b = 0.30$, $A_v = 2000$, $A_b = 200$, $m = 0.5$

TABLE 8.1
Optimal Solutions of n_1 and n_2

n_1	n_2	T	TC_b	TC_v	TC
1	1	0.1367761033	462,112	14,622	476,734
1	2	0.1492229279	459,782	16,550	476,332
1	3	0.1573944795	459,869	17,112	476,981
1	4	0.1641876166	460,536	17,343	477,879
2	1	0.1483985661	460,282	16,258	476,540
2*	2*	0.1619528159	457,034	18,847	475,881
2	3	0.1705693571	456,609	19,766	476,375
2	4	0.1776112502	456,887	20,262	477,149
3	1	0.1561991415	460,587	16,688	477,275
3	2	0.1702076200	456,812	19,644	476,456
3#	3#	0.1789310355	456,120	20,738	476,858
3	4	0.1859845038	456,209	21,353	477,562
4	1	0.1627664426	461,391	16,833	478,224
4	2	0.1770379761	457,210	20,066	477,276
4	3	0.1857767011	456,324	21,283	477,607
4	4	0.1927810989	456,281	21,976	478,257

Note that * indicates the integrated optimal solution of n_1 and n_2, which minimizes TC, and # indicates the buyer's optimal solution of n_1 and n_2, which minimizes TC_b

TABLE 8.2
Comparison of Independent and Integrated Policy

	Independent $n_1^* = 3, n_2^* = 3$	**Integrated** $n_1^* = 2, n_2^* = 2$	**Change in Cost**
Total Cost of Buyers	456,120	457,034	914
Total Cost of Vendor	20,738	18,847	−1891
Total Cost of System	476,858	475,881	−977

In Tables 8.1 and 8.2, the optimal solution is exhibited for independent and integrated decisions. If both the buyers follow independent policy, then ordering policy is $(n_1 = 3, n_2 = 3)$ with a total cost of \$476, 858. If both the buyers agree to join in the integrated system with an ordering policy of $(n_1^* = 2, n_2^* = 2)$, then the total integrated cost is reduced to \$475, 881. The graph of the total cost for independent and integrated inventory policies is shown in Figure 8.2. The buyers' cost increases when both the buyers and the vendors agree to make a joint decision. In the integrated policy, the vendor gain \$1,891 and buyers lose \$914. Since the integrated strategy is beneficial to the vendor, buyers do not agree to the joint decision. To encourage the buyers to cooperate with the integrated system, the vendor should offer the buyers a

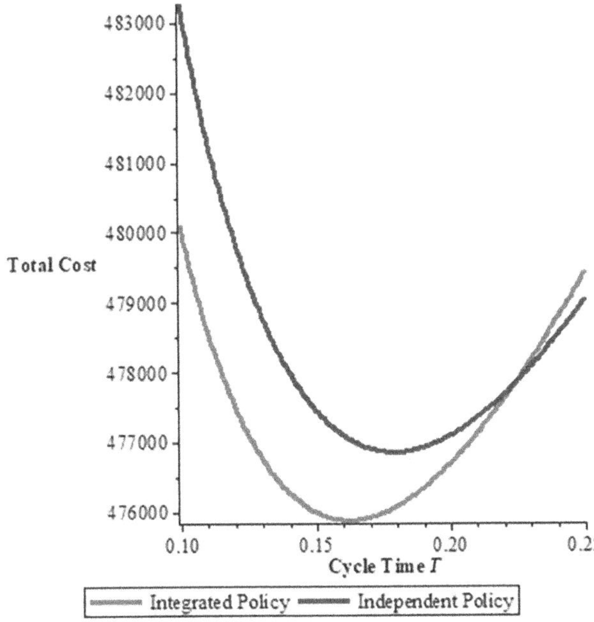

FIGURE 8.2 Total Cost for Integrated versus Independent Vendor-Buyers Inventory System.

permissible delay in payment or some proportion of sharing the extra benefits. This integrated policy reduces the integrated total cost.

Hence the adoption of the integrated policy instead of the independent policy decreases the channel cost.

The results obtained in the numerical analysis indicate the need to perform sensitivity analysis for the model parameter. The aim is to identify the parameters that are more relevant to the performance of the system. The sensitivity analysis is carried out by increasing or decreasing model parameters by 20%, which is shown in Tables 8.3 to 8.8. In Tables 8.3 to 8.8, * indicates the integrated policy, whereas # indicates the independent policy.

From Table 8.3 it is observed that an increase in the constant demand a_i changes the total cost in the range (805–1118).

From Table 8.4 it is seen that an increase in linear demand b_i changes the total cost from 1,050 to 907.

From Table 8.5, the total cost changes from 964 to 991 with an increase in quadratic demand c_i.

From Table 8.6, it is observed that an increase in unit cost c_v and c_b changes the total cost in the range (805–1118).

From Table 8.7, it is seen that an increase in ordering cost A_v and A_b changes the total cost from 720 to 1202.

From Table 8.8, it is observed that an increase in ordering cost I_v and I_b decreases the total cost from 1069 to 896.

TABLE 8.3
Sensitivity Analysis for Constant Demand a_i

a_1	a_2	n_1^*	n_2^*	TC*	$n_1^\#$	$n_2^\#$	TC#	Change in Total Cost
4,800	5,400	2	2	291,603	3	3	292,408	805
6,400	7,200	2	2	383,978	3	3	384,874	896
8,000	9,000	2	2	475,881	3	3	476,858	977
9,600	10,800	2	2	567,448	3	3	568,498	1,050
11,200	12,600	2	2	658,761	3	3	659,879	1,118

TABLE 8.4
Sensitivity Analysis for Rate of Change of Linear Demand b_i

b_1	b_2	n_1^*	n_2^*	TC*	$n_1^\#$	$n_2^\#$	TC#	Change in Total Cost
0.03	0.03	2	2	475,169	3	3	476,219	1,050
0.04	0.04	2	2	475,527	3	3	476,540	1,013
0.05	0.05	2	2	475,881	3	3	476,858	977
0.06	0.06	2	2	476,231	3	3	477,172	941
0.07	0.07	2	2	476,577	3	3	477,484	907

TABLE 8.5
Sensitivity Analysis for the Rate of Change of Quadratic Demand c_i

c_1	c_2	n_1^*	n_2^*	TC*	$n_1^\#$	$n_2^\#$	TC#	Change in Total Cost
0.06	0.06	2	2	476,008	3	3	476,972	964
0.08	0.08	2	2	475,945	3	3	476,915	970
0.1	0.1	2	2	475,881	3	3	476,858	977
0.12	0.12	2	2	475,816	3	3	476,800	984
0.14	0.14	2	2	475,751	3	3	476,742	991

TABLE 8.6
Sensitivity Analysis for Unit Cost c_v and c_b

c_v	c_b	n_1^*	n_2^*	TC*	$n_1^\#$	$n_2^\#$	TC#	Change in Total Cost
6	7.8	2	2	291,603	3	3	292,408	805
8	10.4	2	2	383,978	3	3	384,874	896
10	13	2	2	475,881	3	3	476,858	977
12	15.6	2	2	567,448	3	3	568,498	1050
14	18.2	2	2	658,761	3	3	659,879	1118

TABLE 8.7
Sensitivity Analysis for Ordering Cost A_v and A_b

A_v	A_b	n_1^*	n_2^*	TC*	$n_1^\#$	$n_2^\#$	TC#	Change in Total Cost
1,200	120	2	2	468,123	3	3	468,843	720
1,600	160	2	2	472,238	3	3	473,092	854
2,000	200	2	2	475,881	3	3	476,858	977
2,400	240	2	2	479,188	3	3	480,280	1,092
2,800	280	2	2	482,241	3	3	483,443	1,202

TABLE 8.8
Sensitivity Analysis for Ordering Cost I_v and I_b

I_v	I_b	n_1^*	n_2^*	TC*	$n_1^\#$	$n_2^\#$	TC#	Change in Total Cost
0.09	0.18	2	2	474,307	3	3	475,376	1,069
0.12	0.24	2	2	475,103	3	3	476,125	1,022
0.15	0.3	2	2	475,881	3	3	476,858	977
0.18	0.36	2	2	476,640	3	3	477,576	936
0.21	0.42	2	2	477,384	3	3	478,280	896

8.6 CONCLUSION

In this chapter, a mathematical model for deteriorating items was developed to study an optimal integrated strategy of a single-vendor, multi-buyer system when the demand is trending.

A numerical analysis reveals that the joint policy lowers the cost of an inventory system, even though the total cost of all the buyers increases. To entice the buyers to cooperate, a promotional incentive in terms of trade credit should be offered by the vendor to the buyers, which also helps to maintain long-term contracts between all players in the supply chain.

REFERENCES

Blackburn, J., and Scudder, G. (2009). Supply chain strategies for perishable products: The case of fresh produce. *Production and Operations Management*, 18(2), 129–137. doi:10.1111/j.1937-5956.2009.01016.x.

Chang, C. T., Teng, J. T., and Goyal, S. K. (2010). Optimal replenishment policies for non-instantaneous deteriorating items with stock-dependent demand. *International Journal of Production Economics*, 123(1),62–68. doi:10.1016/j.ijpe.2009.06.042.

Chen, S. C., and Teng, J. T. (2014). Retailer's optimal ordering policy for deteriorating items with maximum lifetime under supplier's trade credit financing. *Applied Mathematical Modelling*, 15–16(38), 4049–4061. doi:10.1016/j.apm.2013.11.056.

Chung, K. J., Cárdenas-Barrón, L. E., and Ting, P. S. (2014). An inventory model with non-instantaneous receipt and exponentially deteriorating items for an integrated three layer supply chain system under two levels of trade credit. *International Journal of Production Economics*, 155, 310–317. doi:10.1016/j.ijpe.2013.12.033.

Fauza G., Amer Y., Lee, S. H., and Prasetyo, H. (2016). An integrated single-vendor multi-buyer production-inventory policy for food products incorporating quality degradation. *International Journal of Production Economics*, 182, 409–417.

Geetha, K. V., and Uthayakumar, R. (2010). Economic design of an inventory policy for non-instantaneous deteriorating items under permissible delay in payments. *Journal of Computational and Applied Mathematics*, 233(10), 2492–2505. doi:10.1016/j.cam.2009.10.031.

Ghare, P. M., and Schrader, G. F. (1963). A model for exponentially decaying inventory. *Journal of Industrial Engineering*, 14(5), 238–243.

Gustavsson, J., Cederberg, C., Sonesson, U., Van Otterdijk, R., and Meybeck, A. (2011). Global food losses and food waste-Extent, Causes and Prevention. *Food and Agriculture Organization of the United Nations, Rome.*

Heldman, D. R. (2011). Kinetic models for food systems. In S. L. Taylor (Ed.), *Food Preservation Process Design* (19–84). San Diego: Academic Press.

Maihami, R., and Nakhai Kamalabadi, I. (2012). Joint pricing and inventory control for non-instantaneous deteriorating items with partial backlogging and time and price dependent demand. *International Journal of Production Economics*, 136(1), 116–122. doi:10.1016/j.ijpe.2011.09.020.

Mis, A. (2003). Influence of the storage of wheat flour on the physical properties of gluten. *International Agrophysics*, 17(2), 71–75.

Ouyang, L. Y., Wu, K. S., and Yang, C. T. (2006). A study on an inventory model for non-instantaneous deteriorating items with permissible delay in payments. *Computers and Industrial Engineering*, 51(4),637–651. doi:10.1016/j.cie.2006.07.012.

Park, K. S. (1983). An integrated production-inventory model for decaying raw materials. *International Journal of System Science*, 14(7), 801–806. doi:10.1080/00207728308926498.

Raafat, F. (1985). A production-inventory model for decaying raw materials and a decaying single finished product system. *International Journal of System Science,* 16(8),1039–1044. doi:10.1080/00207728308926498.

Rau, H., Wu, M. Y., and Wee, H. M. (2004). Deteriorating item inventory model with shortage due to supplier in an integrated supply chain. *International Journal of System Science*, 35(5), 293–303. doi:10.1080/00207720410001714833.

Rong, A., Akkerman, R., and Grunow, M. (2011). An optimization approach for managing fresh food quality throughout the supply chain. *International Journal of Production Economics*, 131(1), 421–429. doi:10.1016/j.ijpe.2009.11.026.

Sarkar, B. (2012). An EOQ model with delay in payments and time varying deterioration rate. *Mathematical and Computer Modelling*, 55(3–4), 367–377.

Shah, N. H., Gor, A. S., and Jhaveri, C. (2011). An integrated inventory policy with deterioration for a single vendor and multiple buyers in supply chain when demand is quadratic. *Revista Investigación Operacional*, 32(2), 93–106.

Shah, N. H., Soni, H. N., and Patel, K. A. (2013). Optimizing inventory and marketing policy for non-instantaneousdeterioratingitemswithgeneralizedtypedeterioration and holding cost rates. *Omega*, 41(2),421–430. doi:10.1016/j.omega.2012.03.002.

Soni, H. N. (2013). Optimal replenishment policies for non-instantaneous deteriorating items with price and stock sensitive demand under permissible delay in payment. *International Journal of Production Economics*, 146(1), 259–268. doi:10.1016/j.ijpe.2013.07.006.

Taleizadeh, A. A., Noori-daryan, M., and Cárdenas-Barrón, L. E. (2015). Joint optimization of price, replenishment frequency, replenishment cycle and production rate in vendor managed inventory system with deteriorating items. *International Journal of Production Economics*, 159, 285–295. doi:10.1016/j.ijpe.2014.09.009

Tsiros, M., and Heilman, C. M., (2005). The effect of expiration dates and perceived risk on purchasing behaviour in grocery store perishable categories. *Journal of Marketing*, 69(2), 114–129. doi:10.1509/jmkg.69.2.114.60762.

Wang, W. C., Teng, J. T., and Lou, K. R. (2014). Seller's optimal credit period and cycle time in a supply chain for deteriorating items with maximum lifetime. *European Journal of Operational Research*, 232(2), 315–321. doi:10.1016/j.ejor.2013.06.027.

Wu, J., Skouri, K., Teng, J. T., and Ouyang, L. Y. (2014). A note on optimal replenishment policies for non-instantaneous deteriorating items with price and stock sensitive demand under permissible delay in payment. *International Journal of Production Economics*, 155, 324–329. doi:10.1016/j.ijpe.2013.12.017.

Wu, K. S., Ouyang, L. Y., and Yang, C. T. (2006). An optimal replenishment policy for non-instantaneous deteriorating items with stock-dependent demand and partial backlogging. *International Journal of Production Economics*, 101(2), 369–384. doi:10.1016/j.ijpe.2005.01.010.

Yang, P. C., and Wee, H. M. (2003). An integrated multi-lot-size production inventory model for deteriorating item. *Computers & Operations Research*, 30(5), 671–682. doi:10.1016/S0305-0548(02)00032-1.

Yu, Y., Wang, Z., and Liang, L. (2012). A vendor managed inventory supply chain with deteriorating raw materials and products. *International Journal of Production Economics*, 136(2), 266–274. doi:10.1016/j.ijpe.2011.11.029.

9 Information Visualization

Perception and Limitations for Data-Driven Designs

Sadia Riaz
S.P. Jain School of Global Management
Dubai, UAE

Arif Mushtaq
City University College of Ajman
Ajman, UAE

Maninder Jeet Kaur
Canadore College
Ontario, Canada

CONTENTS

9.1 INTRODUCTION: BACKGROUND AND DRIVING FORCES

Each second a lot of information and data is generated, which has become an important part of our everyday lives. Big data relates to the wide and diverse collection of rapidly evolving information, including volume of information, its intensity and the context or quality of the data points. Big data often comes from a number of channels and in various formats. It can be captured from publicly shared social networks, forum comments, reviews, personal electronics, cell phones, polls, product purchases, online reviews or check-ins. Social networking alone has produced huge numbers of data, including Facebook, YouTube, Twitter and Instagram [1]. Domo's Data Never Sleeps 5.0 [2] reported the amount of data in numbers generated by different social media applications every minute of the day [1].

- YouTube Videos watch – 4,146,600
- LinkedIn professionals joined – 120
- Instagram photos posts – 46,740
- Snapchat photos shared – 527,760
- Twitter tweets – 456,000

With more than 1.4 trillion months of active social network user interaction, Facebook creates the highest amount of data, where users share more than four million posts per minute – equalling 4,166,667 or up to 250 million posts per hour. More than 2.5 quintillions of data are produced daily from different platforms, and with the advent of the Internet of Things [3], it is estimated that by 2020, every second, 1.7 MB of data will be generated by each person on the earth. Hence the integration of cameras and other sources in smart devices will allow data to be collected and transmitted in a broad variety of dimensions and situations [4–6]. The recent rise of volumes of visual data only confirms that humans by nature are visual creatures. Viewers absorb graphical information to elude processing words or textual information. Visuals are helpful for every type of story, particularly when the narrative becomes complicated [7]. One the one side, visuals may be relevant for the communication of an experimental outcome and unexpected results, as well as provide insight into the underlying structure of the data. One the other side, researchers claim visualization to be an optic gimmick used to generate viewers' interest. Some may even refer to visuals as an icing on the cake with nothing deep or fascinating underneath. This is because the representation of data is loaded with examples that are packed with visuals showing poor data representation. Poor visual data not only compromises the quality of data but also impedes scientific research progress. So with more and more research images joining the news and social media – from climate change to epidemic outbreaks – the scope for misleading visuals is high for impairing a general understanding of science. It confuses readers, as well as deceives data scientists, who are analytical professionals who use technology and social science knowledge to identify trends and manage big data. Data scientists have a range of expertise in technology, such as Hadoop, NoSQL, Python, Spark, R and Java, and use industry knowledge, intellectual insights and cynicism to find solutions to business problems [8]. In practice, there are a lot of challenges for big data processing and analysis. As all data are currently visualized by computers, this leads to difficulties in the extraction of data, followed by its perception and cognition. Those tasks are time-consuming and do not always provide correct or acceptable results. Nonetheless, for data scientists, the dynamic dimensions of big data pose a difficulty because they have to find a way to cleanse and identify problems, reduce their complexity and produce appealing visualizations. Data visualization is the representation of data in a textual, pictorial or graphical presentation by interpreting or analysing data to identify specific patterns, correlations and other perceptive behaviours. These representations are produced with the help of high-end analytical software – known tools for data visualization. Many solutions have already been proposed in this field. Husain et al. [9], for example, presented a range of contemporary and recent visualization methods during their work. The methods proposed are developed for large and commercial data platforms (e.g., IBM and software applications).

Even though data visualization tools also face issues, such as information processing and analysis, they are critical for the process of visualization. It has become evident that the big data era has created massive, complex and heterogeneous databases, making it even more difficult to turn a data-curious user into someone who can interpret and analyse data independently. It has become a challenge as users have little-to-no training and experience in the field of data processing and visualization. Data visualization has, therefore, become a major research problem requiring a range of data storage, querying, indexing, image display, accessibility and personalization problems. This chapter, therefore, discusses various issues and research in the field of data visualization. It is designed to (1) systematically review and synthesize literature on available visualization techniques and discuss applications in sectors, (2) discuss human visual limitations that hinder the decoding of visual data and (3) deliberate data visualization case studies that support human visual limitations in decoding big data effectively.

9.2 VISUALIZATION OF BIG DATA

In this section, existing literature on big data visualization has been synthesized, while prominent visualization techniques and applications for big data in smart cities are discussed.

Data visualization is a modern scientific concept, which in a way is an upgradation of the visual representation of quantitative data that used to be in the form of early mapping techniques, applied thematic analysis, statistical predictions and plotting. In addition, it includes advancements and developments in the fields of other disciplines, such as medicine and science. It is argued to also associate with the rise of 19th-century mathematical reasoning, economic analysis and business data gathering. Likewise, emerging trends and methods, such as imaging and retrieval techniques, innovations in mathematics and statistical research methodologies, data analytics, scientific evaluation and innovation in document preparation have immensely contributed to making data visualization a highly robust and adopted measure in today's time.

9.2.1 Data Analytics and Big Data Visualization

Data analytics is more structured than visualizing information. The word often applies to a set of solutions to evaluate digital information, their perspectives or both. Some common types of data analytics systems include the following: business intelligence software: A business intelligence program may gather data from multiple sources and execute multiple tasks, including packing it for data warehouse storage to data mining, which reveals significant connections in a data set. Online analytical processing (OLAP) utilities: OLAP is a group of database tools designed for fast lookup and analysis. Users who explore OLAP find specific details grouped into "cubes" residing on the back end between the consumer and data warehouse, including metrics such as time, position and product line for various data formats. Advanced analytics: Machine learning methods involve training robotics to perform complex tasks without instruction, and this group includes various artificial intelligence applications. Analytics, irrespective of approaches, help companies answer questions and optimize performance. For example, a data scientist could use an OLAP tool to find a sales report for a specific year and then compare the information as part of an attempt to better understand long-term trends. Big data visualization differs from the previously described processing methods and from conventional visual techniques. It primarily works on feature extraction and geometric simulation to display large-scale data. It exploits the visual senses of the human brain by applying processes to reduce the size of the results [10]. Experts conclude that the vision is our primary sense: 80%–85% of the knowledge that we interpret, understand and process is mediated by this viewpoint. According to the *New York Times* bestselling book *Brain Rules*, a person typically retains 65% of what he or she sees in a picture after three days, compared to just 10% of what has been heard. Because of this limitation, unprecedented opportunities have opened up in the field of data visualization owing to the emergence of big data. Mordor Intelligence [11] predicts that by the end of 2019, visualization demand will be expanded by 9.21% from $4.12 billion in 2014 to $6.40 billion at the annual compound growth rate. Likewise, SAS Institute presents the findings of the International Data Group study in their white paper [12]. This research focuses on the analysis of large-scale data by organizations and reveals that 98% of the most successful companies with big data have used visualization techniques for analysis. The study results highlight the advantages of visualization in terms of faster decision making, stronger ad hoc data analysis, increased collaboration and information sharing within and outside the organization. Therefore, big data visualization refers to the use of more contemporary visualization techniques in describing relationships in information. Scientists of different fields use machine tools/technology to model dynamic phenomena, which cannot be directly observed [13]. They describe concepts such as weather conditions, medical testing or statistical relations with the help of data visualization to demonstrate real-time shifts and images (which go beyond the simple pie, bar and other charts) and illustrate through diagrams (that do not use hundreds of columns and attributes) better visual representation of the results [14]. Hence, in almost all fields of knowledge, data visualization is applied. Comprehensive research and visualization are some of the most efficient ways to distinguish important relationships [13].

Furthermore, typically, graphs, columns and charts are created when companies need to show information and their interrelationships with the help of data analytics. They are visually enhanced by adding a number of colours, words and, sometimes, symbols. However, the main problem with this setup is that it does not do a good job of presenting huge amounts of data. Data visualization uses more immersive, graphical graphics – such as design and animation – to view figures and link data parts [14]. Thus visualization of big data may not be as simple and requires working with different technologies of big data analysis. Although a wide array of new, copyrighted and open sources big data visualization tools are now available, these packages are unique as they tend to provide all or many of the listed features. To meet and exceed the needs of the customer, a certain collection of features should be offered by the big data visualization tools [15]:

- Capacity to process multiple types of incoming data
- Capacity to apply different filters to adjust the results
- Capacity to interact with the data sets during the analysis
- Capacity to connect to other software to receive incoming data or provide input for users

9.2.2 Techniques in Big Data Visualization

The data has no significance in itself, while data interpretations help us drive value. Data visualization applications also promote the interpretation of trends and patterns that can enable critical and informed decisions. We can explain some of the most common methods for visualizing visualize big data. These methods help in choosing the right solution for your scenario. This is, however, a short listing of the several websites, commercial applications and tools for big data visualization popular at present. A company will always have to find the tool that is best suited for it that can turn raw data or information into a number of images and charts that are easy to understand.

- Jupyter – Jupyter is known as a one-stop shop for big data visualization. It is an open-source project that enables, evaluates, accesses and functions together with big data in real time in over a dozen languages to develop software. The device maintains the field of input, and the program executes the code to construct a visually understandable image based on selected design methods. That's just the tip of the iceberg though. You can share Jupyter notebooks with team members to allow internal collaboration and enhance data analysis teamwork. The notebooks can also be uploaded to GitHub or Gitlab to share reports with the public, included in a Kubernetes Docker container and run on any other Jupyter-supported device. Jupyter notebooks now implement kernels for other programming languages, such as Java, Go, C#, Ruby and many others, but it first used Python and R. The ability to communicate with multiple systems, such as Spark, makes Jupyter a complete system capable of handling data from large, data-intensive applications with various input services [15].

- Tableau – Tableau is known as the easiest way to view artificial intelligence (AI), big data and machine learning applications. Tableau is among the big data visualization market leaders, particularly efficient in delivering interactive data visualization for results derived from big data operations, deep learning algorithms and multiple types of AI-driven applications. For Amazon Web Services, MySQL, Hadoop, Teradata and SAP, Tableau can be incorporated, making this software a flexible tool to construct informative graphs and intuitive data representation. It allows C-suite and middle-chain administrators to make grounded decisions based on tableau graphs that are insightful and easy to read.
- Google Charts – Google Charts is a popular tool because of its free availability and secure and efficient Google power integration. Google is now synonymous with innovation, and just as Google Chrome is probably the most popular browser out there, Google Charts is among the best big data visualization tools, not to mention that it's completely free and Google's enthusiastic help. Why is that? Because the data collected through this approach is naturally used to train Google's AI, it's a win-win situation for all parties. Google Charts offers a wide range of types of visualization, from simple pie charts and time series to multi-dimensional interactive matrixes. Adjustment options are numerous, and if deep modification is needed, an extensive support section is available. The tool makes the resulting charts HTML5/scalable vector graphics (SVG) compatible with any browser. Vector markup language support ensures compatibility with older versions of Internet Explorer (IE), and it is possible to port the charts to the latest releases of Android and iOS. Most specifically, Google Charts incorporates data from multiple Google services, such as Google Maps. As a result, interactive charts are generated that ingest data in real time and can be managed using an interactive dashboard.
- D3.js – D3.js is renowned for imaging the big data in any way you want. D3.js stands for data-driven document, a JS library that basically requires real-time interactive big data visualization in any way. It is assumed to be not a tool; therefore, users should have a solid understanding of JavaScript to work with the data and present it in a humanly understandable way. This library transforms the data into SVG and HTML5 formats, but older browsers like IE7 and 8 are unable to exploit the capabilities of D3.js. Data collected from disparate sources such as large-scale data sets are binding in real time with document object model to creating immersive animations (both 2D and 3D) in a very fast way. The architecture of D3 allows users to reuse the codes intensively through a number of add-ons and plug-ins [15].

9.3 APPLICATIONS

In a variety of applications in the big data age, visualization techniques are essential. The volume, speed, heterogeneity and complexity of the data available make exploration and analysis of the data very difficult for humans. Data visualization allows consumers to perform a series of computational activities that, with common data processing tools, are not always feasible [16]. Physics and astronomy are the main domains of applications for data visualization and analysis. Satellites and telescopes generate streams of large and dynamic data every day. Using traditional analytical methods, astronomers can identify noise, patterns and similarities. Visual analytics, on the other hand, enable astronomers to identify unexpected phenomena and conduct many difficult operations which conventional analytical approaches cannot perform. Another field of application is atmospheric sciences, such as meteorology and climate science. High volumes of data are gathered regularly in this area from sensors and satellites. The storage of these data over the years leads to a huge volume of data to be analysed. Visual analysis can help scientists carry out core tasks, such as climate-related analysis and event prediction. However, simulation tools for analysing real-time events, such as earthquakes, explosions, flooding and tsunamis, are being used in this area in a number of scenarios. Visualization methods are used in many bioinformatics projects as well. It is extremely difficult, for example, to interpret the large quantities of biological data provided by DNA sequencers. Visual techniques can help biologists to gain insight and identify important gene areas for experimentation. Likewise, as we are talking about the big data age, technical visualizations are commonly used in the field of business intelligence these days. Finance markets are one area of application in which visual analytics enable the monitoring of markets, the identification of trends and the making of predictions. Besides, market research is also an application area within business intelligence. A vast number of different sources (e.g., consumer perceptions, finance data, social media analytics) are analysed by marketing agencies and home marketing departments. Visual techniques are used to perform tasks, such as identifying trends, emerging market opportunities, influential users and communities, as well as optimizing activities (e.g. product and server troubleshooting), business analysis and development (e.g. churn-rate prediction, optimization of the marketing system).

9.3.1 Visualization Tools for Insurance Risk Processes

This section addresses management mechanisms that are best suited for insurance-related data analysis and visualization to evaluate risk mitigation. At the same time, risk assessments are basic tools for any non-life actuary – they are used to determine how much damage an insurance company can incur. These often benefit from ratings-triggered, step-up bonds tied to volatile interest rate fluctuations, as well as catastrophe bonds in which the size of the voucher pay-off relies on the severity of disasters. A traditional insurance risk model, the so-called aggregate risk model, has two major parameters: an occurrence rate (or incidence) concept and a description of the severity (or value and/or quantity) of an event's gain or loss [17–19]. Incidence and frequency are generally considered cyclical and mutually unrelated. Together, they form the backbone of the practical risk mechanism. Accordingly, both must be adapted to available historical data. All three strategies mentioned in this section agree that while mean excess function, limited value evaluation and likelihood plot are fairly simple techniques, they offer valuable support during the estimation process. In most instances, no theoretical risk conclusions can be drawn based on cyclic time evolution. Nonetheless, this development is essential to practitioners who have to quantify the risk cycle, such as the estimated time to ruin and the likelihood of ruin. It, therefore, requires efficient statistical estimation schemes [20], as well as strong inference methods. This segment discusses four such tools.

1. Probability gates
2. Ruin probability plot
3. Quantile lines
4. Density evolution plot

These four techniques allow an immediate assessment of the model's suitability and the risks of the company on the basis of visual inspection of the plots. As such, they are particularly useful to high-level managers who are interested in an overview of the situation and do not have to examine all the data behind the final result.

9.3.2 Graphical Data Representation in Bankruptcy Analysis

For bankruptcy studies, graphical data representation is important for model selection because the problem is not very transparent, as well as highly non-linear. The simple display of results in a closed form in classical assessment models eliminates the need for graphical tools. While less robust, non-linear and non-parametric models are more visual in comparison. In several stages of corporate default research, simulation methods based on support vector machines (SVM) are applied. The steps that follow include the selection of variables, the calculation of probability of default (PD) and displaying PDs with colour coding for two- and higher-dimensional models. The right colour scheme is chosen at this level to ensure the correct representation of PDs. The findings are translated into PDs as a monotonous and non-parametric regression. SVM learns a non-parametric scoring function, which is in turn non-parametrically converted into PD. Because PDs cannot be viewed directly, alternate display strategies must be sought. The visualization tools [21] are useful to do this.

9.3.3 Reconstruction, Visualization and Analysis in Medical Images

The advances in medical imaging technologies have played an important role in medical diagnosis and procedures by offering anatomical and functional information on human bodies that is difficult to obtain in the absence of these techniques. Such models often generate large quantities of noisy data that require state-of-the-art computation and estimation for accurate image reconstruction and visualization. This segment outlines the latest research in this field and discusses problems to be discussed in future studies. In particular, computational statistics for positron emission tomography (PET), ultrasound images and magnetic resonance images from an image reconstruction perspective, image segmentation and model-driven image analysis are discussed.

Scientists in many fields have been lucky enough to "look at the unseen" through data visualization. It is rather surprising that researchers can view the functions of the brain throughout the human body and interpret functions with state-of-the-art techniques of medical imaging [22]. This may have been hard to imagine at the beginning of the 20th century [22]. Nevertheless, modern medical imaging techniques have significantly contributed to the understanding, diagnosis and treatment in most medical books, precisely elaborating biological processes and diseases inside the human body [23, 24]. As the technology that captures medical images continues to evolve, visualizing and interpreting them has become more difficult because of a massive number of inherent noisy data. Modern computer analytics methods are important for the retrieval of valuable information from different types of modern medical images. The section will now further address computational statistical methods for examinations of medical images. It will include reconstruction, segmentation and vision modelling of images captured by magnetic resonance imaging (MRI), ultrasound and PET.

Computerized tomography (CT) is an important process for visualizing accurate information on the inside of a human body based on observations outside the body. The detailed reproduction of images from the interior of the human body is a technical task. Since the observations detected are indirectly related to the target image, the problem of tomographic reconstruction is an inverse problem, often unplaced or ill conditioned. For example, reconstruction of PET images by maximum probability approximation using an expectation–maximization algorithm [25, 26], the weighted least square estimate and other methods without regularization will generate edge and noise in images. Therefore, computer-based statistical methods should be used to integrate and merge associated and incomplete structural knowledge into other diagnostic modalities, including X-ray, CT, MRI and others [27–29]. Reconstruction challenges become rather complicated for molecular imaging types in postgenomic, genomic and medical studies, such as micro-PET, where reliable and precise imaging at molecular resolution levels require considerable work. Therefore, spatial frequency analyses and segmentation of images are synthesized to model human vision in such cases [30–34]. These studies show that Gabor multichannel filter outputs can reproduce texture feature descriptors to distinguish responses in different directions and frequencies. Using image recognition and statistics, this vision model can be applied to medical image segments. Ultrasound videos are particularly noisy because of inherent noise. Vision model-based methodologies for ultrasound framing are derived from active contour (i.e. snake-balloon model), regional competition and related methods [35–37]. The technique of inverse regression [38] has also been applied to dynamic segment images, such as magnetic resonance [39].

9.3.4 Exploratory Graphics of a Financial Data Set

The first steps of any data analysis are to learn about the research's objectives and the results. In this case, the goal is to estimate the risk of bankruptcy of a company based on its recent financial performance. Many comprehensive prediction models are addressed in the literature on SVM. Viewing methods are used to examine the large data sets, for example, of American corporate accounts available for predicting bankruptcy to understand and evaluate the data set quality. It is an original exploratory study which does not use experience in accounting. The exploratory data analysis (EDA) has been popular within the domain of inferential statistics since the landmark book by Tukey [40]. While everyone understands the importance of EDA, little has been published on the same. Similarly, modern known methods, such as interactive graphics [41], are not as much practically applied. However, for this chapter's exploration and discussions, interactive graphics have been frequently used. The challenge is that while contemplating digital presentations, the presentation and exploratory graphics must be separated. Graphs representing different variables or pairs of variables are often used for describing or recording data distributions. Care must be taken in terms of proportions, aspect ratios, legends and all graphical properties that could influence the performance of the display in communicating information to others. Exploration graphics are distinctive and more likely to be multivariate in nature. Also, aesthetic characteristics of graphics need not be of particular concern as long as data is clearly depicted and makes sense. Presentation graphics are long-lived because they are drawn for review by many people. For instance, Playfair's English business data plots are years old but still informative and insightful [42]. Exploratory graphs can be very short-lived, as they serve the needs of one or two people. A data analyst may analyse a large number of images before discovering any that reveal information, and, after finding that information, the analyst may conclude that it is better presented by a different type of display altogether. This chapter's visual illustrations are a subset of those used in the study; some visuals may not appear "pretty" primarily because they were drawn to do a job. Detailed scales were not included, as the distributional form is always of prime importance. Exploratory analyses are a highly subjective approach, as each analyst can select different graphs or a combination method to capture data insights and reveal information visually. It can only be said with confidence that researchers who do not use graphics to know their data would underestimate the properties of the data they deal with. If nothing else, graphics are extremely useful for data quality assessment and cleansing.

9.4 DATA VISUALIZATION AND HUMAN VISUAL LIMITATIONS

In this section, elements of human cognition factors and visual limitations in data visualization are elaborated on. Data visualization involves translating figures and raw data into visual objects: lines, columns, maps, etc. It attempts to integrate elements of user-friendliness and aesthetics. Such visualizations provide comprehensible raw-to-complex data analysis, making research a strong tool for communication and collaboration [43]. Visual cognition is the psychological branch which combines visual and prior knowledge to create high-level representations in visual scenes and unconsciously agree on content [44]. Visual processing primarily focuses on the recognition and representation of objects and spatial relationships in perception and imagery. Although visual cognition and cognitive vision may sound remarkably similar, they are different. Cognitive vision applies to functional and computational computer-focused vision systems. Vision cognition is the use of partial information in the human visual system to learn from the structure of a visual image. The visualization aims to support data interpretation through the use of the human brain's visual cognition system that recognizes patterns, trends and outliers [45]. Therefore, aspects of human perception and cognition that are considered necessary for effective data visualization are elaborated on in this section. In addition, case studies from real-life scenarios are addressed to highlight the importance of visual cognition for data processing and analysis.

9.4.1 Cognitive Load Factors

When designing visualizations, we must pay attention to human perceptions and cognitive load factors, such as working memory, attention resources, cognitive tunnelling and proceduralization.

- Working Memory and Visual Chunking [46]: People have a minimal working memory to remember new things. And we should avoid creating emotional fatigue in visualization representation.
- Chunking: A short-term brain memory technique that organizes small pieces of information into a coherent whole for fast retrieval. If items can be separated in memory pieces, they can greatly increase performance [47].
- Attention Resources: It has been shown that we have focus limits while doing several things concurrently. Within Daniel Kahneman's ability model, he described concentration as a cognitive effort whose capacities differ based on multiple factors, such as the environment, behaviours and human circumstances. Irrespective of the resource pool, knowledge competes for our attention. Every image has a different coloured photo square [48].
- Cognitive Tunnelling: The mental state in which the brain focuses on the closest things. It doesn't see the remaining environment or other relevant data. However, it is also difficult to direct readers to the best content. Viewers may concentrate on the incredible animation only without attempting to understand the content. This is the same as interactive presentation of results.
- Procedural Knowledge: A collection of items or comments suggesting a cognitive situation and behaviour if the requirement is fulfilled. Looking at the information processing model, we see how long-term memory is created. In John Robert Anderson's paradigm, learning occurs in stages. Declarative knowledge is acquired when high working memory load is essential, resulting in frequent errors. By accumulating facts, declarative knowledge then transforms into procedural knowledge [49].

9.4.2 Visual Encoding Elements

Visualization of data is an ideal tool to compensate for the capacity of the brain to interpret visual information rapidly, as long as you keep in consideration a few human visual requirements.

a. Usually, the brain is much better at identifying and understanding data characteristics and patterns when they are viewed in a well-performed visualization rather than presented in a table.

b. Visual encryption (encoding) refers to the way visual attributes, including colour and shape, can be assigned to various kinds of information. Visual encoding transforms data into a visual representation that brains can usually easily decipher.
c. The majority of data visualization applications have many encoding choices, and the aim of the designer is to find the correct balance, making data simpler and not harder to process.
   i. Simpler Execution: Used to illustrate a specific data set using only one stand-out colour.
   ii. Harder Execution: Different colours for each data set are more difficult to visualize.
d. Preattention features are visual cues, such as colour hue, brightness and saturation, shape, size, position and texture, which are visualized by the brain without deliberate effort.

By carefully selecting which preattention attributes to use and when, data will aid the brain process in an instant. However, using many attributes together can also create complications. Visualizations of data have really been effective when visual encoding elements are objectively incorporated into the presentation of visual data [50].

Data visualization is the encoding of visual data with the help of visual cues. Yet our minds do not treat such visual cues equally. Visual cues tend to have a spectrum or range based on accuracy that is supported by length (for example, a bar chart) and colour bifurcation (such as hue, light and intensity). In addition, for the purpose of better information processing, data presentations can be based on various movements, sometimes up and down the hierarchy of perception [51]. Observing the known visual narrative also sometimes reduces visualization choices for data scientists. For example, dot, line and column charts are ideal to depict changes incurred over a period of time. However, each one highlights different and sometimes multiple perspectives of the story [52]. A line chart is good to reflect the "rate of change," while a comparative analysis is well represented using a column chart. Likewise, line and dot charts are good options to show pattern and the correlation with the particular years to be compared. It is also important to realize the importance of understanding difference between spatial and sequential learning in the context of data visualization. The human brain often reacts to movement and sound in an organized sequence. Therefore, programmers deploy a set of animated GIFs to transmit information in a sequential and repeated way while loops help explain complex concepts by repetition [53]. Designers often make conscious choices to explicitly explain results. For instance, in the case of the data comparison of 58 counties to determine immunization rates in California kindergarten [54], a dot and line graph may have been catastrophic and visibly confusing, while a drop-down approach to display data using perceptual hierarchy (such as colour strength to reflect incomplete immunization percentage) made data visually more presentable and relevant. In this particular case, the detailed comparison has been substituted for an overall trend breakdown analysis. Hence, presentation of data should not make it difficult for the human brain to process information [55]. The correct approach is to use colour that is as natural as possible and supports interpretation seamlessly. The selection of the right colours not only reduces the pressure on the reader's brain; it is also crucial for accessibility and usability development. The use of tools such as ColorBrewer also facilitates colour-blind readers' visualization. Furthermore, while more complex views, such as box plots, may be fine for those who crunch and analyse the data; usually, it is considered complicated by readers at large. It is important, therefore, to include relevant details and contextual information to build a reliable and yet effective information architecture. Annotations are extremely useful to elaborate on and define storytelling in data visualization in addition to guiding readers (through key points). The white space also allows the brain ample time to process information. Finally, drafting and playing with multiple codes and visualizations is highly recommended to check them for the best possible outcomes on different individuals.

9.4.3 Human Visual Limitations

There is also no denying that data visualization has affected the way information is now shared, processed and presented in business meetings. Quantitative data has always been concise and abstract. Even though it may not refer to the physical world, it is still depicted visually – whether it concerns profits, sickness, athletic success or anything else. Important stories exist in our data, and data analytics are a valuable way to learn, interpret and communicate these stories to others. However, we need to find a way to describe the indefinable pattern or information within our data. The translation of the abstract quantitative data into physical attributes of vision (length, position, height, shape and colour) will work only if visual perception and human cognition are understood. In other words, we need to adopt ideas arising from understanding human cognition to interpret visualization outcomes effectively. Likewise, almost everyone can recognize the popularity of viral humour via social media and construct graphs, charts and diagrams based on that. On the other hand, it indicates that people can also now potentially reflect upon poor data visualization. There is a broad collection of data and databases that are so large or complex that conventional data processing systems are insufficient, and this has all been used to describe or refer to big data. With the expansion of big data, the current state is becoming less and less understandable. Analyst Doug Laney presented the 3Vs concept in 2001 [56]. Big data is composed of 3V dimensions, according to Laney: volume (observed data quantity), variety (meaning of data type numbers) and velocity (processing speed or data management). This means that in the near future, people will focus continuously on data visualization to grasp the large amounts of data in terms of its volume, variety and velocity. It means that we need to learn how to best use simulation tools for data visualization and understand how the human brain responds to external stimuli.

The human brain can be a wonderful machine, but only one value can be stored at a time. This is why multi-tasking is considered more detrimental than favourable for information processing and storage. The representation of data often makes it conveniently possible for the brain to obtain and perceive vast quantities of data. The study by Salesforce reveals that most businesses agree that too much data remain unanalysed and nearly half of the test findings are failing. Another study by the Aberdeen Group found that visualized data improved managers' chances to obtain information quickly by 28% [57]. In the Massachusetts Institute of Technology report, it was reported that images and the visual display of data can be processed by the human brain in 13 milliseconds [58]. When visual stimuli enter the retina, information flows into the brain for processing based on shape, texture, light and orientation. Hence visualizations transform intricate information into visual images, which can be better managed by the human brain. If that is the case, then all that is required is to find patterns and triggers – their variants instead of entire data collections.

While data representation uses graphics and images to communicate information, a pioneer in data visualization, Ben Shneiderman [59] said that the object of displaying is not just about pictures but also about providing deep meaningful insights. This means that data becomes much less useful in charts than that which makes data meaningful and relevant for its users. The brain dedicates enormous energy to visually process the information and data. Depending on who you are asking, the energy level will vary between 30% to 60%. Nevertheless, visual processing takes up much more space than the 8% used for data processing by contact or touch or the 2% used during information processing by hearing. The human brain is wired for fast visual processing, a characteristic that helped early humans thrive. The difference between life and death could be seen by subtle gestures, animal patterns or other harmful signs in the world [60]. While we usually don't have to search for danger in our everyday world, our minds have retained the incredible ability to quickly discern patterns and identify obvious and not so apparently visual signals. The brain has such a strong impulse to identify, coordinate and make sense of variations and structures around us that many games are designed to enable people to do just that. Think of *Memory* – a very popular childhood game of the 1970s that would ask kids to find identical objects, memorize from material provided and solve intricate puzzles that focused on teenage or adult memory formation and visual processing skills.

9.5 EXEMPLARY CASE STUDIES

In this section, exemplary case studies are discussed that have used innovative designs that have captured elements of data visualization while encoding human visual limitations.

On a more conventional note, great visualization of data revolves around projecting clarity, application and appropriate use of graphical tools. But these considerations fall short in many ways, primarily because they are able to generate merely acceptable standards of visuals that can be classified as below average or more as a functional view of data. To enhance user-centric, data-driven visualization, a higher level understanding of visualization techniques is required.

A good visualization of data is described as easy to understand, easy to remember and well suited to the data represented. At the same time, visualization of data may be designed to capture the interest of people. There are always some visuals that will gain popularity and establish appealing trends for the future. While for the rest, good design criteria must be developed (for instance, colour, typography, use of space). As stated earlier, human brains have evolved to identify patterns and become attracted to stimuli that do not fall into existing mental models. Forms and colours that are too noticeable can also become patterns. Visual data with unusual or unconventional visual elements is always easier to remember. For example, Napoleon's 1869 march to Russia is excellent and one of the earliest examples of showcasing the power of data visualization [61]. The horizontal axis represents advancement from France (left) to Russia (right), whereas the vibrant bands represent the army's size – the broad beige band indicates its attack; the small black band indicates its withdrawal. His escape. His withdrawal. The fact that its imaginative and transparent is part of the drawing for this kind of picture. It is never recommended to lose simplicity at the altar of innovation when creating visuals to reflect results. Both of these can occur simultaneously. Steve Jobs, who was a designer, entrepreneur and co-founder Apple Computer, said, "Technology alone is not enough—it's technology married with liberal arts, married with the humanities, that yields us the results that make our heart sing" [62]. He was someone supportive of the idea that architecture and technology are inseparably associated, and this philosophy actually performed quite well for him and Apple Inc. Resultantly, we now witness that data visualization is dramatically switching from static to interactive displays. It liberates viewers and users, as they not only have the technological capability or the knowledge to change variables, apply filters or execute certain operations in real time. It gives them the freedom to determine the impact and changes in data representation as they like. Therefore, while creating visualizations, it is significant to ascertain the requirements of those end users, who could be consumers. The following are five carefully selected cases of data visualization designs that can be considered groundbreaking by looking at current standards. After all, outstanding data visualizations stands the test of time—as Minard visualized the march of Napoleon—and can offer insights through visualization years after formation.

9.5.1 The New Yorker: Citi Bike

Citi Bike in New York published a digital map of interactive data showing bike access a few months after New York City unveiled its new bike-sharing service in 2013 [63]. Over a month, the experiment used and collected live data from riding tours. For data visualization, users were given the option to move the red dot on the slider that showcased different times and days of the month. The reported statistics were based on the number of daily trips, along with weather or precipitation that may have obviously affected the cycling pattern of the people. The dots on the map could expand and shrink to suggest the availability of bikes in that area. In addition, dots would also indicate a specified number if hovered over.

Research on Citi Bike is an example of how visualization can make data available and useful in a timely manner. Because it included a newly implemented public program, their insight was widely applicable. Information could have an effect on the lives of thousands of people and, initially, stakeholders found it difficult to comprehend Citi Bike information. However, in the new interactive format and in an extremely relevant time, the New Yorker made data widely available.

9.5.2 *Atlantic*: Healthier Population – Contextualization

Together with Athenahealth, the *Atlantic* has created a comprehensive report on the latest approach to health care and treatment called "People Healthier." This takes the form of a succinct case study using real-time data that updates perceptions by scrolling through the window [64]. What makes this visualization a standout in how data (textual and visual) depicts uniformity and coherence throughout.

The example that follows displays a hue-coded chart of Lowell's ethnic makeup in the area the study was about. Through superimposition, the same hue-coded dots can be applied on major hospitals located in the city. The approach tends to create cohesion and flow in the visualization of information. Without this visualization, meaningful data would be significantly impoverished or lost in translation. Indeed, the in-depth writing preceding the data offers deeper insights and provides an enriched experience to the user for rightfully linking images with textual narration. While the interactive data acquire essential meaning, graphics make writing simpler and appealing. Ultimately, "World Healthier" shows how effective visual data can be created and presented.

9.5.3 Public Web Platform Response – Responsive and Scalable Visualization

It is estimated that Google searches are estimated to be 3.5 billion per day. This means, every nanosecond, a question is being asked, and if there is ever someone who wanted to know what kind, sort or type of questions are being asked through Google, then Answer the Public is the solution [65]. It basically visualizes Google's search inquiries and classifies them into different sub-theme categories. The "data visualization" is based on certain keywords that show what type of questions, queries or expressions are raised by the users. The entire search has been carefully sorted according to question words, such as "how," "when," "where" or "when." The outline of each category is visually appealing, and if branches are visually followed, then they tend to smoothly link to a summarized terrain of sub-questions associated with them. It gives a very clear view of inquirers' thought processes while posting the question. For in-depth analysis, one can search visualizations based on different query trains or keywords (e.g. prepositions, contrasts, related topics). It is a good case example based on using different viewpoints simultaneously and how that is reflective of searchers' mind.

Answer The Public is for internet advertisers who want to understand the requirements of their potential clients. However, considering how user-friendly, versatile and responsive this data is, it can be used by a number of other researchers for a variety of reasons.

9.5.4 SelfieCity – Analysis of New Phenomena

The "selfie" is one of the simplest visual items of the digital generation. It is an image of the photographer's own face, usually very tightly accompanied by a snapshot of its surroundings (because of technological limitations). Nonetheless, this form of visualization masks a wealth of knowledge about social media, society, culture and values. The hidden knowledge in selfies has been researched by a group of researchers through the SelfieCity initiative [66]. As part of this research initiative, researchers have collected arbitrarily 120,000 Instagram pictures from five different cities to understand prominent grouping patterns, including similarities and variations in the ongoing selfie phenomenon. The data was collected based on the identification of patterns in the voice, location and features of the photographer. Findings of the study are demonstrated in the form of different visualizations on the website, where sample pictures are organized alongside posture axes, with its head tilted on the x-axis, while it angles are either positioned upward or downward on the y-axis.

SelfieCity's social data could not provide detailed insight. Therefore, researchers used common analytical techniques to extract available information (e.g. selfies) and made it user-friendly and understandable. Nevertheless, further experiments were made possible by this act of simplification. The age we are in currently is where our perception focuses on the conflict between minds and disciplines; hence, this type of research (and its subject) represents a bold step towards the future of data visualization.

9.5.5 Sitebulb – Internal Link Mapping

It is known that for search engine optimization, a website audit is required to know if the safety of the site and its pages are validated and evaluated. When a website has hundreds or thousands of pages, knowing the layout is incredibly difficult, which makes managing the design of the site for a good user interface impossible. Sitebulb is, therefore, a good case for data visualization [67]. The idea is simple, but it has a huge impact, and the purpose of a simulation is simply to produce something important and relevant. If you analyse the image, the main bubble is the official homepage. Every line and the next green dot show a spoke of the site, like a bike wheel.

9.6 FUTURE RESEARCH AND CHALLENGES

In this section, we will review important challenges and possible solutions related to the future agenda for big data visualization with augmented reality (AR) and virtual reality usage: (1) Application development integration: To operate visualized objects, it is necessary to create a new interactive system for the user. It should support such actions as scaling; navigating in visualized 3D space; selecting sub-spaces, objects, groups of visual elements (flow/path elements) and views; manipulating and placing; planning routes of view; generating, extracting and collecting data (based on the reviewed visualized data). Nevertheless, one of the main issues regarding this direction of development is the fact that implementing effective gestural and voice interaction is not a trivial matter. As for the interaction issue, for appropriate haptic feedback in a magnetic resonance environment, there is a need to create a framework that would allow interaction with intuitive gestures. The disadvantage of hand-tracking input is that there is no tactile feedback. In summary, the interface should be redesigned or reinvented to simplify the user interaction. Software engineers should create new approaches, principles and methods in user interface design to make all instruments easily accessible and intuitive to use. (2) Tracking and recognition system objects and tools have to be traced in virtual space. The position and orientation values of virtual items are dynamic and have to be re-estimated during presentation. Tracking head movement is another significant challenge. It aims to avoid a mismatch of the real view scene and computer-generated objects. This challenge may be solved by using more flexible software platforms. (3) Perception and cognition: Actually, the level of computer operation is high but still not sufficiently effective in comparison to human brain performance even in cases of neural networks. As was mentioned earlier, human perception and cognition have their own characteristics and features, and the consideration of this issue by developers during hardware and interface design for AR is vital. In addition, the user's ability to recognize and understand the data is a central issue.

9.6.1 Human Cognition and Visual Challenges

New technologies and applications have given us exposure to huge amounts of data at a higher level. To extract and collect all the worthwhile information from big data, it is important to apply a level of relatively unprecedented data visualization approaches. And to get a realistic review, a more thorough perspective and an outline of the information, we have to humanize the results.

We live in a world full of data, which exposes us every day to a continuous flow of new knowledge. Human beings produced five exabytes (five billion gigabytes) of data from the beginning of recorded time up until 2003. Every two days, the same sum was generated in 2011. Researchers from the University of California estimated that every day we ingest an average of 34 gigabytes of new information [68]. "Global data," a term referring to the universe of all information ever produced, stored or reproduced, has expanded to an unimaginable size of 33 zettabytes. Thus data visualization is only useful to the extent that it encrypts information in a way that can be interpreted by our eyes and brains. Getting this right is a science rather than an art that can only be done by studying human perception. The purpose is to convert abstract knowledge into visual representations which can be effectively, precisely and sensibly decoded. While there are many types and forms of data, nearly 90% of the data transmitted to the brain are visual. The human brain was actually not designed for texts or numbers. The brain can process a visual image within 13 milliseconds, significantly faster than how much text or numerical information is processed in the brain. The truth is our brains evolved naturally in a world where visual information is key to survival. Our brains adapt quickly and more intuitively to handle visual information, and its implications are also very clear. For example, picture posts get up to 94% more total views compared to their textual counter versions. To sum up, it is obvious that visual data is far more natural and appealing for our brains. As humans, we have the habit of anticipation and the capacity to foresee and prepare for the future, which is crucial to the survival of our species.

9.6.2 New Visualizations, New Challenges

Data visualization is one of the most important ways by which big data can be perceived and made available to most citizens. Data visualization is wayfinding, both literally like the street signs that lead you to a highway and figuratively where abstract elements convey information in colours, scale or location. In either case, if properly aligned, the visual may provide a shorter route to direct decision making and become a device for communicating critical information in all data analysis. Nonetheless, data visualization will provide the right amount of interactivity if real action is to be taken. It must be developed so that it is simple to use, comprehensible, meaningful and usable. It's an ongoing challenge as well because as technology changes, new tools are developed in the hope of improving its utility for industry-wide use.

1. Simplification of results. One of the main benefits of visualization is the potential to capture and allow the use of large amounts of data. For examples, consider real-life basic tests, such as drug toxicity checks, which try to reduce complex systems to "yes" or "no" findings. As the Monder Law Group reported, such tests can be inaccurate.
2. Human limitations to algorithms. This is the largest and most complex potential problem. Every data reduction algorithm is based on human inputs that could be deeply flawed [69]. For example, an algorithm creator may pick out different parts of data which are most important to consider. It does not take all entities or all scenarios into consideration, especially when there are data outliers or unusual circumstances that call for an alternative approach. The dilemma is compounded by the fact that many data display solutions are being applied at a national level; thus, they develop into one-size-fits-all algorithms, which do not meet individual needs.
3. Visual over-dependence. This is more of a customer issue than a developer's problem, but it usually reduces the potential impact of visualization. Once visuals are used to read and interpret data, it is likely that users would start relying on them as input visual modes very quickly. We may, for example, accept their assumptions as absolute truth and never look into the data sets that create such visuals. We may take general conclusions from this, while users may not inform or give critical feedback based on reports or data assessment. The general assumptions can be different and tricky.
4. Visualization's inevitability. There are already hundreds of tools available that can help us understand complex data sets with visual graphics and charts. Furthermore, data visualization is in demand and too trendy to go anywhere soon. Resultantly, researchers and developers are focused more on a quick visualization course that would give them expertise over multiple areas. It may be beneficial in the short run but may lead to shortfalls in visualization inevitability in the future. There are a number of consequences that should be taken into account by businesses to create and for consumers to search for visualized products. The outcome of this practice would be overreliance on visuals that would eventually compound the weaknesses of human errors in the development of algorithms (because businesses want to market as soon as possible).

9.7 CONCLUSION

In this big data era, the traditional ways of presenting data reached a few limitations and traditional data visualization is inadequate to handle big data at this point. In this chapter, we identified some of the methods, techniques and applications of data visualization in the big data era. We also reviewed some related works, current research approaches and related tools. These approaches and tools could provide new ways for visualizing big data. Visualized data can significantly improve the understanding of the preselected information for an average user. In fact, people start to explore the world using visual abilities beginning at birth. Images are often easier to perceive in comparison to text. In the modern world, we can see a clear evolution towards visual data representation and imagery experience. Moreover, visualization software becomes ubiquitous and publicly available for ordinary users. As a result, visual objects are widely distributed from social media to scientific papers, and thus the role of visualization while working with large amounts of data should be reconsidered. Hence there's no stopping the development of data visualization, and under no conditions is it being argued that it should be stopped. However, it is essential to develop it the right way because it can be an extraordinary tool in countless different areas, but collectively, we need to be aware of the potential problems and biggest obstacles data visualization will need to overcome.

REFERENCES

1. Manyika J, Chui M, Brown B, Bughin J, Dobbs R, Roxburgh C, and Byers AH (2011) Big Data: The Next Frontier for Innovation, Competition, and Productivity. June Progress Report. McKinsey Global Institute.
2. DOMO – Data Never Sleeps 5.0. Accessed 14 Feb 2020, from https://www.domo.com/learn/data-never-sleeps-5.
3. IGSR (2019) The International Genome Sample Resource, Supporting Open Human Variation Data. Accessed 14 Feb 2020, from https://www.internationalgenome.org/.
4. Via M, Gignoux C, and Burchard EG (2010) The 1000 Genomes Project: New Opportunities for Research and Social Challenges. *Genome Med*, Vol. 2(3).
5. Internet Archive (2015) Internet Archive Wayback Machine. Accessed 14 Feb 2020, from http://archive.org/web/web.php.
6. Nielsen J (2015) Comparing Content in Web Archives: Differences between the Danish Archive Netarkivet and Internet Archive. In *Two-Day Conference at Aarhus University*, Denmark.
7. The Lemur Project (2015) The ClueWeb09 Dataset. Accessed 14 Feb 2020, from http://lemurproject.org/clueweb09.php/.
8. Russom P (2013) Managing Big Data. TDWI Best Practices Report, TDWI Research.
9. Husain SS, Kalinin A, Truong A, and Dinov ID (2015) SOCR Data Dashboard: An Integrated Big Data Archive Mashing Medicare, Labor, Census and Econometric Information. *J Big Data*. Vol. 2(1), pp. 1–18.
10. Thompson D, Levine JA, Bennett JC, and Bremer PT, Gyulassy A, Pascucci V, and Pébay PP (2011) Analysis of Large-Scale Scalar Data Using Hixels. In *Proceedings of Symposium on Large Data Analysis and Visualization (LDAV), IEEE 2011*. pp 23–30.
11. Report (2014) Data Visualization Applications Market Future of Decision Making Trends, Forecasts and the Challengers (2014–2019). Mordor Intelligence.
12. SAS (2013) Data Visualization: Making Big Data Approachable and Valuable. Market Pulse: White Paper.

13. Sciforce (2019) Best Data and Big Data Visualization Techniques. Accessed 14 Feb 2020,fromhttps://medium.com/sciforce/best-data-and-big-data-visualization-techniques-e07b897751dd.
14. Techopedia, Big Data Visualization. Accessed 14 Feb 2020, from https://www.techopedia.com/definition/28988/big-data-visualization.
15. Towards Data Science (2018) Top 4 Popular Big Data Visualization Tools. Accessed 14 Feb 2020, from https://towardsdatascience.com/top-4-popular-big-data-visualization-tools-4ee945fe207d.
16. Keim DA, Kohlhammer J, Ellis GP, and Mansmann F (2010) Mastering the Information Age – Solving Problems with Visual Analytics. Eurographics Association.
17. Klugman SA, Panjer HH, and Willmot GE (1998) *Loss Models: From Data to Decisions*. Wiley, New York.
18. Panjer HH and Willmot GE (1992) *Insurance Risk Models*. Society of Actuaries, Chicago, IL
19. Teugels J and Sundt B (2004) *Encyclopedia of Actuarial Science*. Wiley, Chichester, UK.
20. Burnecki K, Härdle W, and Weron R (2004) Simulation of Risk Processes. In J Teugels, B Sundt (Eds), *Encyclopedia of Actuarial Science*, Wiley, Chichester, UK.
21. Chun-houh C, Wolfgang H, and Antony U (2008) Handbook of Data Visualization, Springer-Verlag Berlin Heidelberg. Accessed 14 Feb 2020, from https://haralick.org/DV/Handbook_of_Data_Visualization.pdf.
22. Kevles BH (1997) *Naked to the Bone: Medical Imaging in the Twentieth Century*. Rutgers University Press, Piscataway, NJ.
23. Suetens P (2002) *Fundamentals of Medical Imaging*. Cambridge University Press, Cambridge.
24. Prince JL and Links J (2005) *Medical Imaging Signals and Systems*. Prentice Hall, Upper Saddle River, NJ.
25. Shepp LA and Vardi Y (1982) Maximum Likelihood Reconstruction for Emission Tomography. *IEEE Trans Med Imag*, Vol. 1(2):113–122
26. Vardi Y, Shepp LA, and Kaufman L (1985) A Statistical Model for Positron Emission Tomography. *J Am Stat Association*, Vol. 80(389): 8–20.
27. Lu HHS, Chen CM, and Yang IH (1998) Cross-Reference Weighted Least Square Estimates for Positron Emission Tomography. *IEEE Trans Med Imag*, Vol. 17(1): 1–8.
28. Lu HHS and Tseng WJ (1997) On Accelerated Cross-Reference Maximum Likelihood Estimates for Positron Emission Tomography. *Proc IEEE Nucl Sci Symp*, Vol. 2: 1484–1488
29. Tu KY, Chen TB, Lu, HHS, Liu RS, Chen KL, Chen CM, and Chen JC (2001) Empirical Studies of Cross-Reference Maximum Likelihood Estimate Reconstruction for Positron Emission Tomography. *Biomed Eng – Appl Basis Commun*, Vol. 13: 1–7.
30. Malik J and Perona P (1990) Preattentive Texture Discrimination with Early Vision Mechanisms. *J Opt Soc Am A.*, Vol. 7(9): 1632–1643.
31. Jain AK and Farrokhnia F (1991) Unsupervised Texture Segmentation Using Gabor filters. *Pattern Recogn*, Vol. 24(12): 1167–1186.
32. Dunn D, Higgins WE, and Wakeley J (1994) Texture Segmentation Using 3-D Gabor Elementary Functions. *IEEE Trans Pattern Anal Mach Intell*, Vol. 16: 130–149.
33. Tan TN (1995) Texture Edge Detection by Modelling Visual Cortical Channels. *Pattern Recogn*, Vol. 28(9): 1283–1298.
34. Zhu SC, Wu Y, and Mumford DB (1998) Filter, Random Field, and Maximum Entropy (FRAME): Towards a Unified Theory for Texture Modelling. *Int J Comp Vis*, Vol. 27(2): 107–126.
35. Chen CM, Lu HHS, and Hsiao AT (2001) A Dual Snake Model of High Penetrability for Ultrasound Image Boundary Extraction. *Ultrasound Med Biol*, Vol. 27(12): 1651–1665.

36. Chen CM, Lu HHS, and Han KC (2001) A Textural Approach Based on Gabor Functions for Texture Edge Detection in Ultrasound Images. *Ultrasound Med Biol*, Vol. 26(2): 273–285.
37. Chen CM and Lu HHS (2001) An Adaptive Snake Model for Ultrasound Image Segmentation: Modified Trimmed Mean Filter, Ramp Integration and Adaptive Weighting Parameters. *Ultrason Imag*. Vol. 22: 214–236.
38. Li KC (2000) *High Dimensional Data Analysis via the SIR/PHD Approach*. Lecture Notes, Department of Statistics, UCLA, Los Angeles, CA (http://www.stat.ucla.edu/~kcli/sir PHD.pdf).
39. Wu HM and Lu HHS (2004) Supervised Motion Segementation by Spatialfrequential Analysis and Dynamic Sliced Inverse Regression. *Stat Sinica*.
40. Tukey JW (1977) *Exploratory Data Analysis*. Addison-Wesley, Reading, MA. http://theta.edu.pl/wp-content/uploads/2012/10/exploratorydataanalysis_tukey.pdf
41. Unwin A, Hawkins G, Hofmann H, and Siegl B (1996) Interactive Graphics for Data Sets with Missing Values – MANET. *J Comput Grap Stat*, Vol. 5(2): 113–122.
42. Playfair W (1786) Commercial and Political Atlas: Representing, by Copper-Plate Charts, the Progress of the Commerce, Revenues, Expenditure, and Debts of England, during the Whole of the Eighteenth Century, Corry, London. Re-published in Wainer H and Spence I (eds), *The Commercial and Political Atlas and Statistical Breviary*, Cambridge University Press. https://www.christies.com/lotfinder/Lot/playfair-william-1759-1823-the-commercial-and-5388575-details.aspx
43. Lee TY, Jones C, Chen BY, and Ma KL (2010) Visualizing Data Trend and Relation for Exploring Knowledge, in *Proceedings of the IEEE Pacific Visualization Poster*.
44. Posner M (1980) The Psychology of Science, *The American Journal of Psychology*, Vol. 93(4): 728–730. doi:10.2307/1422380
45. Edward S and Jeffrey H (2010) Narrative Visualization: Telling Stories with Data Manuscript. Accessed 14 Feb 2020, from http://vis.stanford.edu/files/2010-Narrative-InfoVis.pdf.
46. George AM (1956) The Magical Number Seven, Plus or Minus Two: Some Limits on Our Capacity for Processing Information. *Psychological Review*, Vol. 63: 81–97.
47. Zhang AX, Verou L, and Karger D (2017) Wikum: Bridging Discussion Forums and Wikis Using Recursive Summarization.
48. Haroz S and Whitney D (2012) How Capacity Limits of Attention Influence Information Visualization Effectiveness. *IEEE Trans Vis Comput Grap*, Vol. 18(12): 2402–2410.
49. Hancock PA and Kim JW (2010) Cognitive Modeling of Performance Response Capacity Under Time Pressure (No. TR-2010–0907). University of Central Florida Orlando Dept of Psychology.
50. Rieber LP (1995) A Historical Review of Visualization in Human Cognition. *ETR&D*, Vol. 43: 45–56. doi:10.1007/BF02300481.
51. Powers WT (1973) *Behavior: The Control of Perception*. http://www.pctresources.com/Other/Reviews/BCP_book.pdf
52. Finke RA (1990) *Creative Imagery: Discoveries and Inventions in Visualization*. Hillsdale, NJ: Lawrence Erlbaum Associates.
53. Robertson G, Fernandez R, Fisher D, Lee B, and Stasko J (2008) Effectiveness of Animation in Trend Visualization, *IEEE Transactions on Visualization and Computer Graphics*, Vol. 14(6): 1325–1332.
54. Kindergarten Immunization Assessment (2018) – Executive Summary California Department of Public Health. Accessed 14 Feb 2020, from Immunization Branch https://www.cdph.ca.gov/Programs/CID/DCDC/CDPH%20Document%20Library/Immunization/2017-2018KindergartenSummaryReport.pdf.
55. Tufte ER (1983) *The Visual Display of Quantitative Information*. Cheshire, Connecticut: Graphics Press.

56. META Group (2001) Application Delivery Strategies. Accessed 14 Feb 2020, from https://blogs.gartner.com/doug-laney/files/2012/01/ad949-3D-Data-Management-Controlling-Data-Volume-Velocity-and-Variety.pdf.
57. David W (2013) Visualization: Set Your Analytics Users Free. Accessed 14 Feb 2020, from https://www.tableau.com/sites/default/files/media/8604-ra-business-intelligence-analytics.pdf.
58. Ioannides AA, Liu L, Theofilou D et al. (2000) Real Time Processing of Affective and Cognitive Stimuli in the Human Brain Extracted from MEG Signals. *Brain Topogr*, Vol. 13: 11–19.
59. Shneiderman B (1996) The Eyes Have It: A Task by Data Type Taxonomy for Information Visualizations, *Proceedings 1996 IEEE Symposium on Visual Languages*, Boulder, CO, USA, pp. 336–343. doi:10.1109/VL.1996.545307.
60. Rolls ET (1999) *The Brain and Emotion*, Oxford University Press Inc., New York.
61. Jones J (2019) Napoleon's Disastrous Invasion of Russia Detailed in an 1869 Data Visualization: It's Been Called "the Best Statistical Graphic Ever Drawn." Accessed 14 Feb 2020, from http://www.openculture.com/2019/07/napoleons-disastrous-invasion-of-russia-explained-in-an-1869-data-visualization.html.
62. Lehrer J (2011) Steve Jobs: Technology Alone is Not Enough. Accessed 14 Feb 2020, from https://www.newyorker.com/news/news-desk/steve-jobs-technology-alone-is-not-enough.
63. Guerriero M (2013) Interactive: A Month of Citi Bike. Accessed 14 Feb 2020, from https://www.newyorker.com/news/news-desk/interactive-a-month-of-citi-bike.
64. The Atlantic (2019) Population Healthier. Accessed 14 Feb 2020, from https://www.theatlantic.com/sponsored/athenahealth/population-healthier/598/.
65. CoverageBook (2019) Answer The Public, Is Built by the Team Behind. Accessed 14 Feb 2020, from https://answerthepublic.com/.
66. SELFIECITY (2014) Investigating the Style of Self-Portraits (*Selfies*) in Five Cities Across the World. Accessed 14 Feb 2020, from http://selfiecity.net/.
67. Sitebulb (2019) Data Visualization. Accessed 14 Feb 2020 https://sitebulb.com/.
68. Bohn R and Short J (2012) Measuring Consumer Information. *Int J Comm*, Vol. 980–1000: 1932–8036/20120980.
69. Ouyang X, Wong WH, Johnson VE, Hu X and Chen CT (1994) Incorporation of Correlated Structural Images in PET Image Reconstruction. *IEEE Trans Med Imaging*.

10 IoT, Big Data, and Analytics

Challenges and Opportunities

Priyanka Vashisht
The NorthCap University
Gurugram, Haryana, India

Vijay Kumar
Department of Mathematics, AIAS, Amity University
Uttar Pradesh, Noida, India

Meghna Sharma
The NorthCap University
Gurugram, Haryana, India

CONTENTS

10.1 INTRODUCTION: BACKGROUND AND DRIVING FORCES

The rapid convergence and advancement of technology in digital electronics, wireless communication, and machine-to-machine (M2M) technologies have resulted in the materialization of the Internet of Things (IoT). With the miniaturization of IoT devices, a huge amount of data have been produced over the past decade. According to a report published by Cisco in 2011, the number of Internet-connected devices has

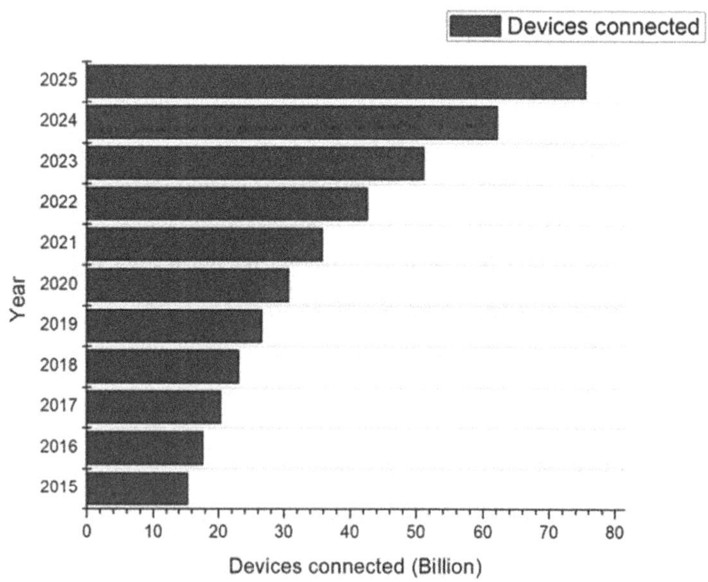

FIGURE 10.1 IoT Connected Devices [2].

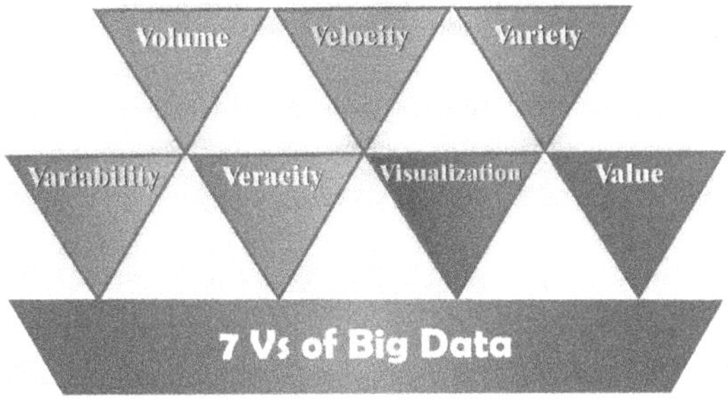

FIGURE 10.2 7 Vs of Big Data.

surpassed the number of human beings across the globe [1]. In another report, it is depicted that the amount of Internet-connected devices is likely to increase fivefold from 15.41 billion in 2015 to 75.44 billion by 2025, as shown in Figure 10.1 [2]. The advent of a huge amount of data produced by IoT devices simply reflects how the growth of big data relies entirely on IoT [3]. To illustrate the challenges of big data, Gartner categorized it according to three characteristics: (i) volume, (ii) velocity, and (iii) variety or the 3Vs. Volume signifies the amount/size of data produced on a daily basis. Velocity indicates the growing speed at which data is produced and the intensity at which the processing of data can be done. Variety refers to a kind of data produced by IoT devices, such as unstructured, structured, and semistructured. Most of the data produced by Internet-enabled devices are unstructured. The 3Vs certainly provide insight into the almost unenvisionable range of data and the rapid pace at which these enormous data sets are produced. Variety in the data hides the actual depth and complexity of big data. In 2013, Mark van Rijmenam [4] proposed four more Vs to further recognize the extremely complex characteristics of big data. The other Vs are variability, veracity, visualization, and value, as shown in Figure 10.2. Variability points to accountability and availability. Veracity deals with the accuracy of data, specially where data is automated during decision making. After processing the data, visibility of data is crucial, as it presents the information in a more readable form, which can be accessed easily. Data is significant if it is incorporated and processed effectively, which generates value that has the ability to formulate valuable, proficient, and correct decision making regarding the opportunities and threats of the organization [4].

An IoT-based infrastructure consists of a large number of sensors and devices that exchange a substantial amount of data over the Internet. In 2014, the Distributed Event-Based Systems (DEBS) Grand Challenge [5], 40 smart houses generated four billion events over a month [6], whereas in 2011, the United Kingdom census revealed that if 26.4 million households converted to smart houses, they are projected to generate 2.64 quadrillion (short scale) of data every month [7], which is a huge volume of data. Also, real-time data generated from sources, such as intelligent transportation systems [8, 9], weather forecasting, and health care, are produced with high-velocity data streams. Diverse IoT data devices collect unlike forms of data, which consequentially lead to a variety of data, although the diverse data produced by IoT devices are not valuable until they are analyzed.

Analytics is a discipline or technique by which complex things are observed and analyzed to extract something meaningful [10, 11]. When this technique is applied to data, it mines some meaningful information and knowledge. The analysis of big data of IoT helps in revealing trends, hidden relationships, unknown patterns, and novel information [12].

IoT has become the dominant source of big data [13], also the insight provided by the analytics compel the researchers to comprehend the practice of deriving analytics insight from the IoT. This can be justified by an argument by Akoka [14] (i.e., "data of IoT is useful only when analyzed"). It has been noticed that data analytics with a focus on IoT applications has, to the best of our knowledge, not been explored much. The rest of this chapter is organized as follows: Section 10.2 discusses the taxonomy of analytics, which is further subdivided based on (i) their application and (ii) the stakeholders during the deployment of IoT. In Section 10.3, various platforms are considered that facilitate the analysis of IoT data. Section 10.4 provides the major application domains of IoT and how the data are analyzed in these fields. Based on previous sections, Section 10.5 discusses the various challenges of IoT big data using analytics. Finally, in Section 10.6, the conclusion and future directions will be provided.

10.2 TAXONOMY OF ANALYTICS

Analytics has advanced over a period of time from managing uncomplicated raw data analysis of data sets to complex analytics with the intention of ensuring that enterprises with competitive advantage [15, 16] analytics are characterized by three key phases: value, difficulty, and intelligence [25]. In this section, the analytics are further classified based on two parameters: (i) different applications and (ii) stakeholders involved.

10.2.1 Classification of Analytics Based on Different Applications

Over the years, various researchers have provided the taxonomy of analytics and their application in different domains of IoT. Bertolucci et al. [17] classified analytics as descriptive, predictive, and prescriptive; however, Gartner [18, 19] added one more category – namely, diagnostic analytics. Corcoran et al. [20] suggested another supplementary classification called discovery analytics. This taxonomy allows researchers to establish the intent of the analytics and to correlate the concept of IoT deployment as it is often articulated in research road maps, as well as improved business outcomes.

10.2.1.1 Descriptive Analytics

In this technique, the data is assessed to answer the questions, "What happened? What is happening?" The data is assessed manually and characterized by conventional business intelligence (BI). The outcome of descriptive analytics can be visualized, such as line graphs, bar charts, pie charts, tables, or generated narratives [7].

10.2.1.2 Diagnostic Analytics

In this process, the root cause of the problem is analyzed. This method facilitates an answer to the question, "Why something happened?" It is categorized by using certain practices, such as data mining, drill down, data discovery, and correlations. [7]

10.2.1.3 Discovery in Analytics

It is a process of identifying outliers and patterns by visually exploring the data or applying guided sophisticated analytics. Visualization is a crucial feature in data discovery, which exploits the pattern recognition capabilities of the brain to digest information. However, guided advanced analytics offers statistical information of data using some tools, which users can use for more refined pattern-oriented data analysis. The discovery of accurate data is an iterative procedure that does not require extensive upfront model creation [7].

10.2.1.4 Predictive Analytics

Predictive analytics mines the information from existing data to establish some patterns so as to forecast future conclusions and trends. Predictive analytics predicts "what might happen in the future." It uses historical data and knowledge to forecast future outcomes [21] and offers the means by which the quality and reliability of these predictions can be assessed [22].

10.2.1.5 Prescriptive Analytics

This is a statistical procedure used to provide recommendations and formulate conclusions by using computational conclusions of mathematical models. It tries to answer the question, "What should I do about what has happened or is likely to happen?" Prescriptive analytics represents the best action to perform on the knowledge acquired from gaining insight from data in a timely manner [23], while considering the scope of uncertainty. This form of analytics has the ability to produce optimal results while at the same time considering answering "what if" questions for analytics to determine the best solutions. A summary of various analytics is presented in Table 10.1 in terms of their feature, usage, and popularity.

Descriptive analytics offers analysts key metrics and business measures. While diagnostic analytics helps in identifying the root cause of the problem [24, 25], both diagnostic and descriptive analytics facilitate the creation of a narrative of the past. Predictive analytics forecasts the possibility of certain occurrences that are likely to happen in the future, while prescriptive analytics helps in determining the best action an organization/individual can take. Prescriptive and predictive analytics assist in

TABLE 10.1
Taxonomy of Analytics

	Descriptive	Diagnostic	Discovery	Predictive	Prescriptive
Real Time Data Analytics	✓			✓	✓
Spatial Analytics	✓	✓	✓	✓	✓
Time Series Analytics	✓			✓	✓
Business Analytics	✓			✓	✓
Visual Analytics	✓	✓	✓	✓	✓

framing the future. In comparison to descriptive and predictive analytics, prescriptive analytics is not explored much [26]. Recently, many researchers are focusing on predictive and prescriptive analytics [27]. Predictive and prescriptive analytics use the statistical or optimization models for decision making well before the results, which leads to better business [26, 28]. New information and communication technologies, such as real-time streaming, sensor data, and IoT, have strengthened predictive analytics by providing businesses with probabilistic recommendations that facilitate effective decision making.

The three phases of business analytics with respect to time are illustrated in Figure 10.3. The main focus of descriptive analytics is to establish what is happening in the present by congregating and analyzing information that could direct the origin of the event. This type of analytics is able to identify patterns that indicate a problem or a prospect for a business, whereas predictive analytics is intelligent enough to forecast whether an event will occur, the expected time of the occurrence, and the major cause of the incidence of the event, as well as the major cause of its occurrence. According to [27], predictive analytics has significantly contributed to business values, as shown in Figure 10.3. The capabilities of predictive analytics can be well exploited in collaboration with prescriptive analytics for making the best decision ahead of time. The outcomes of the predictive analytics act as an input to prescriptive analytics for generating proactive decisions, but the time interval between the prediction and proactive decision leads to inevitably business value losses. Because of the time gap, it is not possible to use the maximum capabilities of predictive analytics without prescriptive analytics, especially when dealing with real-time or IoT data. However, it is very crucial to detect the occurrence of an event in a timely manner using descriptive analytics and an appropriate prediction of emerging events using predictive analytics to avoid the potential loss of business values.

10.2.2 Classification of Analytics Based on Different Stakeholders

Further, the classification of data analytics can be done based on stakeholders engaged in the deployment of IoT investment. The relationship of the types of analytics is shown in Table 10.2. Stackholder-based big IoT data analytics is discussed next.

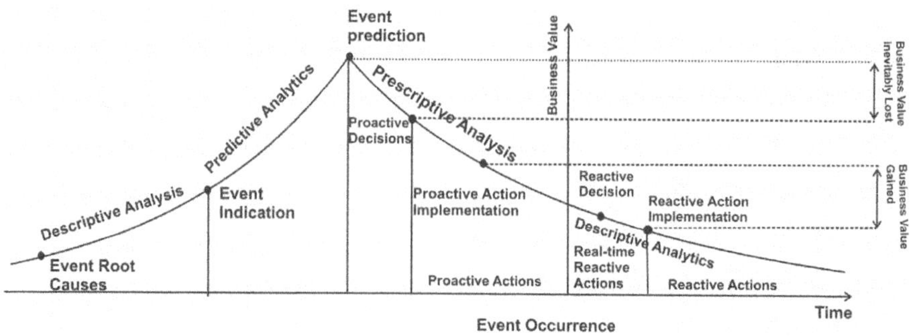

FIGURE 10.3 Phases of Business Values [27].

TABLE 10.2
Comparsion of four types of Analytics

	Type of Question address	Tools used	Use AI & ML	Application	Popularity
Descriptive	What happened? / What is happening?	• Data Aggegration • Data Mining	No	• standard report generation • basic spreadsheet functions • Vertical and horizontal analysis of financial statements	Used by almost all Organizations
Diagnostic	W*hy* something happened?	• Optimize • CrazyEgg	Sometimes	• Variance analysis • Dashboard	Used by many organizations
Predictive	might happen in the future?	• Statistical Models • Simulation	Yes	• Health care • Collection Analysis • Fraud Detection • Risk Management and many more	Used by smaller but growing grop of organizations
Prescriptive	what action needs to be taken for what has happened or is expected to happen	• Optimization Model • Heuristics	Yes	• Oilfield equipment Mangement • Pricing • Healthcare • Resource Development	Not yet Widespread

10.2.2.1 Real-Time Data Streaming Analytics

In this type of analytics, processing is performed on real-time data gathered from various IoT devices. The data is analyzed where immediate action needs to be taken, such as traffic analysis and financial transactions.

Ed-daoudy and Maalmi [29] proposed a novel architecture of the analytical system for predicting the health status of a person in real time. A distributed model is implemented where the streaming of real-time data events are ingested in Spark. A standard decision tree (c4.5) algorithm is modified to support parallel and distributed computing, as well as more scalable using Spark. Also, analytics is applied to real data from distributed sources for a variety of diseases to predict the health status of a person. The model is evaluated on parameters, such as execution time and throughput, to mark the effectiveness of a proposed system. Experimental results predict that the system has the capacity to effectively analyze and predict real-time big IoT data for various diseases. Manogaran and Lopez [30] implemented an architecture that integrates technologies, such as Apache Flume, HBase, and Pig, to gather and accumulate enormous amounts of sensor data in the Amazon Web Service. For training in the logistic regression model using the Cleveland heart disease database, sensor-enabled wearable devices were used to capture the data of patients, such as blood sugar level, blood pressure, and heart rate, to forecast the status of heart wellness. Predictive analytics was used by the proposed model that efficiently classified the heart disease. The accuracy of the training data and the validation sample for the model was 81.99% and 81.52%, respectively. The study in [31] was intended to gather information from the sensors of smart cities and reform that information into data forms, such as Resource Description Framework (RDF) and JavaScript Object Notation (JSON), through data modeling. A case study based on weather data set was used to obtain the results in the forms of RDF, XML, and JSON. XML assured the interoperability of the proposed method. RDF helped the semantic representation of interlinked data that can supplementary link and recycle with other linked data in the cloud. Using this technique, data sustainability for both deduction and analytics can be assured in real time.

10.2.2.2 Spatial Analytics

In this method of analytics, geographical locations or patterns are used to establish spatial association among physical objects. Locality-based applications in IoT domains are smart parking, smart barcode readers, etc. Limkar and Jha [32] proposed a scheme to compute IoT-generated encrypted spatial data. A novel technique – namely, homomorphic encryption, was used to guarantee the privacy of data. Indexing was implemented using the R tree and its variants for encrypted data. A scalable and efficient method was also proposed in real-time encrypted spatial data, with parallel construction of the R tree and its variants. A range of spatial queries was fired on encrypted spatial data. The proposed scheme was efficient, compatible, and scalable with spatial data generated by IoT devices. The authors claimed that the rigorous experiment was performed on real-time spatial data sets to verify the efficiency of the system. Lee et al. [33] investigated the various opportunities and challenges that come along with geospatial data. Various case studies were executed to illustrate the significance and advantages of analytics on geospatial big data, including health care, revenue increase, urban planning, etc. An emerging platform has also been considered for sharing information collected from geospatial big data to track human movement via some sensor device. The authors contributed their research mainly on interactive analytics for dynamic data. The online analytical processing engine and spatial complex event processing (CEP) were modified by the KAIST team for further enhancing the benefits of the proposed work. The proposed work deals with the complication faced by IoT in the discovery field service [34]. Here analysis of missing data is done to find the outlier in the data generated by the sensors. If the missing data is not handled properly, then it is difficult to analyze the correlation between spatial and temporal data, which leads to inaccuracy in the results. The proposed method decreases the complexity in handling the outlier data by taking care of missing value in the data. Experimental results show considerable improvement in response time in contrast with the conventional method of dealing with missing values.

10.2.2.3 Time-Series Analytics

Here the data and events analytics are time based to expose the related trends and their patterns. Applications such as health monitoring and weather forecasting include time series that help to find a "long-term change in the mean level" [35]. For storing data collected for sensors, a number of sensor databases are developed. Bader et al. [36] evaluated 12 well-known time series sensor databases with 27 decisive factors. These factors are categorized into six groups (i.e. (i) function, (ii) clustering, (iii) granularity, (iv) tagging of data and long-term storage, (v) interfaces and extensibility, and (vi) license and support). Authors claim that their work is extensive, repeatable, and an open-source benchmark for enterprise readiness. Databases that meet all the requirements of time-based big IoT data are KairosDB (2018), InfluxDB (2018b), and MonentDB [33]. KairosDB uses the distributed NoSQL database management system [37], and for storing the data, it uses Apache Cassandra. Three column families are used for storing time series–based sensor data – namely, Row Key Index Column Family, Data Points Column Family, and String Index Column Family [38, 39]. A data compression technique was used in this database during the writing process [40]. In InfluxDB, the logical grouping of a database is done for storing time series–based IoT data. The logical group of InfluxDB is known as measurements. The identification of each time series is done using a unique tab in the measurement. InfluxDB supports query language similar to SQL [41]. Here the values and their time stamps are compressed and stored independently using some encodings scheme, which is dependent on the type of data and its characteristics. Storing data autonomously permits the same encoding scheme to be used for all time stamps, while allowing diverse encodings for unlike field types.

10.2.2.4 Business Analytics

Business analytics is a discipline used to determine the insight of data by means of refined statistical, mathematical, machine learning, and network science techniques, along with the diversity of data and expert knowledge to perform accurate and timely decision making [42]. D'Hondt et al. [43] implemented a prototype in collaboration with existing technologies in business process management (BPM) and distributed analytics. Here the reverse engineering methodology is applied where instead of sending data for analytics of deal ownership, heterogeneity, and size of data, analytics are brought to data in distributed form. The dynamic deployment of service repository into the IoT edge automation of containerized micro-services is done using the BPM engine. The flexibility of this approach is seen while processing the data, where huge volumes of unprocessed data are left at its origin.

10.2.2.5 Visual Analytics

This is an interactive visualization through data analytics techniques "for an effective understanding, reasoning and decision making on the basis of very large and complex data sets" [44]. Visual analytics can contribute to not only describing and diagnosing what happened but also help users to discover new insights. Zhang et al. [45] applied visual analytics in the health-care domain for answering questions such as, "What is the distribution of pregnancy age?" Researchers analyzed data based on the supposition that two disease patterns caused by "fever" and "diarrhea" are not linked and discovered through detecting the delayed outbreak of the two diseases. Maciejewski et al. [46] suggested a model for spatiotemporal data, where data hot spots are being identified for a section of space at a particular instance. Hot spots are the points where the frequency of the occurrence of some events is high. Researchers proposed a tool kit that helps analysts with results in which they get a combined statistical and spatiotemporal view. The proposed method simulates the spatiotemporal event by grouping a seasonal tendency decomposition and kernel density estimation. The data smoothing process is responsible for seasonal tendency decoposition and event distribution accountable for kernel density estimation. Yue et al. [47] explicitly concentrated on predictive analytics. They proposed a reasoning mechanism – namely, Resource-Bounded Information (RESIN) – that has the capacity to forecast future inclination with time-series analysis. This is an agent-based approach in which an artificial intelligence–based smart agent incorporates with classification and visualization tools to identify an abstract representation of the event that could principally enhance the accuracy of the proposed system. Riensche et al. [48] suggested a methodology and an architecture that support the expansion of games by using predictive analytics. The game takes input data from the events generated during execution, which are analyzed to acquire knowledge using social and predictive models in such a way that the game is always engaging. A prototype has been developed such that while executing the test versions of games, it always uses the predictive analysis to make them more dynamic and adaptive. Moreover, the flexibilty in the architecture allows for opportunities for future extension and growth of the proposed prototype. Huang et al. [50] converted a huge amount of data into functional data using visual analytics techniques that were applied to complex network procedures. Functional data are categorized into various risk levels and then visualized as a color-contoured map. This helps the application to improve decision making in real time for managing network operations. Using the color-contoured map, the network analyst is able to forecast the possible failures at the network level and can take preventive measures in response to undesirable situations. The proposed system facilitates the constant supervision of network performance and takes timely action so the big IoT data can be analyzed. As the constant flow of data from numerous sources is increasing intensively, plenty of sophisticated and extremely scalable big data analytics platforms have been introduced over a period of time [49]. In the next section, a few of the platforms that support big IoT data analytics are discussed.

10.3 ANALYTICS PLATFORMS FOR IOT BIG DATA

In this section, the study of various analytics platforms is conducted to help in analyzing big data produced by IoT devices. The services of cloudlets [51], mobile-edge computing [52, 53], and fog computing [54] are explored to analyze the big data produced by IoT devices so that the analytics can be carried out in the vicinity of the origin of the data.

10.3.1 Apache Hadoop

Hadoop [3, 55] is able to store and process large amounts of data produced by IoT devices. As the interconnected devices are increasing, the data produced by these devices also rises exponentially. Hadoop provides reliable solutions to such data by scaling up based on the volume of data produced by IoT devices. There is no performance issue because of some unexpected events, as the nodes used for analysis are distributed across clusters. Hadoop provides high availability and fault tolerance, which are required for the dynamic environments, such as clouds. Researchers claim that for the efficient storage and analysis of IoT big data, Hadoop offers the best solution.

10.3.2 1010data

1010data [56] is a cloud-based platform that primarily deals with semi-structured data, such as IoT big data. It offers an interactive analytical capability with the ease of implementation. In addition to all the standard analytical and financial functions, it offers two advanced functions to broaden the analytical capabilities of the system. The two analytics functions are time-series analysis and text analysis. The time-series analysis looks for possibly related events in a data set on the basis of some sequence or time. The text data is analyzed based on some correlations, such as tweets, on a particular event in text-based analysis. On the 1010data platform, the big data is analyzed using a (i) high-performance analytical database, (ii) subsets of data by means of interactive analytical software, and (ii) an interactive analytical database [57].

10.3.3 Cloudera Data Hub

Cloudera introduced a Hadoop-based framework for big data processing and analytics called the Enterprise Data Hub [58]. The framework administers a huge number of geographically distributed interconnected devices and a gateway for enhanced security [59]. It provides end-to-end solutions for IoT devices with reduced cost and increases the speed of deployment of IoT use cases. Business rules and sophisticated analytics models can be deployed by the core platform, as well as the edge, taking benefits from past data and machine learning and facilitating BI at the edge.

10.3.4 SAP HANA

SAP HANA [60] is a cloud-based platform that supports intelligent and efficient IoT data analytics, along with consistent operational workloads. A variety of solutions are supported by SAP to accommodate big unstructured data generated by IoT devices. Hive is used by HANA to access big data, while Sybase IQ is exploited by SAP for creating a columnar Database Management System (DBMS). A built-in analytic library is also provided by HANA, which consists of functions, such as spatial processing, support for R programming, and text analytics. Apart from the flexibility provided by HANA for storage, manipulation, and analysis of data, it also offers low latency and better throughput.

10.3.5 HP HAVEn

HP launched the Hadoop Autonomy Vertica Enterprise (HAVEn) [61] platform that administers and analyzes enterprise data through a comprehensive suite of tools. The data can be of any form (i.e, unstructured, semistructured, or structured). The key components of HP HAVEn are a platform and an ecosystem. The platform offered by HP HAVEn is further divided into three components: (i) data collectors, (ii) engines, and (iii) applications. The data collector consists of a connector that helps to collect data from various sources. Autonomy and ArcSight provide the connectors to HAVEn for searching and exploring data. HAVEn is built on various engines that support specific tasks, such as the Vertica engine, which is well suited for large databases and provides scalability, along with parallel processing, of feature results into accelerated data analytics. HAVEn supports any number of IoT applications. The ecosystem of HAVEn is a collection of resources, such as hardware, software, and services. With all the resources HP HAVEn provides, only intelligence, technology, vision, and insight, it also supports the transformation of businesses [62].

10.3.6 Hortonworks

Hortonworks [63] provides a management and analytics tool for big data produced by IoT devices. The Hortonworks Data Platform is a freeware, open-source software distribution based on Hadoop that focuses on the enhancement of Hive. It is a container-based service platform that supports multiple executive versions of an application. The multiversion supports the enhancement of a service without intervening with the previous version. Along with enhancing the storage capabilities, it also manages the workload of the system very efficiently. Hortronworks consists of TensorFlow, a container that, along with Graphics Processing Unit (GPU) pooling, carries the designing, training, and building of training data for deep learning models [64].

10.3.7 Pivotal Big Data Suite

The Pivotal Big Data Suite (Pivotal BDS) [65], is a cloud-based data platform that is sustained on event-driven architecture. It allows an organization to transform existing data architecture, provides deep insights into the data with sophisticated analytics, and offers more dimension to analytic applications. Pivot BDS provides three crucial solutions known as Pivotal HDB, Pivotal Greenplum, and Pivotal GemFire, all of which are carried under one license. Pivotal HDB is a Hadoop-based analytical database that combines robust ANSI SQL compliance with massively parallel processing (MPP)–based analytics. It permits ad hoc query processing and analytics of data stored in Hadoop Distributed File System (HDFS) by means of SQL syntax and other related tools. Pivotal Greenplum is a well-equipped MPP analytical database used for fast analytics on huge volumes of data. It provides a database with built-in machine learning techniques through Apache MADlib (incubating) for optimized query processing. Pivotal GemFire is intended to support low-latency, high-volume transactional and functional applications. In spite of its many benefits, Pivotal BDS is still in its formative years, and its open acceptance is hindered by several unresolved issues [66].

10.3.8 Infobright

Infobright offers IoT and M2M application developers with efficient and scalable analytics [67]. It can store and analyze huge volumes of data produced by various IoT applications [68]. Infobright has embed capability for scalable storage, fast data insight, and composite ad hoc analysis of data produced by IoT devices. Infobright enables strength for the ThingWorx IoT platform by providing liberty to M2M and IoT application developers to transmit data to a highly dedicated analytics engine. It has a high data skipping and compression ratio, which makes Infobright an appropriate tool for big IoT data. The data-skipping expertise and columnar fashion of Infobright guarantees that suitable data will be picked for every query execution. Automatic indexing of this tool avoids the hassle of partitioning and tuning out data. Even with all of these advanced features and support for diverse data, optimal execution of all queries is not possible through the Infobright optimizer.

10.3.9 MapR Edge

MapR Edge [69] helps with realizing enterprises' full capabilities of their IoT devices by addressing issues, such as limited bandwidth and constrained physical space. Here collection, processing, and analysis of IoT data are done in the vicinity of their source. MapR Edge offers the safe processing of data at the local level, whereas at the global level, fast aggregation of data insights is provided. Moreover, it merges in real-time analysis of data with functional decision making at the edge. MapR Edge also ensures the security of data by the encryption of the wire used for communicating data between the main data center and the edge. MapR Edge holds standards such as HDFS API or POSIX for accessing files, SQL for query processing, Apache Kafka™ API for event streams, and HBase and OJAI API for NoSQL database. Better business solutions are provided by pushing the intelligence gained through data analysis back to the edge [70].

Big data platforms facilitate quick, economical, and enhanced solutions across various application domains of IoT. In the forthcoming section, an investigation of various application domains of big IoT data analytics has been done.

10.4 APPLICATION DOMAIN OF BIG IOT DATA ANALYTICS

Big data technology offers efficient processing and storage of large and varied data services in an IoT environment, although analytics allow enterprises to make advanced and better decisions. Major sources of big data are IoT applications such as health care, transport, and smart cities. In this section, the roles of analytics in several IoT domains are discussed.

10.4.1 Health Care

In the last few decades, a huge volume of data has been generated by the health-care sector. A large amount of data leads to challenges in extracting useful information. The valuable information extracted from health-care databases helps in predicting pandemic situations in advance and finding timely cures. In addition, health-care professionals can analyze a large number of patients through data analytics. Moreover, data analytics can also be used by health-care insurance companies while making policies.

Paul et al. [71] proposed an IoT-based framework for a smart and healthy township using predictive analytics. An improved Hadoop foundation with HBase provided the proposed framework with big data analytics and MapReduce. The framework restrained the requirement of both static and dynamic air pollution monitoring. The framework is based on citizen science concepts in which the air pollution data is collected from sensors and predicts the finding so that the community can be free from airborne diseases, such as asthma.

Chen et al. [72], proposed a three-tiered architecture that includes three platforms – namely, sensing and gateway, resource and administration, and open application platform. On the first layer, all the Internet-enabled devices are connected to the IoT network layer. They present the idea of IoT development from the China perspective: software interfaces, data formats, and modularization of hardware for IoT gateways and terminals. Common interface modules are responsible for collecting physical interfaces of diverse sensors into a standard interface. IoT gateways and terminals should be supported by such applications or software that provide self-configuration and self-adaption. At tier two, the network and service layer are composed of backbone networks and resource administration platforms. This layer supports IoT applications using backbone networks, such as the Internet, 3G, and 4G, along with significant network connectivity control functions. The lowest level is closer to where the IoT application developer can develop some application using some Application Programming Interface (API). Chen et al. mentioned that in "remote medical monitoring," data is gathered from several sensor-like devices positioned at the patient's body, and after processing the data, useful suggestions can be provided. Islam et al. [73] surveyed various health-care technologies, architecture, and applications based on IoT infrastructure. Researchers also studied a variety of IoT privacy and security features, along with threat models, attack classifications, and requirements for security in view of health-care infrastructure. In addition, a secure and intelligent model was proposed that deals with the security challenges countered by IoT devices. Islam et al. also explored how IoT-based, health-care systems can assist different age groups while monitoring their health. Yang et al. [74], implemented a home mobile health-care system (mHealth) prototype for wheelchair users. They suggested an architecture that uses the capabilities of wireless technology to collect data such as the electrocardiogram (ECG), blood pressure, and heart rate for sending and receiving instructions from the source and sinks nodes. Also, the statuses of living and nonliving environments were monitored to analyze the safety of the wheelchair users. An advanced remote interactive system was also enabled so that users could intelligently interact with smart home devices. The proposed protocol system was a people-centric sensing device that monitors human activities in an efficient way and interacts with the living environment. In [75] Amendola et al., the researchers discussed the present radio frequency identification (RFID) technology used in various smart devices, particularly wearable tags. Numerous existing opportunities are explored at the application level, such as wearable devices that discover the movement of body parts, such as legs and arms. They also gathered and analyzed multichannel data to understand human behavior in varying situations, such as exposure to power and hygienic conventions. Amendola et al. also identified a few challenges and possible trends for the future.

Mukherjee et al. [76] surveyed the application of data analytics in health-care systems. Also, in [7] authors discussed the role of analytics in several applications across numerous fields, particularly in IoT systems. Because of the exponential growth of IoT devices and increasing prospects of infrastructure development, the future of analytics has become more prominent. A methodical review of analytics and its applications in the IoT system were discussed. A layered taxonomy was presented that characterized and classified analytics based on its abilities and application. Finally, some trade-offs for analytics in IoT were identified.

10.4.2 Transport

The transport industry can revolutionize with the help of data analytics. Through analytics, the authorities can manage transport and logistics efficiently and effectively. The analysis of data helps in acquiring knowledge, such as the peak time of traffic and causes of road mishaps; it also supports authorities in taking necessary measures, such as optimizing route plans during peak time, imposing rules for road safety, and enhancing end-to-end user experiences in terms of delivery times and managing parking spaces.

Killeen et al. [77] proposed architecture for IoT that used predictive analytics. An approach called improved consensus self-organized model (ICOSMO) was proposed that uses a semi-supervised machine learning technique. ICOSMO is an enhancement of consensus self-organized model (COSMO) [78], which is proposed for improving the sensor feature selection of the COSMO approach [78], which is a sensor-feature selection approach. The COSMO approach helps in diagnosing defective buses and deviates them from the rest of the bus fleet, which improves the transportation system by handling the faults in an efficient way. Liu et al. [79] studied several technologies and applications that assist video analytics and intellectual systems. To effectively plan the traffic flow, identification of road mishaps and the imposition of traffic rules for safety measure video analytics can be applied effectively

[88]. An additional application of smart vehicles is to assist the driver. Danner et al. [80] proposed a framework called precedent-aware classification (PAC) in which decision making is done heuristically. PAC helps in detecting vehicle and pedestrian data in the absence of dedicated hardware. It uses historical data of former trips and their corresponding routes for reducing computational time, which results in meeting the tight time constraints of embedded systems in which a conventional system is likely to fail. Jara et al. [81] used analytics for smart cities by correlating data gathered from IoT devices and temperature that allowed interpolating and predicting complex human behavior. The visual analytics were applied to the data collected from the SmartSantander smart city testbed [82] based on the Poisson distribution to interpolate and predict traffic density. In addition, a data analytics model has been suggested for the open-source data analytics platform Konstanz Information Miner (KNIME). Liebig et al. [83] provided a prediction model for traffic flow in the areas with a limited degree of sensor coverage. It uses a spatial and temporal conditional random field [84] for estimating future patterns of traffic flow. Moreover, Gaussian process regression is used to gather information from unobserved areas. The proposed model suggested an optimum route for every trip to avoid traffic congestion. The cost for an optimal trip is calculated using the A* search algorithm [85], which accelerates the processing of finding routes using contraction hierarchies. The researcher claimed that the model is promising, as it uses both real-time data and continuous streaming backbones. He et al. [86] proposed a cloud-based multilayered platform that collaboratively uses IoT and cloud technologies. A smart parking platform was introduced that facilities safe driving by using logistic regression and the Naïve Bayes decision. It has developed an intelligent parking system with features such as the location of a vehicle, availability of a parking spot, and reservation information. This smart parking system was modeled based on birth-death stochastic that predicts and optimizes the availability of parking spots.

10.4.3 Living

Chianese et al. [87] proposed an integrated approach that collectively uses big data, BI, and IoT to exhibit a platform for cultural behavior analysis. The analytics is applied to data such as unstructured or structured, multiple domain vocabularies, geospatial and social network data, classifiers and ontologies, and multimedia. Classification of movement in a museum was done for proximity evaluation. In this model, data gathered from the sensors was enriched with semantics knowledge of social and cultural tourism to analyze cultural activities through visualization using an associative model. Razip et al. [88] designed a system based on mobile visual analytics tools. The visualization in mobile tools, along with human abilities, were used to identify certain patterns for better public safety and proper law enforcement. The proposed system helped in the timely identification of high-risk zones by using an interactive risk assessment tool. Furthermore, various military applications and public safety use video analytics for tracking movement and recognizing intruders or targets. In [89], the authors discussed use cases on public safety and their probable solutions using big data analytics. They argued that when video analytics are applied to a large volume of visual inputs captured from various sources, suspicious activities can be flagged very quickly and efficiently. Even in crowded places, the identification of potential targets can be done by recognition of directional movement and faces using multiple cameras and intelligent video analytics. The analytics can proactively bring suspicious people and activities to users' attention, leading to the timely activation of alarms and facility lockdown. Ploennigs et al. [90] proposed the architecture and approach that demonstrated the use of semantic sensors, web technologies, and reasoning for the automation of temperature monitoring using analytics. The proposed system is able to detect irregularity in the building temperature, such as high occupancy in a room and cooling system break down. The Building Automation and Control Systems [91] can automatically develop rules for diagnosis and behavior using the semantic approach. The experimental results show that power consumption of buildings is 40% of the world's energy usage and, hence, assures sustainability. Mukherjee et al. [92] proposed a fast and economical real-time evaluation of high dimensional data. The algorithm provided insight into real-time data gathered from wearable IoT devices to identify human activity and then generalized the findings to other high-dimensional data. A ranking system was also proposed for multivariate data. The significance of the proposed algorithms in identifying irregularity and determining data patterns from IoT and streaming data form a basis for intelligent devices, such as activity trackers and monitors.

10.4.4 Environment

Schnizler et al. [93] proposed a disaster detection mechanism that works on heterogeneous streams of data collected through a sensor device. An agent-based intelligent sensor method was used to detect the outliers by preprocessing the low-level events. The outliers then went through the round table that mapped several heterogeneous data to familiar event ontology. The intelligent agent helped with mapping the event through feedback loops, which might engage adjusting parameters so that agents can do mapping or crowdsourcing. The CEP [94] engine processes the homogeneous stream of data to reconstruct the disaster situation using aggregation and clustering of semantic data so that prediction, as well as simulation of disaster, can be done to estimate the damage caused by it. Xu et al. [95] suggested another disaster detection method focusing on urban disaster. Visual results were produced by applying analytics on media events from multimodal micro-blog posts such as text, images, or videos. A real-time story of a metropolitan crisis was analyzed using spatiotemporal data helped to improve the situational responsiveness of the disaster management team. Mukherjee [96] used condor as a wind forecasting application. Data gathered from wind speed sensors within wind turbines were composed from doing parallel data analytics. "Black-box" style program execution was used on historical data for prediction. Constrained Application Protocol [97] was used for analytics, which has a positive effect on overall system performance. This application plays a crucial role in energy provisioning and planning. Ghosh et al. [98] implemented an intelligent energy system called the Plug Load Energy Management Solution. It uses intelligent plugs, along with data analysis, to dynamically supervise energy policy. The weighted moving average model identifies energy consumption patterns of an appliance that results in an opportunity to save energy. Based on the usage and requirements, the new policies are proposed that trigger some events to further lower the amount of electricity bills. Researchers proved that the proposed system significantly saved energy, nearly 60%, and, ultimately, led to the reduction of carbon footprints. Marinakis [99] proposed an energy-efficient, high-level, data-driven architecture for smart buildings. The data produced by such buildings were exchanged, managed, and processed using this model. An algorithm based on artificial intelligence, along with distributed ledger technology, incorporated cross-domain data. The model helped with solving advance problems, such as energy usage prediction, geoclustering, and multicriteria evaluation of external factors. The model allowed for effective and consistent policy making and supported the establishment and utilization of novel energy-efficient service by using a wide variety of data for a well-organized function of buildings. The analytics provides visual outputs and a multicriteria support system, along with BI. Alonso et al. [100] used a rule-based expert system and machine learning for personalized suggestions by analyzing data gathered from smart homes. The proposed system assisted a customer with using energy more efficiently. Additional suggestions, such as cheaper options and anomalies owing to some devices, are predicted by processing data collected from other smart homes. Ahmed [101] exploited big data tools and techniques, such as MapReduce and Apache Hadoop ecosystem tools, for collecting, processing, and analyzing data produced by IoT devices and provided insight that enhanced the efficiency of energy consumption. Furthermore, a model was proposed for the classification of energy patterns of buildings. There are seasonal shifts in this classification owing to changes in patterns because of more detailed analysis, such as appliance-specific data, which helps to predict future energy usage.

10.4.5 Industry

Vargheese et al. [102] proposed an automated system that improves the shopping experience of customers by enhancing the on-shelf availability (OSA) of commodities. In addition, a five-level algorithm was proposed that processed data from a range of IoT devices to formulate certain decisions. The stages were detection, quantification, verification, replacement, identification, and resolution. Detection of the deviation of the availability of a product was taken care of at level one. For detecting the variability, the planogram compliance models were used. On the second level, the relative size of the box and location of products was used to identify the number of remaining products on the shelf, whereas in verification, additional IoT elements were used to ascertain that false alerts should not be raised. For a store manager to reconcile the inventory, a spatial map was generated by using visual analysis of data or by using RFID-based location maps for tracking misplaced data. Finally, the resolution stage incorporated the alliance between the store manager and automated systems to refill the shelf. Further, few parameters, such as local demand, current weather, events, and endorsement information were analyzed with existing OSA to forecast the product demand and model consumer activities, which act as feedback/input into the system. Nechifor et al. [103] implemented a medium-level, IoT-based infrastructure for developing a logistic-based application. An emulator was used for analyzing real-time data in the cold chain monitoring [104] process. Sensors were used for monitoring the transportation conditions, such as humidity and temperature, of the goods so as to automatically activate the alarms when the acceptance criteria of goods were no longer maintained. Moreover, the proposed framework was used to predict the delay in routes and suggest an alternative route so as to reduce the amount of misplaced and damaged perishable products during transportation. Verdouw et al. [105] and Robak et al. [106] designed a framework that supported virtualization in the IoT. They explored the supply chain management system with a physical flow from unprocessed stuff to the end product with a collective objective. The intelligent analysis and reporting of the virtual supply chain system were supported at the highest level. When data mining, BI and predictive analytics were applied to fourth party logistics and floriculture, which showed advanced warning in the case of any unforeseen behavior or outliers in the system. This helped in predicting the consequences when the product reached its destination. Predictive analytics was used, which forecast the changes at the destination by analyzing data collected from sensors and identifying if there were any pattern deviations during the course of the transaction. Moreover, real-time data analysis facilitated instant control of actuators so that inconvenience such as unsuitable truck conditions for consumable products or non-availability of products could be rectified. Kamilaris et al. [107] proposed a Agri-IoT framework that uses the CEP engine to determine major events on semantic data streams generated from sensors using two smart farming scenarios. In the first scenario, the productiveness of a cow was identified based on dairy farm conditions, such as temperature, fodder, and cleanliness, to recommend the best time for insemination. In the second scenario, an adaptive approach was used to control the soil condition for the cultivation of crops. Agri-IoT supports large data analytics, along with event detection, and guarantees smooth interoperability among processes, IoT devices, operations, services, farmers, and other pertinent stakeholders, as well as online information resources and data streams accessible on the web. With the evolution of IoT, various domains of organizations have become smarter, such as smart machinery, smart homes, smart cards, smart cities, intelligent health-care systems, and more. Still, some irregularities and drawbacks exist, which are discussed in the next section.

10.5 CHALLENGE OF BIG IOT DATA ANALYTICS

With the advent of IoT, the prospects of enterprises and end users have increased manyfold in various application fields. Still, there are various challenges that have not been addressed. The following challenges have been identified based on the previous studies.

10.5.1 Privacy

As big data analytics techniques are widely used on IoT data, privacy has become a major issue while mining the data. Moreover, service-level agreements (SLA) are not very user oriented and do not ensure safety in regard to users' personal information. Existing security architecture is not very compatible when applied to IoT systems, as it is intended to deal with current setups in which the heterogeneity of different kinds of machines and varied forms of data coexist. The logical association among the IoT devices require a cheap and easy-to-implement M2M-oriented solution to assure privacy and security of individual data [108]. Moreover, the existing solutions for privacy are specifically designed for static databases, while IoT data is very dynamic in nature. Therefore, the deployment of these solutions is not easy for dynamic data [108]. Therefore, while assembling the SLAs conditions, legal and regulatory technicalities should be considered.

10.5.2 Visualization

While applying analytics on big data generated by IoT devices, visualization plays a significant role when enormous amounts of data are generated. Since the data is large and heterogeneous with multiple dimensions, this makes it difficult for conducting data visualization on big IoT data. Providing solutions for visualizations that support the indexing framework for data with multiple dimensions is a tricky job. To overcome the issue of multidimensionality and complexity of IoT data, various dimensionality reduction techniques have been introduced [109, 110]. However, the proposed techniques are not suitable for every data form that is presented. In addition, the locality of data is also very crucial to gain useful information in an efficient manner [108]. Today, the majority of visualization tools used for big IoT data analytics exhibit bad results in terms of response time, functionality, and scalability. During visual analytics, the uncertainties of IoT data impose a big challenge [111].

10.5.3 Scalability

Another challenge of big IoT data analytics is scalability. Scalability is defined as the capability of a machine that can automatically adapt according to changing requirements and the environment of a device [112]. Before scalability is provided to IoT-based architecture, various factors need to be considered, such as filtration and processing of data at the edge and managing analytics applications at the edge, as well as cloud and security. Enterprises require a mechanism with the help of which it can handle large amounts of data from different devices at once. There does not exist such a mechanism that handles distributed data and process it at once. Moreover, real-time data streaming in certain applications involves parallel processing, which will increase the overhead of maintaining information regarding the processes that are distributed across the network [112].

10.5.4 Data Integration

Data produced by IoT devices are not uniform and hence increase the level of complexity needed to handle it. The diversity in the design and pattern of such data is widespread and often composed of hundreds of unstructured forms or pseudo-structured formation [113]. To understand the data produced by IoT devices, all the structured and unstructured data must be processed. Investigating a particular type of data can limit the potential insight of the actual data. Currently, enterprise data warehouses focus only on structured data, as incorporating different data types is a complex job in the integration of diverse applications [114]. Consequently, a solution needs to be considered to facilitate the storage and analysis of unstructured data. The solution must be simple and without any technical limitations. Implementing any solution for analytics that has technical constraints will limit the ability of analysis and reduce the potential for possible value creation.

10.6 CONCLUSION

With the advancement of new-age communication devices, such as sensors and smartphones, analysts have the opportunity to extract valuable data so as to anticipate the future trends of some analytics tools. The tools in turn use predictive analytics on big enterprise data with some organizational value to benefit organizations. In this work, the focus was primarily on IoT data and analytics applied to data produced by these devices. First, the correlation between big data analysis and IoT data were identified. Then the classification of analytics is done at two levels. The categorization of analytics is done based on their application to IoT data and the stakeholders at the time of deployment of the IoT. Then a relationship is determined between two categorizations. Various platforms that facilitate IoT and analytics services were also discussed. This chapter has illustrated how analytics play a crucial role in various applications across the domains. Finally, a few trade-offs for analytics in the IoT were discussed that could shape research directions going forward.

REFERENCES

1. Cisco. The internet of things: How the next evolution of the internet is changing everything. 2011. https://www.cisco.com/c/dam/en_us/about/ac79/docs/innov/IoT_IBSG_0411FINAL.pdf
2. Alam, T. "A reliable communication framework and its use in internet of things (IoT)." *International Journal of Scientific Research in Computer Science, Engineering and Information Technology (IJSRCSEIT)*, vol. 3, no. 5, pp. 450–456, May 2018.
3. Ahmed, E., Yaqoob, I., Hashem, I.A.T., Khan, I., Ahmed, A.I.A., Imran, M. and Vasilakos, A.V. "The role of big data analytics in Internet of Things." *Computer Networks*, vol. 129, pp. 459-471, 2017.
4. Mark, V. R. "Why the 3v's are not sufficient to describe big data." *BigData Startups*, 2013.
5. Holger, Z. and Z. Jerzak. "The DEBS 2014 grand challenge." *Proc. of the 8th ACM International Conference on Distributed Event-based Systems, DEBS*, vol. 14, pp. 10–1145, July 2014.
6. Fernandez, R. C.,Weidlich, M., Pietzuch, P. and Gal, A. "Scalable stateful stream processing for smart grids." *Proceedings of the 8th ACM International Conference on Distributed Event-Based Systems*, pp. 276–281, 2014.
7. Siow, E., Tiropanis, T. and Hall, W. "Analytics for the internet of things: A survey." *ACM Computing Surveys (CSUR)*, vol. 51, no. 4, pp. 1–36, 2018.
8. O'Hara, N., Slot, M., Marinescu, D., Čurn, J., Yang, D., Asplund, M., Bouroche, M., Clarke, S. and Cahill, V. "MDDSVsim: an integrated traffic simulation platform for autonomous vehicle research." *Proc. of the International Workshop on Vehicular Traffic Management for Smart Cities*, 2012.
9. van Nunen, E., Kwakkernaat, M.R., Ploeg, J. and Netten, B.D. "Cooperative competition for future mobility." *IEEE Transactions on Intelligent Transportation Systems*, vol. 13, no. 3, pp. 1018–1025, 2012.

10. Sivarajah, U., Kamal, M.M., Irani, Z. and Weerakkody, V. "Critical analysis of big data challenges and analytical methods." *Journal of Business Research*, vol. 70, pp. 263–286, 2017.
11. Hopkins, J., Hawking, P., Hopkins, J. and Hawking, P. "big data analytics and IoT in logistics: A case study." *The International Journal of Logistics Management*, 2018.
12. Golchha, N. "Big data-the information revolution." *Int. J. Adv. Res*, vol. 1, no. 12, pp. 791–794, 2015.
13. Chen, M., Mao, S., Zhang, Y. and Leung, V.C. "*Big data: related technologies, challenges and future prospects*" (vol. 96). Heidelberg: Springer, 2014.
14. Akoka, J., Comyn-Wattiau, I. and Laoufi, N. "Research on big data–a systematic mapping study." *Computer Standards & Interfaces*, vol. 54, pp. 105–115, 2017.
15. Mikalef, P., Pappas, I. O., Krogstie, J. and Giannakos, M. "Big data analytics capabilities: A systematic literature review and research agenda." *Information Systems and e-Business Management*, vol. 16, no. 3, pp. 547–578, 2018.
16. Krumeich, J., Werth, D. and Loos, P. "Prescriptive control of business processes." *Business & Information Systems Engineering*, vol. 58, no. 4, pp. 261–280, 2016.
17. Bertolucci, J. "Big data analytics: Descriptive vs. predictive vs. prescriptive." *Information Week*, 2013.
18. Kart, L. *Advancing analytics: Technical report*. Gartner Inc., 2012.
19. Chandler, N., Hostmann, B., Rayner, N. and Herschel, N. *Gartner's business analytics framework: Technical report*. Gartner Inc., 2011. http://www.gartner.com/imagesrv/summits/docs/na/business-intelligence/gartners
20. Corcoran, M. "The five types of analytics," technical report. *Information Builders*, pp. 68–69, 2012. http://www.informationbuilders.co.uk/sites/www.informationbuilders.com/files/intl/co.uk/presentations/four
21. Hair Jr., J. F. "Knowledge creation in marketing: The role of predictive analytics." *European Business Review*, vol. 19, pp. 303–315, 2007.
22. Shmueli, G. and Koppius, O. R. "Predictive analytics in information systems research." *MIS Quarterly*, pp. 553–572, 2011.
23. Basu, A.T.A.N.U. "Five pillars of prescriptive analytics success." *Analytics magazine*, pp. 8–12, 2013.
24. Soltanpoor, R. and Sellis, T. "September. Prescriptive analytics for big data." *Australasian Database Conference*, Springer: Cham, pp. 245–256, 2016.
25. Lepenioti, K., Bousdekis, A., Apostolou, D. and Mentzas, G. "Prescriptive analytics: Literature review and research challenges." *International Journal of Information Management*, vol. 50, pp. 57–70, 2020.
26. Hagerty, J. (2017). Planning guide for data and analytics. *Gartner Inc (2017)*: 13. https://www.gartner.com/binaries/content/assets/events/keywords/catalyst/catus8/2017_planning_guide_for_data_analytics.pdf.
27. Larson, D. and Chang, V. "A review and future direction of agile, business intelligence, analytics and data science." *International Journal of Information Management*, vol. 36, no. 5, pp. 700–710, 2016.
28. den Hertog, D. and Postek, K. "Bridging the gap between predictive and prescriptive analytics-new optimization methodology needed", technical report. *Tilburg University*, Netherlands, 2016.
29. Ed-daoudy, A. and Maalmi, K. "A new Internet of Things architecture for real-time prediction of various diseases using machine learning on big data environment." *Journal of Big Data*, vol. 6, no. 1, pp. 104, 2019.
30. Manogaran, G. and Lopez, D. "Health data analytics using scalable logistic regression with stochastic gradient descent." *International Journal of Advanced Intelligence Paradigms*, vol. 10, no. 1–2, pp. 118–132, 2018.

31. Malik, K.R., Sam, Y., Hussain, M. and Abuarqoub, A. "A methodology for real-time data sustainability in smart city: Towards inferencing and analytics for big-data." *Sustainable Cities and Society*, vol. 39, pp. 548–556, 2018.
32. Limkar, S. V. and Jha, R. K. "Computing over encrypted spatial data generated by IoT." *Telecommunication Systems*, vol. 70, no. 2, pp. 193–229, 2018.
33. Lee, J. G. and Kang, M. "Geospatial big data: challenges and opportunities." *Big Data Research*, vol. 2, no. 2, pp. 74–81, 2015.
34. Balakrishnan, S. M. "Aspect-oriented modeling of spatial data interpolation for estimating missing data in Internet of Things (IoT) service discovery." *Journal of Testing and Evaluation*, vol. 47, no. 6, pp. 4004–4016, 2019.
35. Chatfield, C. and Xing, H. *The analysis of time series: An introduction with R.* CRC Press, New York, 2019.
36. Bader, A., Kopp, O. and Falkenthal, M. "Survey and comparison of open source time series databases." *Datenbanksysteme für Business, Technologie und Web (BTW 2017) – Workshopband*, 2017.
37. Idreos, S., Groffen, F., Nes, N., Manegold, S., Mullender, S. and Kersten, M. "Monetdb: Two decades of research in column-oriented database." *IEEE Data Engineering Bulletin*, 2012.
38. Lakshman, A. and Malik, P. "Cassandra: a decentralized structured storage system." *ACM SIGOPS Operating Systems Review*, vol. 44, no. 2, pp. 35–40, 2010.
39. Burdack, M., Rössle, M. and Kübler, R. "A concept of an in-memory database for IoT sensor data." *Athens Journal of Sciences*, vol. 5, no. 4, pp. 355–374, 2018.
40. Kairos D. B. "Cassandra schema," https://kairosdb.github.io/docs/build/html/CassandraSchema.html
41. Salmaan, S. and Amjad, D. "INFO-H415–advanced databases project: Time Series Database with InfluxDB." Ecole polytechnique de Bruxelles – ULB, Brussels 2018–2019.
42. Delen, D. and Ram, S. "Research challenges and opportunities in business analytics." *Journal of Business Analytics*, vol. 1, no. 1, pp. 2–12, 2018.
43. d'Hondt, T., Wilbik, A., Grefen, P., Ludwig, H., Baracaldo, N. and Anwar, A. "Using BPM technology to deploy and manage distributed analytics in collaborative IoT-driven business scenarios." *Proc. of the 9th International Conference on the Internet of Things*, pp. 1–8, 2018.
44. Icke, I. and Sklar, E. "Visual analytics: A multifaceted overview." *Faces*, pp. 1–23, 2009.
45. Xu, Z., Zhang, R., Kotagiri, R. and Parampalli, U. "An adaptive algorithm for online time series segmentation with error bound guarantee." *Proc. of the 15th International Conference on Extending Database Technology*, pp. 192–203, 2012.
46. Maciejewski, R., Hafen, R., Rudolph, S., Larew, S.G., Mitchell, M.A., Cleveland, W. S. and Ebert, D. S. "Forecasting hotspots—A predictive analytics approach." *IEEE Transactions on Visualization and Computer Graphics*, 17(4), pp. 440–453, 2010.
47. Yue, J., Raja, A., Liu, D., Wang, X. and Ribarsky, W. "A blackboard-based approach towards predictive analytics." *AAAI Spring Symposium: Technosocial Predictive Analytics*, vol. 154, 2009.
48. Riensche, R. M., Paulson, P. R., Danielson, G., Unwin, S. D., Butner, S., Miller, S., Franklin, L. and Zuljevic, N. "Serious gaming for predictive analytics." *AAAI Spring Symposium: Technosocial Predictive Analytics*, pp. 108–113, 2009.
49. Poornima, S. and Pushpalatha, M. "A survey of predictive analytics using big data with data mining." *International Journal of Bioinformatics Research and Applications*, vol. 14, no. 3, pp. 269–282, 2018.
50. Huang, Z., Wong, P. C., Mackey, P., Chen, Y., Ma, J., Schneider, K. and Greitzer, F. L. "Managing complex network operation with predictive analytics." *AAAI Spring Symposium: Technosocial Predictive Analytics*, pp. 59–65, 2009.

51. Shaukat, U., Ahmed, E., Anwar, Z. and Xia, F. "Cloudlet deployment in local wireless networks: Motivation, architectures, applications, and open challenges." *Journal of Network and Computer Applications*, vol. 62, pp. 18–40, 2016.
52. Ahmed, E., and Rehmani, M.H. "Mobile edge computing: opportunities, solutions, and challenges." *Future Generation Computer Systems*, vol. 70, pp. 59–63, 2017.
53. Peng, K., Leung, V., Xu, X., Zheng, L., Wang, J. and Huang, Q. "A survey on mobile edge computing: Focusing on service adoption and provision." *Wireless Communications and Mobile Computing*, 2018.
54. Siow, E., Tiropanis, T. and Hall, W. "Ewya: An interoperable fog computing infrastructure with RDF stream processing." *Lecture Notes in Computer Science*, vol. 10673, Springer: Cham, pp. 245–265, 2017. doi:10.1007/978-3-319-70284-120.
55. Manikandan, S. G. and Ravi, S. "Big data analysis using Apache Hadoop." *2014 International Conference on IT Convergence and Security (ICITCS)*, pp. 1–4. IEEE, 2014. doi: 10.1109/ICITCS.2014.7021746
56. Morabito, V. "Managing change for big data driven innovation." In *Big Data and Analytics*, pp. 125–153. Springer: Cham, 2015. https://doi.org/10.1007/978-3-319-10665-6_7
57. 1010data: https://www.ciosummits.com/media/pdf/solution_spotlight/1010_bloor-group.pdf
58. Bhardwaj, A., Bhattacherjee, S., Chavan, A., Deshpande, A., Elmore, A. J., Madden, S. and Parameswaran, A. G. "Datahub: Collaborative data science & data set version management at scale." arXiv preprint arXiv:1409.0798, 2014.
59. Cloudera. *An enterprise, end-to-end, open source architecture for IoT.* Joint Venture of Red Hat, Eurotech, and Cloudera. https://www.cloudera.com/solutions/gallery/red-hat-eurotech-end-to-end-open-source-architecture-for-iot.html
60. Färber, F., Cha, S. K., Primsch, J., Bornhövd, C., Sigg, S., Lehner, W., Färber, F., Cha, S. K., Primsch, J., Bornhövd, C. Sigg, S. and Lehner, W. "SAP HANA database: Data management for modern business applications." *ACM Sigmod Record*, vol. 40, no. 4, pp. 45–51, 2012.
61. Burke, S. "Hp haven big data platform is gaining partner momentum." *CRN [online]* http://www.crn.com/news/applications-os 240161649, 2013.
62. HP HAVEn. *Profite from big data: Turn disconnected intelligence.* Hewlett-Packard Development Company, August 2013. https://www.officeproductnews.net/sites/default/files/ProfitfromBigData.pdf
63. Hortonworks. *Harton data platform (HDP) 3.0: Faster, smarter and hybrid data.* Hartonworks Inc., 2011–2018. https://www.cloudera.com/content/dam/www/marketing/resources/datasheets/hdp-datasheet.pdf.landing.html
64. Dhayne, H., Haque, R., Kilany, R. and Taher, Y. "In search of big medical data integration solutions—a comprehensive survey." *IEEE Access*, vol. 7, pp. 91265–91290, 2019.
65. Zhuang, Y., Wang, Y., Shao, J., Chen, L., Lu, W., Sun, J., Wei, B. and Wu, J. "D-Ocean: An unstructured data management system for data ocean environment." *Frontiers of Computer Science*, vol. 10, no. 2, pp. 353–369, 2016.
66. Saggi, M. K. and Jain, S. "A survey towards an integration of big data analytics to big insights for value-creation." *Information Processing & Management*, vol. 54, no. 5, pp. 758–790, 2018.
67. Ray, P. P. "A survey of IoT cloud platforms." *Future Computing and Informatics Journal*, vol. 1, no. 1–2, pp. 35–46, 2016.
68. Slezak, D., Synak, P., Wrăłblewski, J. and Toppin, G. "Infobright analytic database engine using rough sets and granular computing." *2010 IEEE International Conference on Granular Computing*, San Jose, CA, pp. 432–437, 2010 doi: 10.1109/GrC.2010.177.
69. MapR. [Online]. https://mapr.com
70. Lalitha, B. "Recover the missing data in IoT by Edge Analytics." *i-Manager's Journal on Software Engineering*, vol. 13, no. 2, pp. 25, 2018.

71. Paul, S. T., Raimond, K. and Kanaga, G. M. "An IoT-enabled hadoop-based data analytics and prediction framework for a pollution-free smart-township and an asthma-free generation." In J. Peter, A. Alavi and B. Javadi (eds.), *Advances in Big Data and Cloud Computing*, vol. 250, pp. 577–587, Springer: Singapore, 2019. doi:10.1007/978-981-13-1882-5_51
72. Chen, S., Xu, H., Liu, D., Hu, B. and Wang, H. "A vision of IoT: Applications, challenges, and opportunities with China perspective." *IEEE Internet of Things Journal*, vol. 1, no. 4, pp. 349–359, 2014.
73. Islam, S. R., Kwak, D., Kabir, M.H., Hossain, M. and Kwak, K. S. "The Internet of Things for health care: A comprehensive survey." *IEEE Access*, vol. 3, pp. 678–708, 2015.
74. Yang, L., Li, W., Ge, Y., Fu, X., Gravina, R. and Fortino, G. "People-centric service for mHealth of wheelchair users in smart cities." In G. Fortino and P. Trunfio (eds.), *Internet of Things based on Smart Objects*. Internet of Things (Technology, Communications and Computing), pp. 163–179, Springer: Cham, 2014. doi:10.1007/978-3-319-00491-4_9
75. Amendola, S., Lodato, R., Manzari, S., Occhiuzzi, C. and Marrocco, G. "RFID technology for IoT-based personal healthcare in smart spaces." *IEEE Internet of Things Journal*, IEEE, vol. 1, no. 2, pp. 144–152, 2014.
76. Mukherjee, A., Pal, A. and Misra, P. "Data analytics in ubiquitous sensor-based health information systems." *2012 Sixth International Conference on Next Generation Mobile Applications, Services and Technologies*, Paris, pp. 193–198, 2012. doi:10.1109/NGMAST.2012.39
77. Killeen, P., Ding, B., Kiringa, I. and Yeap, T. "IoT-based predictive maintenance for fleet management." *Procedia Computer Science*, vol. 151, pp. 607–613, 2019.
78. Rögnvaldsson, T., Nowaczyk, S., Byttner, S., Prytz, R. and Svensson, M. "Self-monitoring for maintenance of vehicle fleets." *Data mining and Knowledge Discovery*, vol. 32, no. 2, pp. 344–384, 2018.
79. Liu, H., Chen, S. and Kubota, N. "Intelligent video systems and analytics: A survey." *IEEE Transactions on Industrial Informatics*, vol. 9, no. 3, pp. 1222–1233, 2013.
80. Danner, J., Wills, L., Ruiz, E. M. and Lerner, L. W. "Rapid precedent-aware pedestrian and car classification on constrained IoT platforms." *2016 14th ACM/IEEE Symposium on Embedded Systems for Real-Time Multimedia (ESTIMedia)*, Pittsburgh, PA, pp. 1–8, 2016.
81. Jara, A. J., Genoud, D. and Bocchi, Y. "Big data for smart cities with KNIME a real experience in the SmartSantander test bed." *Software: Practice and Experience*, vol. 45, no. 8, pp. 1145–1160, 2015.
82. Sanchez, L., Galache, J. A., Gutierrez, V., Hernandez, J. M., Bernat, J., Gluhak, A. and Garcia, T. "Smartsantander: The meeting point between future internet research and experimentation and the smart cities." *2011 Future Network & Mobile Summit*, Warsaw, pp. 1–8, 2011.
83. Liebig, T., Piatkowski, N., Bockermann, C. and Morik, K. "March. Predictive trip planning-smart routing in smart cities." In *EDBT/ICDT Workshops*, pp. 331–338, 2014.
84. Lafferty, J., McCallum, A. and Pereira, F.C. "Conditional random fields: Probabilistic models for segmenting and labeling sequence data." *Proceedings of the 18th International Conference on Machine Learning 2001 (ICML 2001)*, pp. 282–289, 2001.
85. Hart, P.E., Nilsson, N. J. and Raphael, B. "A formal basis for the heuristic determination of minimum cost paths." *IEEE transactions on Systems Science and Cybernetics*, vol. 4, no. 2, pp. 100–107, 1968.
86. He, W., Yan, G. and Da Xu, L. "Developing vehicular data cloud services.". *The IoT environment IEEE transactions on industrial informatics*, vol. 10, no. 2, pp. 1587–1595, 2014.
87. Chianese, A., Marulli, F., Piccialli, F., Benedusi, P. and Jung, J. E. "An associative engines based approach supporting collaborative analytics in the internet of cultural things." *Future Generation Computer Systems*, vol. 66, pp. 187–198, 2017

88. Razip, A. M., Malik, A., Afzal, S., Potrawski, M., Maciejewski, R., Jang, Y., Elmqvist, N. and Ebert, D. S. "March. A mobile visual analytics approach for law enforcement situation awareness." In *2014 IEEE Pacific Visualization Symposium*, pp. 169–176, 2014.
89. Gimenez, R., Fuentes, D., Martin, E., Gimenez, D., Pertejo, J., Tsekeridou, S., Gavazzi, R., Carabaño, M. and Virgos, S. "The safety transformation in the future internet domain." In *The Future Internet Assembly*, pp. 190–200 Springer: Berlin, Heidelberg, 2012.
90. Ploennigs, J., Schumann, A. and Lécué, F. "Adapting semantic sensor networks for smart building diagnosis." In *International Semantic Web Conference*, pp. 308–323, Springer: Cham, 2014.
91. Aste, N., Manfren, M. and Marenzi, G. "Building automation and control systems and performance optimization: A framework for analysis." *Renewable and Sustainable Energy Reviews*, vol. 75, pp. 313–330, 2017.
92. Mukherjee, U. K. and Chatterjee, S. "Fast algorithm for computing weighted projection quantiles and data depth for high-dimensional large data clouds." In *2014 IEEE International Conference on Big Data (Big Data)*, pp. 64–71, 2014.
93. Schnizler, F., Liebig, T., Marmor, S., Souto, G., Bothe, S. and Stange, H. "Heterogeneous stream processing for disaster detection and alarming." *2014 IEEE International Conference on Big Data (Big Data)*, Washington, DC, pp. 914–923, 2014. doi: 10.1109/BigData.2014.7004323
94. Luckham, D. "The power of events: An Introduction to Complex Event Processing in Distributed Enterprise Systems." In N. Bassiliades, G. Governatori, and A. Paschke (eds.), *Rule Representation, Interchange and Reasoning on the Web: International Symposium, Addison-Wesley*, vol. 5321, p. 3, Springer: Berlin, Heidelberg, 2008. doi:10.1007/978-3-540-88808-6_2
95. Xu, Z., Liu, Y., Zhang, H., Luo, X., Mei, L. and Hu, C. "Building the multi-modal storytelling of urban emergency events based on crowdsensing of social media analytics." *Mobile Networks and Applications*, vol. 22, no. 2, pp. 218–227, 2017.
96. Mukherjee, A., Dey, S., Paul, H.S. and Das, B. "Utilising condor for data parallel analytics in an IoT context—An experience report." In *2013 IEEE 9th International Conference on Wireless and Mobile Computing, Networking and Communications (WiMob)*, pp. 325–331, IEEE, 2013.
97. Aigner, W., Miksch, S., Müller, W., Schumann, H. and Tominski, C. "Visual methods for analyzing time-oriented data." *IEEE Transactions on Visualization and Computer Graphics*, vol. 14, no. 1, pp. 47–60, 2008. doi: 10.1109/TVCG.2007.70415
98. Ghosh, A., Patil, K. A. and Vuppala, S. K. "PLEMS: Plug load energy management solution for enterprises." In *2013 IEEE 27th International Conference on Advanced Information Networking and Applications (AINA)*, pp. 25–32, 2013.
99. Marinakis, V. "Big data for energy management and energy-efficient buildings" *Energies*, vol. 13, no. 7, pp. 1555, 2020.
100. González Alonso, I. and Rodríguez Fernández, M. "A holistic approach to energy efficiency systems through consumption management and big data analytics." *International Journal on Advances in Software*, vol. 6, 2013.
101. Ahmed, H.. 2014. "Applying big data analytics for energy efficiency. Masters Thesis. Aalto University, 2014.
102. Vargheese, R. and Dahir, H. "An IoT/IoE enabled architecture framework for precision on shelf availability: Enhancing proactive shopper experience." *IEEE International Conference on Big Data (Big Data)*, pp. 21–26, 2014.
103. Nechifor, S., Petrescu, A., Damian, D., Puiu, D. and Târnaucă, B. "Predictive analytics based on CEP for logistic of sensitive goods." *International Conference on Optimization of Electrical and Electronic Equipment (OPTIM)*, pp. 817–822, 2014.

104. Abad, E., Palacio, F., Nuin, M., De Zarate, A. G., Juarros, A., Gómez, J. M. and Marco, S. "RFID smart tag for traceability and cold chain monitoring of foods: Demonstration in an intercontinental fresh fish logistic chain." *Journal of Food Engineering*, vol. 93, no. 4, pp. 394–399, 2009.
105. Verdouw, C.N., Beulens, A.J.M. and Van Der Vorst, J.G.A.J. "Virtualisation of floricultural supply chains: A review from an Internet of Things perspective." *Computers and Electronics in Agriculture*, vol. 99, pp. 160–175, 2013.
106. Robak, S., Franczyk, B. and Robak, M. "Applying big data and linked data concepts in supply chains management." *Federated Conference on Computer Science and Information Systems*, pp. 1215–1221, 2013.
107. Kamilaris, A., Gao, F., Prenafeta-Boldu, F. X. and Ali, M. I. "December. Agri-IoT: A semantic framework for Internet of Things-enabled smart farming applications." *IEEE 3rd World Forum on Internet of Things (WF-IoT)*, pp. 442–447, 2016.
108. Marjani, M., Nasaruddin, F., Gani, A., Karim, A., Hashem, I.A.T., Siddiqa, A. and Yaqoob, I. "Big IoT data analytics: Architecture, opportunities, and open research challenges." *IEEE Access*, vol. 5, pp. 5247–5261, 2017.
109. Wang, L., Wang, G. and Alexander, C. A. "Big data and visualization: Methods, challenges and technology progress." *Digital Technologies*, vol. 1, no. 1, pp. 33–38, 2015.
110. Azar, A. T. and Hassanien, A. E. "Dimensionality reduction of medical big data using neural-fuzzy classifier." *Soft Computing*, vol. 19, no. 4, pp. 1115–1127, 2015.
111. Chen, C. P. and Zhang, C. Y. "Data-intensive applications, challenges, techniques and technologies: A survey on big data." *Information Sciences*, vol. 275, pp. 314–347, 2014.
112. Gupta, A., Christie, R. and Manjula, P. R. "Scalability in Internet of Things: Features, techniques and research challenges." *Int. J. Comput. Intell. Res*, vol. 13, no. 7, pp. 1617–1627, 2017.
113. Acharjya, D. P. and Ahmed, K. "A survey on big data analytics: challenges, open research issues and tools." *International Journal of Advanced Computer Science and Applications*, vol. 7, no. 2, pp. 511–518, 2016.
114. Ma'ayan, A., Rouillard, A. D., Clark, N. R., Wang, Z., Duan, Q. and Kou, Y. "Lean Big Data integration in systems biology and systems pharmacology." *Trends in Pharmacological Sciences*, vol. 35, no. 9, pp. 450–460, 2014.

11 Multiple-Criteria Decision Analysis Using VLSI Global Routing

Subhrapratim Nath
Jadavpur University
Kolkata, West Bengal, India

Rana Majumdar
MSIT, TIG
Kolkata, India

CONTENTS

11.1 INTRODUCTION

It is evident from empirical study that minimizing energy along with miniaturizing switch area usage are two significant design objectives in very large scale integration (VLSI) global routing architecture policy when it comes to Moore's law, which vehemently states, "The number of transistors per square inch on integrated circuits had doubled every year." To combat such a real-time situation optimization in terms of energy and space, it is inevitable to set a VLSI chip toward progressive functionalities. Now for real-time application, the word totally independent doesn't satisfy its relevance when one makes a practical approach toward designing from scratch, which involves the visualization of several upcoming problems institutively in the case of VLSI routing to enhance the efficiency of VLSI circuits, a number of issues need to be focused on, including time management, space management, reduction in power dissipation, etc. Considering space management, one needs to concentrate on objectives such as wire-length minimization, as execution time for the efficiency of an algorithm in VLSI routing has a direct relationship with area usage and wire length. Now to minimize wire length in the early phase yields better results, and there comes the concept of bend reduction to encourage smooth flow and to avoid unnecessary path-related issues. Beyond that, it also saves power and cost. Delay reduction, an important part of time management, is only an effect of the wire-length minimization, as data will flow in a superfluous manner through a path of shorter length free of hindrances, thereby reducing any further delays in output. Previous work on space management includes the STAI Route framework, which led to improper utilization of the routing area, specifically in higher routing layers with fewer routing blockages, as the lack of placement of standard cells does not facilitate any routing of their interconnections. Interestingly, similar fallacies also arose in time management before the delay model, as the previous outputs of an input were the stated bits of the finite-state machine before the arrival of the synchronized clock concept, which was the earlier version before the arrival of the delay model as shown in Figure 11.1. To achieve enhanced performance, the objectives become multiples for constructing a decision tree by applying operational research, thereby incorporating a certain number of conflicting criteria, such as cost, time management, space management, and other internal and external issues. Multiple-criteria decision making (MCDM) is a subdiscipline of operational research, confirming it's dynamicity with the emergence of different technical and nontechnical parameters; hence, it claims the fact that optimization, when done with a multiple objectives performance, can be enhanced at an appreciable rate, as with the present situation, it possesses the capability to adjust with live changes. The concept was first whimsical in 1979 when

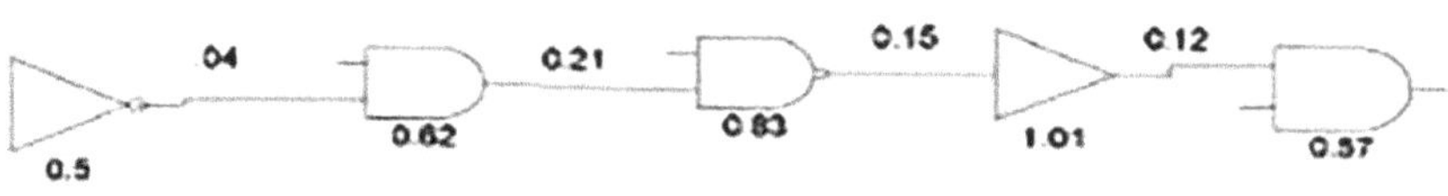

FIGURE 11.1 Delay Model (Source: http://3.bp.blogspot.com/delay3.bmp).

Stanley Zionts popularized the idea for an entrepreneurial audience. It is based on the concept of metaheuristics, which involves an iterative master strategy that guides and modifies the operations of subordinate heuristics by combining intelligently different concepts for exploring and exploiting the search space applicable to a large number of problems. A large number of metaheuristic algorithms, such as invasive weed optimization (IWO) and PSO-MU, have been deployed to achieve the result after focusing on multiple issues and influencing the result, which justifies the term in the case of the VLSI routing multi-objective. In a more generalized sense here, the authors model the preferences of a decision maker over a set of decision alternatives and explore more diverse ways of comparing them than in data analysis (i.e., from a broader perspective, the sole objective of MCDM or multiple-criteria decision analysis [MCDA]) to generate a set of solutions, which shows exemplary competence in computing the results of a particular class of practical problems in certain modules, although the result is purely nondeterministic because of ever-changing constraints parameters. Here the authors assume the case of the evaluation of a solution deploying multiple criteria with growth is better in each criterion. However, it is unlikely to have a single deterministic solution. Typically, performance of a particular solution varies criteria-wise. Finding a way of trading off between criteria is one of the main aims in the MCDM approach. Statistically, the MCDM problem associates with arguments as [2]

$$\text{“max”}\ q;\ \text{subject to}\ \mathbf{q} \in \mathbf{Q};$$

where $\mathbf{q}$ represents the vector of k criterion functions (objective functions) and $\mathbf{Q}$ stands for the feasible set, $\mathbf{Q} \subseteq \mathbf{R}k$.

If Q is defined explicitly (by a set of alternatives), the resulting problem will be in multiple-criteria flora. If Q is defined implicitly (by a set of constraints), the resulting problem represents a multiple-criteria design problem. The quotation marks that are deployed to act as an indicator representing the maximization of a vector are not well-defined mathematical operations.

11.2 PROBLEM CONCEPTUALIZATION AND RELATED STUDY

11.2.1 VLSI Global Routing

Global routing in VLSI design serves as the base of any discrete optimization problems in computational theory when it comes to architectural paradigms. From this point of view, only the physical layout of a circuit is realized by getting suitable information about its functional description and specifications. Because of the exponential increase in complexity of integrated circuits, computer-aided design tools have been instrumental in this design process. Global routing has threefold objectives, which include the minimization of total overflow, wire length, and running time, meaning it perfectly maps with the idea of MCDM, as both global routing and MCDA has multi-objectives to satisfy promptly in addition to which it requires careful decisional analysis to achieve its desired outcome. A deep insight into the concept speaks of a few introductory steps to be set; one of them includes region assignment, which vividly describes the urgency of assigning routing regions to each net with the collaboration of two internal factors:

(1) Timing budgets of net, and
(2) Routing congestion of the region.

The additional maiden phase accounts for the assignment of a pin to prepare the detailed routing stage for each region. Coming to the core-programming part, there are two distinct tactics that can be used based on the circumstantial demand, which are the sequential approach and the concurrent approach. The former routes the net once and the order of the result bears a proportionate relationship with criticality, estimated, wire length, etc. The latter is basically an improvement over the former, as it eliminates the evolving shortcomings of the net ordering problem and takes into account all the nets simultaneously. The resulting formulation of the concurrent approach is an integer program. Here the authors are formulating a problem statement of wire-length minimization, bend reduction, and VLSI delay formulation with interactions and collaborations with decisional routes that lay their base upon their structural and behavioral competencies.

11.2.2 Wire-Length Minimization

Wire-length minimization is one of the prime aspects among three objectives, as previously mentioned. It involves miniaturizing the length of the wire to minimize resource consumption as the net with longer wire length will lead to the exhaustion of more routing resources, and the resources are limited in nature. The objective of wire-length minimization can be achieved by finding an optimal solution for VLSI physical design components, such as partitioning and floor planning. These two parameters, again, have a direct influence on criteria, such as power, cost, and clock speed. The idea behind the technical implementation is to have an evolutionary algorithm that incorporates one or more local search phases within its evolutionary cycle to obtain the minimum wire length by reducing delay in partitioning and by reducing area in floor planning. The idea of simulated annealing is also involved, which accounts for a promising result, as it shows improvement in the solution at each discrete stage by comparing the better of the current and the previous solution. The idea is to partition the cell using simulated annealing so as to minimize the estimated wire length. Either a cell is chosen randomly and placed in a random location on the chip, or two cells are selected randomly and interchanged. The performance of the algorithm was observed as the ratio of displacements to interchanges. But it's undoubtedly the genetic algorithm that deals with the final issue of ultimate minimization of wire length, as fitness function has a lot to do with the minimization. Basically, the fitness function is calculated based on the area of the module (i.e., Total Area = $\sum An$).

The sum of the weighting factors equals one. The complete fitness function is

$$G_f = \frac{A_a}{A_m} > 1, \frac{A_a}{A_m} \leq 1, \frac{A_a}{A_m} + A_f \tag{11.1}$$

11.2.3 Bend Reduction

Bend reduction enhances a dimension and is critical to execute, as it directly helps in global routing by minimizing delay with a decreasing wire length in the initial phase only in the subsequent later phase, and there lies the competency of the existing algorithm to execute the task. The central idea is to diminish the number of bends with a twofold objective to get rid of excessive manufacturing and excess power dissipation. The fitness function of the bend reduction is given by [3]

$$\text{Fitness} = \text{L}(\text{Ts}) + \beta * \text{n}, \tag{11.2}$$

where the term n represents the number of bends and the parameter β portrays the weighting factor that is associated with the bends' quantity. For the purpose of reducing the number of bends, which is one of the key factors of chip manufacturability, an edge-vertex encoding strategy combined with edge transformation is also helpful.

11.2.4 VLSI Delay Formulation

VLSI circuit suffers two major sorts of delay – namely, cell delay, which collaborates timing delay between an input pin and output pin of a cell, and cell delay information. The latter is called the net delay, which consists of characteristics of the driver cell, along with load characteristics of the cell, and RC value of the net. These are the few factors that need consideration to calculate delay. Alternatively, delay can be computed in real-time circumstances by implementing pre-layout or post-layout or during sign-off timing.

As realized from Figure 11.2 for the previous logic circuit,

$$\text{delay} = 0.5 + 0.04 + 0.62 + 0.21 + 0.83 + 0.15 + 1.01 + 0.12 + 0.57 = 4.05\text{ns}$$

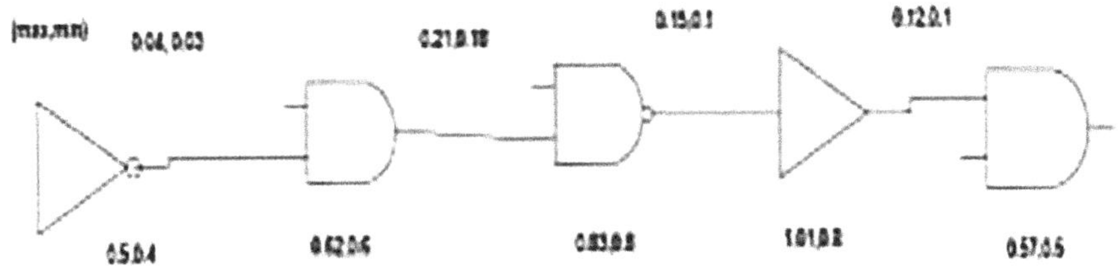

FIGURE 11.2 Logic Circuit (Source: http://3.bp.blogspot.com/delay1.bmp).

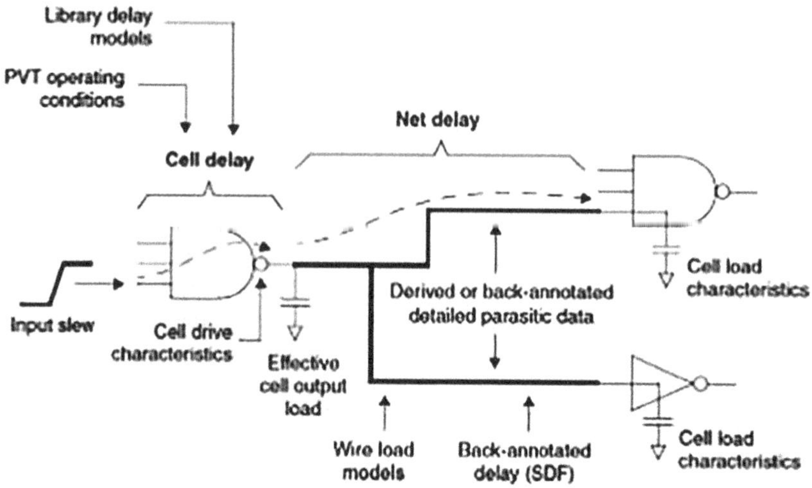

FIGURE 11.3 Logic Circuit for Delay Maximization and Minimization (Source: http://3.bp.blogspot.com/delay2.bmp).

Considering Figure 11.3, max delay and min delay are computed as

$$\text{Delay}\left(\text{max}\right) = 0.5 + 0.04 + 0.62 + 0.21 + 0.83 + 0.15 + 1.01 + 0.12 + 0.57 = 4.05\text{ns},$$

$$\text{Delay}\left(\text{min}\right) = 0.4 + 0.03 + 0.6 + 0.18 + 0.8 + 0.1 + 0.8 + 0.1 + 0.5 = 3.51\text{ns}.$$

There also exist other parameters that directly affect the delay library setup time, including the library delay model, external delay, cell load characteristics, etc. Variation of any of the parameters will affect the delay. Few of them possess the properties of mutual exclusiveness, and that calls for the effect of only one parameter at a time, which is to be considered during delay formulation. So as per their value, classification is done based on the fall and rise delay of the entire cell in the max and min bucket.

11.2.5 Resistor-Capacitor (RC) Circuit Network Delay Model

A distributed RC network is an effective and popular option for exhibiting interconnections of a VLSI circuit. The interconnect delay from source to sink can be estimated by the dint of the Elmore delay formulation. In the following approach, an individual segment of the routing path is simulated by a π-model RC circuit, and the total delay starting from source to sink is determined by dint of iteration over all the individual wire segments along that particular path followed by calculation of the delays of the individual segments in accordance with the formulation [4].

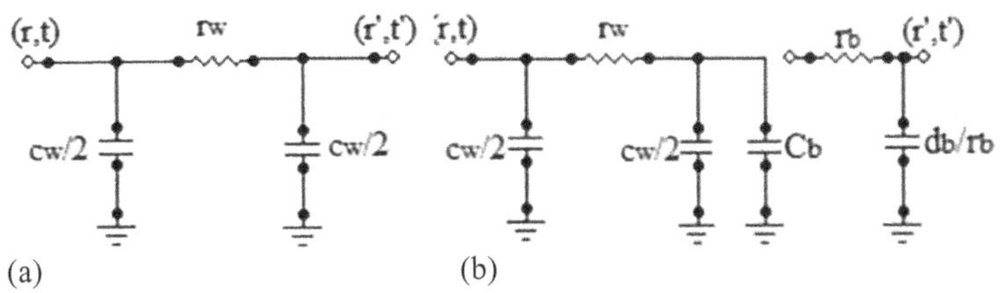

FIGURE 11.4 RC Delay Model (Source http://3.bp.blogspot.com/delay4.bmp)

Figure 11.4(a) unveils the wire segment π-model with resistance *r* and capacitance *c*. The same segment, with a buffer inserted at the proper position with the sole objective to minimize the interconnect delay, is displayed in Figure 11.4(b) with intrinsic buffer delay *db*, buffer output resistance *rb*, and buffer input capacitance *cb*. Basically, in this type of model Elmore delay calculation, where range is from source to sink, plays an instrumental role in computing the delay pair.

$$r' = r_w + r \tag{11.3}$$

$$t' = \left(r + r_w / 2\right) c_w + t \tag{11.4}$$

In the equations, rw and cw represent the resistance and capacitance of the wire segment, respectively. In the second figure, a segment includes wire with buffer.

$$r' = r_b \tag{11.5}$$

$$t' = r\left(c_w + c_b\right) + r_w\left(c_w / 2 + c_b\right) + d_b + t \tag{11.6}$$

11.3 MAPPING ROUTING PROBLEM IN GRAPH THEORY

11.3.1 Grid-Graph Model for Wire length-Minimization

This is the phase where, by minimizing the overall wire length of interconnected terminals or pins, attempts have been made to discover the rectilinear minimal Steiner tree (RMST) problem of the graph [5]. Here we see a graph model G = (V, E) representing a routing region layout where region is sliced into a number of unit square cells, as shown in Figure 11.5. Each cell is represented by vertex a vi and the edge eij, interlacing the two adjacent lying vertices vi and vj. The vertices correspond to the nodes, and edges correspond to the routing paths of the layout. To output the solution of the routing problem for a multi-terminate net obtaining the Steiner tree of optimal cost in the graph, one of the greatest challenges, as well as a serious form of threats, for the delay model is obstacles or hindrances in the path. This parameter directly influences the output of the delay model, and, in fact, can be called a kind of "time deterministic step."

0	0	0	0	0	0	0	0
0	0	0	0	0	0	0	0
0	0	2	2	2	2	0	0
0	0	2	1	1	1	0	0
1	0	2	1	1	1	0	1
0	0	2	1	1	1	0	0
0	0	2	2	2	2	0	0
0	0	0	0	0	0	0	0
0	0	0	0	0	0	0	0

(a) Color-Coded Representation

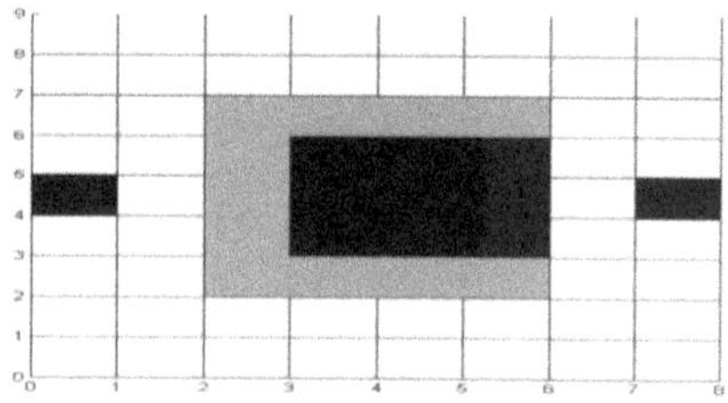

(b) MatrixRepresentation

FIGURE 11.5 Graph Model for Color-Coded Representation and Matrix Representation.

As mentioned in Figure 11.5, the source and sink have been colored dark, buffer obstacles have been colored gray, and wire obstacles have been colored dark. Basically, the vertex locations with corresponding values are given by

V (r; c) = z [5], where z is the value of vertex V located at row r and column c. z = 1 for wire obstacle, z = 2 for buffer obstacle, and z = 0 for locations amenable to both routing and buffer placement [5]. The key idea consists of the joint application of wire-length minimization with the buffer-insertion algorithm to derive the buffered path in the routing area with the minimum time delay between source *s* and sink *t*. In the absence of a buffer, the routing process gets momentum as now the routing is completed in two phases, which includes discovering the shortest path between source and sink followed by performing buffer insertion, as required along the length of that path. In the presence of a buffer, the entire routing phase experiences an appreciable setback in terms of accuracy. However, the buffered minimum delay path does not coincide with any of the shortest wire length between source and sink. Possible solution accounts can be overcome by the simultaneous application of wire-length minimization and buffer insertion during the routing process (i.e., directly making a comparison among the interconnect delays of different buffered paths while the search for the optimal path is still in progress).

11.3.2 RSMT

Rectilinear Steiner minimum tree (RSMT) is literally counted as one of the global routing techniques in VLSI design, deploying the spanning of a tree; the tree spans a given set of pins by reaching individuals either in a vertical way or horizontally [6]. The problem by nature is nondeterministic polynomial (NP) time and runs in polynomial time, and it is derived from the minimum spanning tree (MST) of classical graph theory. For a given set of vertices, the MST is formed by interconnecting them where the optimum solution is the sum of the weights of all edges in the tree. In the Steiner tree, extra intermediate vertices are added to the existing tree to reduce the total length of the MST. The new nodes, which are introduced to decrease the total cost of interconnection, are known as Steiner points, and the tree is known as the Steiner tree.

11.3.3 RSMT for Obstacle Avoidance

In any RSMT constructions, obstacles are prevalent, although attempts have been made to eliminate the hindrance factor. The prime application of RSMT is to project all possible approaches vital to compute obstacle-avoiding rectilinear minimal Steiner tree (OARMST). According to the standard definition, "Given a set P of n number of input points, a new set of points S called a set of Steiner points need to be found such that the length of the path spanning the points P U S is minimum." The major challenge in RSMT is to overcome the obstacle that by nature is a rectangular block, which may further possess common edges or common vertex points. Generally, for the initial resolution of the problem, an empty list of obstacles is generated followed by the introduction of the obstacles in the obstacle list leading to a refining of previously generated OARMST. Iterations are performed to detect obstacle overlapping. Sometimes a slight variation in the problem statement is detected, contradicting the fact that all the edges have to be avoided over an obstacle but to have a constraint so that the maximum length of edge/s is strictly within some threshold named slew constraint. This is established from the fact that wires are given allowance over the blockages, but the case is negative for buffers or repeaters, which in turn limits the length of the wire allowed over an obstacle. The minimum total path length of the interconnected terminal nodes in a VLSI layout becomes the cost of RMST, which involves rigorous calculation by summing together each segment's length, which is expressed as

$$\mathrm{L}(Ts) = 12\Sigma(|xi - xj| + |yi - yj|)\ i \neq j \tag{11.7}$$

where for a tree Ts, the Cartesian coordinate of a terminal point Pi is considered (xi, yi).

For the formation of a rectilinear tree (Figure 11.6), the path from one coordinate point 1 to another 2, as portrayed in Figure 11.5, will alternate between the line segment P and follow two different approaches. The x-option represents an approach where 1 follows a vertical path down to reach the pseudo-Steiner point as the former step, followed by the horizontal path toward the left to reach the end point 2, and the

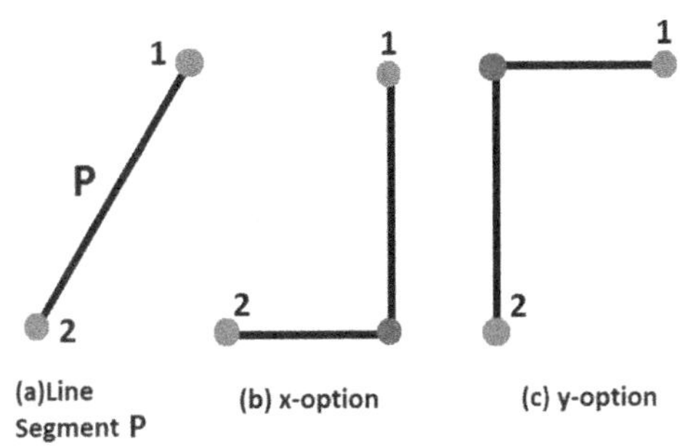

FIGURE 11.6 Rectilinear Tree.

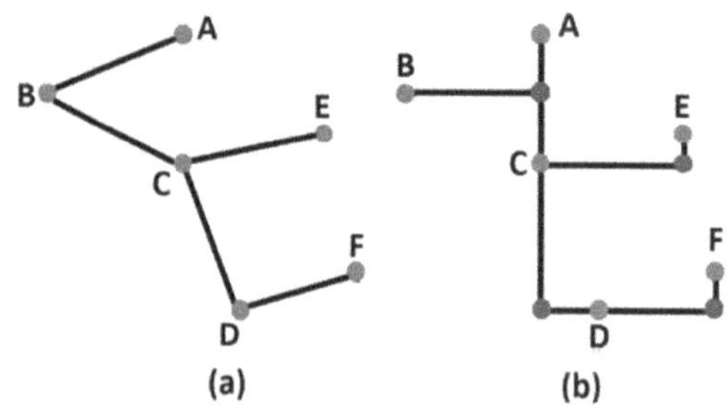

FIGURE 11.7 RMST Bend Reduction.

y-option reflects another approach where 1 follows a horizontal path toward left to reach the pseudo-Steiner point as the first step, accompanied by the vertical path down to reach the end point 2.

11.3.4 RMST for Bend Reduction

RMST, when applied for bend reductions, and if a spanning tree, as shown next, is considered by making utilization of the x-choice and y-choice, the corresponding rectilinear Steiner tree can be formed. In this work, the candidate rectilinear Steiner tree is portrayed as a collection of spanning tree edges, each of which is extended to be used as the global best arrangement in the PSO calculation to recognize local minima that is even more supreme, and the cycle is iterated in a loop. Along these lines, the global minima is found out by the genetic-based discrete particle swarm optimization (GDPSO) [5] calculation by continuously finding out better local minima, which in turn reduces the RMST cost together with reducing numbers of bends as depicted in Figure 11.7.

11.4 METAHEURISTICS

Metaheuristics signify a higher-level procedure or heuristic designed to obtain a heuristic partial search algorithm that may output a competent solution to an optimization problem, especially with incomplete or imperfect information or limited computation capacity. Metaheuristics ensure the dynamic usability of the selected algorithm on the basis of a few assumptions to solve an optimization problem. Compared to optimization algorithms and iterative methods, metaheuristics eliminate the possibility of obtaining a globally optimal solution that can be found for some class of problems. There exist several algorithms, such as gradient descent, PSO-MU, and IWO, integrating metaheuristic as the key perception of their working principles.

11.4.1 PSO

The PSO model consists of a swarm of n particles following an iterative approach in a D-dimensional problem space to encourage searching for a better solution. The mathematical approach involves considering the particles representing the position vector xI = (x1, x2, x3, … ,xn) and velocities represented by the velocity vector vI = (v1,v2, v3, … , vn). In the continuous search process, an individual's particle positional change is based on the position of the neighboring particle and its own previous experience. Each iteration time t leads to the updating of position and velocity by following some standard mathematical equations:

$$\begin{aligned} &\underset{V_1}{\rightarrow}(t) = \vec{V_1}(t-1) + \varphi_1 r_1 (p_1 - \underset{X_1}{\rightarrow}(t-1) + \varphi_2 r_2 (p_{g.} \underset{X_1}{\rightarrow}(t-1) \\ &\underset{X_1}{\rightarrow}(t) = \underset{X_1}{\rightarrow}(t-1) + \underset{V_1}{\rightarrow}(t) \end{aligned} \tag{11.8}$$

Here positive constants $\varphi 1$ and $\varphi 2$ are acceleration coefficients initialized to 2. r1 and r2 are random variables whose values are in between 0 and 1. The maximum velocity is clamped to a value of vmax that is used to provide assistance to the particle effectively in the search space. Pi and pg are the particles' previous best and group's previous best position. The convergence to the optimal solution by PSO is achieved by updating the velocity and position vector.

11.4.2 PSO-MU

This is basically the mutant version of the existing PSO algorithm. The algorithm basically bears the concept of the selection- and fitness-based distribution approaches of the genetic algorithm (i.e., from the technical point of view, swarms possessing a high degree of fitness have the chance for survival following the concept of Charles Darwin's "survival of the fittest"). Here the authors presented the modified PSO algorithm based on supporting the new concept. The application of a reproductive system is deployed in the context of altering the position vector of randomly selected particles in individual iteration. The agent's position follows a replacement with high fitness value via the selection of some bits of the position vectors of the agents from its prior output. This approach consists of an algorithm, which performs an initial exhaustion that initiates with the selection of swarms from the existing generation in the first phase. The swarms with high fitness probability get the chance to strictly obey the probability factor [5, 6].

$$\frac{f_j}{\sum_{j=1}^{N} f_j} \tag{11.9}$$

where N is the population size.

The high fitness factor is extracted from the selected pool followed by the generation of a mutant for the second phase. This enhanced knowledge of high fitness property is induced in the position vector Xi, t + 1 = Xi, t + Vi, and t + i for the emergence of a new generation of swarms resulting from mutation in PSO.

11.4.3 Gradient Descent Algorithm

Gradient descent algorithm proves its applicability and competency in resolving the wire-length optimization issue by bend reduction. The working principles involve applying a first-order iterative optimization algorithm to obtain the minimum of a function. To find a local minimum of a function using gradient descent, one takes steps proportional to the negative of the gradient (or approximate gradient) of the function at the current point. On the contrary, for the positive of the gradient, it is "gradient ascent." Mathematical interpretation reflects the detrimental behavior of multivariable function F(x) in the negative direction of a specified point at a quick rate provided that the function F (x) F(x) is defined and differentiable in a neighborhood of that point. if

$$a_{n+1} = a_n - \gamma \Delta F\left(a_n\right) \tag{11.10}$$

$F(a_n) \geq F(a_{n+1})$. In other words, the term $\gamma \triangle F(a_n)$ is subtracted purposely to move against the gradient toward the minimum. With this vision, one commences with a guess x_0 for a local minimum of F and considers $x_0, x_1, x_2, \ldots$ such that

$$\begin{aligned} x_{n+1} &= x_n - \gamma_n \Delta F\left(x_n\right),\ n \geq 0 \\ F\left(x_0\right) &\geq F\left(x_1\right) \geq \ldots., \end{aligned} \tag{11.11}$$

In case the of gradient descent, an expansive advance size will lead to the quick convergence of the approximation to the local minimum at iteration k (i.e., X(k) toward the base), whereas once in the neighborhood of the closest local minima, oscillations take place as X(k) overshoots the base. Further, a little advanced size will result in a leisurely convergence toward the base awaiting the exactness of the ultimate outcome. Since at a local minimum, the slope remains zero, and a small advancement will eventually be noticed when a base is drawn closer. With a certain number of assumptions on the function F, ∇F Lipschitz and particular choices of γ

$$\gamma_n = \frac{\left(x_n - x_{n-1}\right)^T [\nabla F\left(x_n\right) - \nabla F\left(x_{n-1}\right)}{\| \nabla F\left(x_n\right) - \nabla F\left(x_{n-1}\right) \|^2}, \tag{11.12}$$

convergence to a local minimum can be guaranteed. When the function F is convex, eventually, all local minima are also presuming global minimum nature, so it can be inferred from the specified case that gradient descent possesses the capability to converge to the global solution (see Figure 11.8) [7].

11.4.4 IWO

IWO [8] is a continuous, stochastic numerical algorithm by nature inspired from weed colonization [9]. The algorithm has set up a benchmark because of its

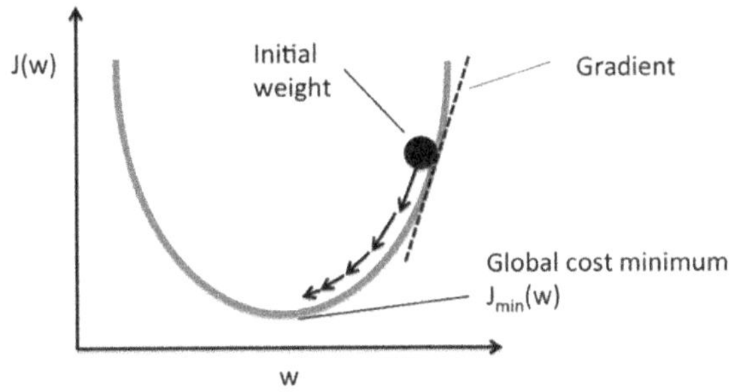

FIGURE 11.8 Gradient Descent.

exceptional competencies in exploration and dissimilarity in properties and took an exceptional place for solving continuous optimization problems. In formal IWO, there is a uniform focus on the search space. DIWO finds special application in discrete combinatorial problems, such as a time-cost, trade-off problem and cooperative multiple task assignment of unmanned aerial vehicles.

The idea involves random spreading of a population of initial seeds n_o over the search space n_o. Plants or the agents in the colony are taken as the initial weed population with the

$$p = \left(\sum_{a=1}^{n} \mathrm{I} e_a - e_{a+2} \mathrm{I} \right) + x_{s_i} - e_{n-1} + x_{s_o} - e_1 \tag{11.13}$$

The individuals, after the growth phase, are given allowance to reproduce new seeds linearly depending on their own, the lowest, and the highest fitness of the colony maximum (S max) and minimum (S min) number of seeds that are the predefined parameters of the algorithm and adjusted according to structure of the problem.

$$S_{plant} = S_{\min} + \left[f_{plant} \times \frac{S_{\max} - S_{\min}}{f_{\max} - f_{\min}} \right] \tag{11.14}$$

This phase involves the random scattering of generated seeds with a normal distribution over the search space. The mean of distribution is equal to the location of the parent plant, but standard deviation (SD), σ, will be reduced from a specified initial value, σ initial, to the final value, σ final.

Competitive exclusion is instrumental when the maximum number of colony members inclusive of both plants and seeds is reached. The fitness of all the members is computed followed by sorting by rank until the population size doesn't violate the maximum limit; finally, the elimination of lower fitness plant weak solutions from the colony takes place. The algorithm ends upon reaching the maximum number of iterations. Because of competitive exclusion, only the plants with the best fitness in the colony survive to this stage, leading to the achievement of an optimal solution. The optimal solution with best fitness is the minimum overall wire length of the interconnected nodes from source to sink in the VLSI problem space.

11.5 PROCESS ADOPTED FOR MODELING THE MCDA APPROACH

PSO in the field of wire-length minimization and bend reduction [10]: Looking from the MCDA perspective, wire-length minimization, along with bend reduction, forms two of the many important objectives in the routing phase that contributes to the expertness of the IC. Henceforth it is indispensable to incorporate the necessary factors or parameters within the swarm intelligence itself to direct the workings considering the newly taken objectives, which will influence the output In a distinct way reflecting a mark improvement from accuracy of result to the algorithmic complexity in a remarkable way. In this case, the resulting algorithm is called gradient descent algorithm, which fulfills the minimization objectives with PSO acting as the base. The algorithm is as follows:

Step 1: *The initial populations set, as well as the various other parameters, are initialized.*

Step 2: *For every iteration, the pseudo-gradient is computed at the time of initialization of the particles.*

Step 3: *Using the equation $L(Ts\ )+\beta^*n$, every particle's fitness value is computed. And for each particle, the personal optimal solution, together with the population's global optimal solution is initialized.*

Step 4: *The velocity and position of a particle having inertia weight w is updated using $w = [3 - exp(-s/200) + (R8*D)2] - 1$, acceleration constants c1 and c2 are updated using linear decreasing and linear increasing function.*

Step 5: *For every particle, the fitness value is recomputed using the equation $L(Ts)+\beta^*n$, and at the same time, its personal optimal solution is updated.*

Step 6: *For the entire population, the global optimal outcome is recomputed.*

Step 7: *After completion of the total number of iterations, the best position, as well as the best fitness of the swarm, is achieved and then, using the coordination of the global best position, the RMST cost id reckoned.*

Step 8: *A comparison is made between the best cost acquired from all the swarms and the swarm's global best cost, and if it is found that the Steiner best cost has a larger value, then it is brought up to date with the swarm's global best cost.*

Step 9: *The ending condition (when the utmost number of iterations is reached or a good enough position is attained) is examined, and if fulfilled, the code gets terminated, and if not, it goes to step 4.*

The IWO algorithm for VLSI routing optimization is basically based on chaotic theory, which if in excess in a routing path and can lead to network congestion and even blockage. Now, previously, wire-length minimization by bend reduction helped a lot in delay reduction [11] beside space management. Here special care is also taken, as any form of network congestion may directly lead to delay. Hence the algorithms aim to optimize the function wire length, along with minimizations and bend reduction. Here the object is to delay reduction directly by avoiding congestion [11].

```
Procedure IWO(crop field(N_N); itermax)
Generate initial population of weeds W
Place buffers randomly for the population agents
For iter = 1 to itermax do
Calculate the maximum and minimum fitness in the colony
using equations (1) to (5) [Marked in IWO sections of
metaheuristics]
For each weed Wi 2 W do
Compute the number of seeds Si for each weed WI
Randomly distribute the seeds following normal
distribution function around the parent plant Wiusing (7)
Place the buffers randomly for the seeds
Add the number of seeds to the solution Wsol
If Wsol + W > pmax then
Sort the seeds based on the fitness of each weed Wi 2 W
Eliminate the worst fit seeds so that Wsol + W_pmax
End if
End for
End for
End procedure
```

11.6 EXPERIMENTAL RESULTS AND INFERENCES

11.6.1 Experiment-A

This experiment involves bend reduction using the gradient descent algorithm [12]. A test bed is created where data sets of 10, 20, 25, 50, 100, and 500 terminal nodes are generated randomly in a 200 x 200 2D grid. The test bed is applied in addition to applying it to the existing DPSO_BRRA algorithm and the GeoSteiner 3.1 benchmark. All experiments were conducted in a system with 4.0 GHz Intel i7 quad core processor and 16GB RAM. To test our proposed algorithm gradient-based PSO against DPSO_BRRA and the GeoSteiner benchmark considering the inertia weight, we initialize the controlling parameters before executing the optimizer algorithm [12] given in Table 11.1.

The inertia weight (w) is varied, where s is the population size, D is the dimension size, and R is relative quality of each solution normalized to [0,1].

$$w = [3 - \exp(-s/200) + (R8 * D)2] \tag{11.15}$$

TABLE 11.1
Initialization Parameter

PARAMETERS	ALGORITHMS	
	GDPSO	DPSO_BRRA
MAX_ITER	60	60
C1	Decreased linearly from 2.0 to 1.49	Decreased linearly from 0.9 to 0.2
C2	Increases linearly from 1.49 to 2.0	Increased linearly from 0.4 to 0.9
w	According to (7)	Decreased linearly from 0.9 to 0.1
Population size	100	100
Bend coefficient	50	50
Gradient step size (γ)		0.1

TABLE 11.2
Comparisons Over RMST Cost (Wire Length) and Bend Count

Instance	GPSO		DPSO_BRRA		GeoSteiner 3.1	
Terminal Points	Wire Length	Bend Count	Wire Length	Bend Count	Wire Length	Bend Count
10	350	7	350	7	350	7
20	593	16	595	16	582	17
25	628	22	631	21	609	24
50	820	42	825	45	798	48
100	1,702	79	1,712	84	1,680	89
500	87,344	490	87,822	496	85,311	509

Here the authors compared the proposed algorithm with the GeoSteiner method where the weighting factor α was set as 200. The comparison of routing results between DPSO_BRRA, GeoSteiner method, and the outcome is shown in Table 11.2.

Table 11.2 reveals that for ten terminal nodes, our proposed GDPSO algorithm bears a resemblance to that of the GeoSteiner benchmark and the existing algorithm with the lowest minimum wire-length cost of 350 and minimum bend count of 7. The values confirm the reliability of GDPSO for outputting the best for a data set of greater terminal nodes when contrasted with the existing DPSO_BRRA but with the most extreme deviation from the benchmark value. The conclusion reflects the competent manageability of RSMT problem graphs with safety and with exemplary competence, thereby satisfying the wire-length minimization issue to a great extent, along with minimum bend count, by using our proposed algorithm.

Routing experiences a serious threat in terms of vias, which is instrumental for a proportionate relationship between the power dissipation and delay. Hence the primary goals instigate us to miniaturize the number of bends besides the formation of the RSMT. Special efforts have been made to limit the wiring length, which is not only useful for via diminishment but also increases the reliance of the routing phase. From the experimental outcome, it is inferable that the proposed algorithm comes out with flying colors in regard to the RSMT issue and can converge at a fast pace. Moreover, the proposed algorithm miniaturizes the quantity of bends following a smooth convergence to the global minimum. A comparison has been drawn between the performance of the hybridized technique (GDPSO) with the already existing metaheuristic approach, such as the DPSO_BRRA and GeoSteiner benchmark, and minute investigations highlight the GDPSO technique as a better alternative approach for apparent cost reduction, along with bend reduction in the VLSI regime.

11.7 EXPERIMENT B

For this experiment, the objective is to measure the performances of the IWO algorithm. The simulation results of this study are compared with the previously proposed approaches, PSO and FA, for solving the VLSI routing problem. The grid-graph of size 22_17 is taken as the test case. To compare the IWO algorithm with the previously described PSO and FA, the same parameter values are used for wire, buffer, source, and sink. The effectiveness of IWO can, therefore, be evaluated with the experimental results by considering the previous results as a benchmark for a test graph of size 22_17.

For the test graph, the parameters are set as follows: source resistance is 140, load capacitance at the sink is 0:002 pF, the wire segment has a resistance of 58 and a capacitance of 0:042 pF, and the buffer segment has an input capacitance of 0:002 pF, an output resistance of 140, and an intrinsic delay of 40 ps.

The optimal solution of 521:73 ps for the problem statement is realized from Table 11.3. In the process, IWO converged faster than FA by a significant fringe and nearly at the same time as PSO.

It is instrumental to convey that more optimization, the more accurate the results, marking a total improvement in the overall architectural design because of overall improvement in the back-end logic. Observational parameters for the PSO-CG approach for global routing and IWO also support the following factors. For example, in the case of IWO, the iterations are much less (i.e. only ten iterations from Table 11.4 of experimental analysis of IWO), and so is the number of computations. The overall effect of this is delayed reduction by an appreciable amount, along with the ease of convergence to the global optimum in comparison to other algorithms,

TABLE 11.3
Comparison of PSO, FA from Previous Research and Modified IWO

Least No. of Iterations	PSO	FA	IWO
Average No. of Iterations	26	31	25
Convergence to Global Optimum		79	68
Optimal Solution Found	521:73 ps	521:73 ps	521:73 ps

TABLE 11.4
Initialize of Parameters

Number of agents q	PSO	FA	IWO
Number of iterations t	-	10	10
Number of computations	400	400	400
Maximum number of seeds smax		5	3
Maximum number of seeds smax			150
Standard deviations			[800, 1]
Nonlinear modulation index N			3

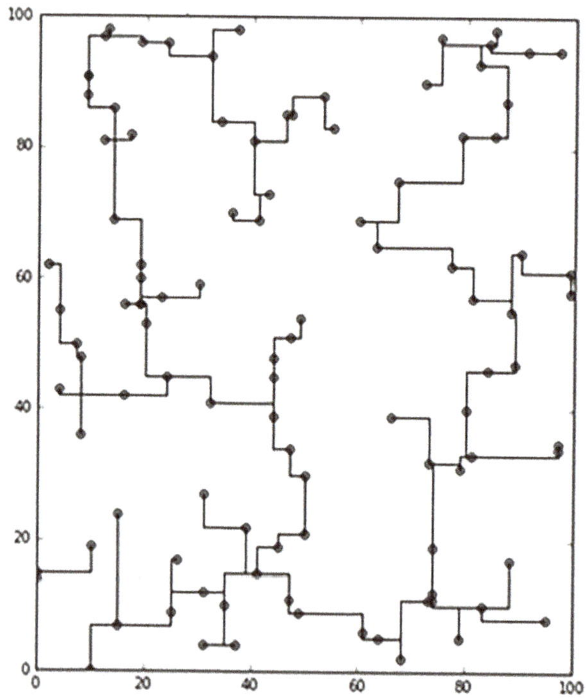

FIGURE 11.9 GDPSO for 100 Nodes.

such as FA and standard PSO. A similar theory holds true for wire-length minimization and bend reduction, where reduction in bend count and wire length results in fetching the best data sets as output. In this context, the competency of GDPSO for outputting the best data set of greater terminal nodes in comparison to the existing DPSO_BRRA but with the most extreme deviation from the benchmark value has been proved already (Figure 11.9). So it can be inferred that when all the optimizing factors associated with time, space, power, and cost are incorporated as multi-objective for MCDA, the result would be near perfect for the MCDA concept when applied in terms of VLSI global routing. This shows the success of the analysis.

11.8 CONCLUSION

The key idea in the MCDA approach is to clearly define the problem or dilemma that faces the decision maker(s). This may comprise a situation that requires a one-off decision or one that requires multiple, ongoing decisions. The MCDA model approach is suitable when an intuitive approach is not appropriate, for example, because the decision maker perceived the enormity and complexity of the decision intuitively involving multiple numbers of conflicting objectives or multiple stakeholders with diverse views. Often, there is a desire for a formal procedure so that the decision-making process can be made open and transparent and is seen to be fair. If the problem is one that can be structured as involving a collection of alternatives that can be tested against several criteria, then the MCDA approach may be suitable. Such alternatives may be alternative choices, actions, strategies, or alternative units. From the technical perspective, it's the competency of different metaheuristic algorithms that accounts for back-end effectiveness.

REFERENCES

1. Ghosh S, Nath S, Sarkar S K. 2015. A Novel Approach to Discrete Particle Swarm Optimization for Efficient Routing in VLSI Design. *4th International Conference on Reliability, Infocom Technologies and Optimization (ICRITO) (Trends and Future Directions)*, pp. 1–4. doi:10.1109/ICRITO.2015.7359375
2. Juneja S, Saraswat P, Singh K, Sharma J, Majumdar R, Chowdhary S. 2019. Travelling Salesman Problem Optimization Using Genetic Algorithm. *Amity International Conference on Artificial Intelligence (AICAI)*, Dubai, United Arab Emirates, pp. 264–268, doi:10.1109/AICAI.2019.8701246.
3. Srivastava A, Singh B, Chabra A, Majumdar R. 2017. Application and Use of MCDM Technique in Software Industry. *International Conference on Infocom Technologies and Unmanned Systems (Trends and Future Directions) (ICTUS)*, Dubai, United Arab Emirates, pp. 487–491, doi:10.1109/ICTUS.2017.8286058.
4. Batra A, Shukla A, Thakur S, Majumdar R. 2012. Survey of Routing Protocols for Mobile Ad Hoc Networks. *IOSR Journal of Computer Engineering (IOSRJCE)*, vol. 8, no. 1, pp. 34–40.
5. Eberhar R C, Kennedy J. 1995. A New Optimizer Using Particle Swarm Theory. *Proceedings of 6th International Symposium on Micro Machine and Human Science*, Nagoya, Japan, pp. 39–43.
6. Julstrom B A, Antoniades A. 2004. Two Hybrid Evolutionary Algorithms for the Rectilinear Steiner Arborescence Problem. *Proceedings of the 2004 ACM Symposium on Applied Computing*, Nicosia, Cyprus, pp. 980–984.
7. Nath S, Ghosh S, Sarkar, S K. 2015. A Novel Approach to Discrete Particle Swarm Optimization for Efficient Routing in VLSI Design. *IEEE Proceedings of Fourth International Conference on Reliability, Infocom Technologies and Optimization.*
8. Zhou H, Wong D F, Liu I-M, Aziz S. 2000. Simultaneous Routing and Buffer Insertion with Restrictions on Buffer Locations. *IEEE Transactions on Computer-Aided Design of Integrated Circuits and Systems*, vol. 19, no. 7, pp. 819–824.
9. Mohamed K H, Shaikh-Husin N. 2008. Simultaneous Routing and Buffer Insertion Algorithm for Interconnect Delay Optimization in VLSI Layout Design. *Proceeding of IEEE International Conference on Microelectronics*, Sharjah, United Arab Emirates.

10. Karimkashi S, Kishk A A. 2010. Invasive Weed Optimization and Its Features in Electromagnetics. *IEEE Transactions on Antennas and Propagation*, vol. 58, no. 4, pp. 1269–1278.
11. Nath S, Chakravarty A K, Ghosh S, Sarkar S K. 2017. Invasive Weed Optimization Approach to VLSI Routing. *Proceeding of IEEE International Conference on Devices for Integrated Circuit (DevIC)*, Kalyani, India.
12. Genggeng L, Guolong C, Wenzhong G, Zhen C. 2011. DPSO-Based Rectilinear Steiner Minimal Tree Construction Considering Bend Reduction. *IEEE Proceedings of Seventh International Conference on Natural Computation*.

12 Application of IoT in Water Supply Management

Reshu Agarwal and Adarsh Dixit
Amity Institute of Information Technology
Amity University
Uttar Pradesh, Noida, India

Shylaja Vinaykumar Karatangi
G. L. Bajaj Institute of Technology and Management
Greater Noida, Uttar Pradesh, India

CONTENTS

12.1 INTRODUCTION

In today's world, as we keep on moving forward with technology and innovative techniques, we continue to find things that optimize our daily tasks and make them easy and robust. We keep making our tasks easy to handle and find ways that help us in making a specific task easier to complete. The same goes for our water supply management as it does for our work. To make it easier for people to reduce the wastage of water in a manner that is within reach has become a model for building a strong water management system

The Internet of Things (IoT) is an idea that mirrors an associated set of anybody, anything, whenever, wherever, any assistance, and any system (Pradnya & Ghewari, 2017; Sharma & Tiwari, 2016). The IoT is a megatrend in cutting-edge advancements that can affect the entire business range and be thought of as the interconnection of interestingly recognizable keen items and gadgets inside the present web framework with expanded advantages. It is composed of two words: "Internet" and "things." It is a methodology/technology in which live monitoring, processing, and accessing of systems or things is done via Internet for unique IoT devices. It collects data from various devices, and, further, this data is sent to various servers for data processing. IoT acts as a dynamic global network for establishing protocols to identify the virtual and physical devices, parameters, specifications, and smart interfaces that are integrated in the network to establish communication between the system and the user. Alhamoud et al. (2014) presented the structure for acknowledging vitality effective savvy homes dependent on remote sensor systems and human movement discovery. Their work depends on the possibility that the greater the part the client exercises at home is identified with a lot of electric machines, which are important for playing out these exercises. In this way, they demonstrate how it is conceivable to identify the client's present movement by observing his or her fine-grained apparatus level energy consumption. Zanella et al. (2014) proposed a system specifically focusing on an urban IoT system. Martinez, Llavori, Cabo, and Pedersen (2007) proposed a six-layered architecture by combining web services, radio frequency identification (RFID) and wireless sensor network (WSN), whereas a five-level architecture was proposed by Zhang, Sun, and Cheng (2012) based on a telecommunication management network. Lin et al. (2017) proposed an essential and basic engineering composed of three layers, which are perception, network, and application. Spillages and channel blasts are regular disappointments in urban water circulation systems; when there is a significant rush in the fundamental pipeline, a total shutdown of the water appropriation process becomes imminent (Babovic, Drécourt, Keijzer, & Friss Hansen, 2002; Misiunas et al., 2005). Joseph et al. (2018) proposed a model based on IoT devices that assist in controlling and planning the utilization of water. Further, Trina, Mohring, and Andersen (2017) proposed a model for an active decision for fruit trees and grasses so as to make water management intelligent. George, Bezdedeanu, Vasilescu, and Suciu (2017) proposed a model for water management that uses IoT and cloud computing. Anjana et al. (2015) proposed a smart metering technique to automate water-meter reading. The proposed technique decreases the meter handling manual problem and decreases reading and billing issues. Further, Mukta et al. (2019) proposed a system to measure water quality. It helps in continuous measurement of water condition based on four physical parameters i.e., temperature, pH, electric conductivity and turbidity properties. Wadekar et al. (2016) proposed an IoT gadget that helps to oversee and design the use of water. Sensors set in the tank constantly illuminate the water level at the current time. This data will be refreshed on the cloud and using an Android application, the client can view the water level on a smartphone in anyplace that is associated with the Internet. Machado, Júnior, Silva, and Martins (2019) presented a board framework using the microcontroller ZR16S08 as the IoT arrangement for water conveyance backing and misfortunes anticipated. Rajurkar, Prabaharan, and Muthulakshmi (2017) proposed a model for limiting and minimizing the usage of water. Moreover, the usage of water in every season is predicted. Mo, Ying, and Chen (2012) proposed a model for a water-quality measuring system with the help of numerous sensors and a microcontroller. Xiaocong, Jiao, and Shaohong (2015) proposed a coordinated framework dependent on IoT for checking water assets and management. It consists of three layers for observing water data, constant data transmission, and water management. Gupta et al. (2018) proposed a water management model for housing societies. This model helps the users understand the importance of using water judiciously and equips them with the knowledge of the functions of the water management system. Ye, Liang, Zhao, and Jiang (2016) proposed an engineering of a smart grid dependent on IoT. This model consists of smart detecting, wise reenactment, insightful determination, clever early admonition, keen guideline, wise removal, and savvy control. Günther et al. (2015) proposed a water distribution system that serves as an example for implementing a smart water network solution. Further, Kamienski et al. (2019) proposed a model for IoT-based smart water management for an accurate water system in agriculture. The significant disadvantage of this model is that it required exceptionally structured arrangements and the rebuilding of certain segments to provide higher versatility. Hadipour, Derakhshandeh, and Shiran (2019) proposed an imaginative method of multi-intelligent control framework (MICS) of a water pump and a pump station, which is essentially planned, set up, and used in a rural area. Results dependent on this model showed that 60% of water can be spared by using MICS by means of IoT. Zidan, Maree, and Samhan (2018) proposed a model for remote real-time monitoring and controlling systems for water treatment. The proposed framework screens the convergence of chlorine and controls the dosing pump to keep maintain chlorine levels with as little human mediation as conceivable. Parameswari and Moses (2018) proposed a model for checking drinking water quality by using remote sensors. The significant commitment of this model was to build up a safe procedure for a modern observing arrangement of water quality dependent on remote sensors. Progressively, the inserted framework relies on secure IoT that is arranged using WSN. Further, the world's urban populace has developed quickly from 1.019 billion in 1960 to 4.117 billion in 2017 (Butler et al., 2014). It is assessed that the population will arrive at 9.7 billion by 2050 (Tao et al., 2016). The over-the-top population development will cause critical water issues, such as water deficiency and water quality corruption in urban territories (Lee, Sarp, Jeon, & Kim, 2015). Water supply management (WSM) is a significant exploration region that influences the monetary development of our nation (Sonaje & Joshi, 2015). WSM has two issues: first is the checking of the status of water levels in tanks, and the second is the notification of the water supply in pipelines. It helps to avoid wasting water. As indicated by the report from the World Health Organization in 2017, around 2.1 billion individuals around the globe need safe drinking water (Mounce, Machell, & Boxall, 2012; Nguyen et al., 2018). So there is a need to guarantee that the wastage of water can be decreased by using IoT.

This chapter is based on an Android application that creates ease for corporate workers and senior citizens for monitoring and controlling real-time data collected by sensors. This chapter consists of IoT technology, water supply methodology, IoT-enabling technologies, Arduino systems, applications of IoT in the field of water supply systems, and their benefits in daily life.

Further, at the end of this chapter, the future scope of the system is presented. The model uses a smart sensing system, modules, and components that are being implemented in the working application. The application consists of an algorithm and flowchart of the data captured by the sensors' system and output.

12.2 WATER MANAGEMENT STRATEGIES

The primary goal of smart water management is to use water supplies efficiently and sustainably and to recycle them. An increasing population, growing environmental issues, and strain on the food and agriculture sectors make water a much more important asset.

Water management strategies and practices address the following aims:

1. Reduce wastewater, which is used in large amounts in areas such as manufacturing, irrigation, and power generation. This means the implementation of high-tech activities, such as precision planting, intelligent irrigation, and water metering, in real time.
2. Improve the quality of the water and avoid hazardous waste contamination and environmental pollutants, such as acidification. Companies use sensor technology for real-time monitoring and control to enhance and preserve water quality.
3. Boost the quality of water facilities, such as water tanks, treatment plants, wastewater collection centers and distribution mains. Using IoT and asset management data solutions, businesses can keep abreast of important measurements, such as water pressure, temperature, and velocity, conduct predictive maintenance, and prevent breakage and downtimes.
4. Using smart water management systems fitted with leakage and moisture sensors to enforce leakage control. Eliminating the harm caused by yearly leakage and leakage management is vital to saving water supplies and budgets.
5. Practice monitoring of consumption to maximize and keep under control the use of water supplies at various levels – in households, industries, regions, or the entire world.

12.3 IOT TECHNOLOGY

A standard IoT device is connected by wired or wireless networks and consists of different interfaces for communication. The IoT system requires the following:

- Interface for connecting to the Internet
- Interface for text, visual, or audio representation
- Interface for connecting Internet
- Interface for memory
- Interface for storage

Some IoT devices include the following:

- Smartwatches
- Smart home monitoring

- Smart home appliances
- Smart vehicles/cars
- Smart health and care devices
- Smart agricultural system

IoT-enabling technologies include the following:

- Embedded systems and protocol-enabling communication
- Security architectures and protocols
- Cloud computing and big data analytics
- Web services and search engines
- WSNs

12.4 IOT APPLICATIONS IN WATER MANAGEMENT

1. Smart irrigation: This is an IoT-based automated device for analyzing soil moisture and the environment for irrigation processes in framing to maintain the water level.
2. Water system integrity: This is an IoT-based automated device for water distribution monitoring, meter reading, water supply mapping, and water-quality reading by sensors and quality checkers. To increase the security of user data, systems, and networks, data is transferred to a framework for operations.
3. Smart water monitoring: This is an IoT-based automated application for smart monitoring and controlling of the water supply in a local area for distribution and within quality standards.
4. Rain and stormwater management: This is an IoT-based device for the automated collection of rainwater in storage tanks from roofs and drainage pipes for future use. It senses the climate or weather and then the sensor detects the data and sends it to a user as a notification/alert.
5. Smart gardening: This is an IoT-based device that is used for self-watering and nutrients for plants. This sensor detects the light, temperature, water, moisture, soil, etc., and the application reminds the user per the need.
6. Aquaculture system: This is an IoT-based device that is used to monitor the water quality for aquatic culturing and for aquarium environment monitoring. It helps the user to recycle or change the water of farms and aquariums to increase productivity.

12.5 BENEFITS OF IOT IN WATER MANAGEMENT

The following are the benefits of IoT in water management:

1. The IoT system easily stores and manages data collected from sensors that can be integrated with cloud computing services and cloud storage, etc., which can be accessed, controlled, and monitored from any place by the user or client.

2. The IoT system in water management increases the accuracy and smart treatment of the water supply.
3. The controlling and monitoring cost reduces and increases profitability and sustainability.
4. The IoT system increases the efficiency level of the usage of energy consumption, treatment level, supply, etc.
5. The IoT system contributes to the protection of the environment and saves water from being wasted.
6. The IoT system collects the data from installed systems and sensors, which can be used to make decisions and operations based on the weather, climate, and spontaneous changes in the surroundings.
7. Risk reduction occurs through up-to-date data collection, which helps users to understand the environment and expected problems.
8. Livestock monitoring of the storage of water tanks and access to all information related to water supply.
9. High data collection frequency to check the water measures of water stock to maintain based on the need.

12.6 FACTORS AFFECTING THE SMART WATER SYSTEM

1. There are various factors that can be used in determining the quality of water, which majorly consist of the property of water and its application. Although there are various factors that can be used in determining the quality of water, monitoring all of them is a tedious task and would cause an unnecessary workload. Furthermore, it would also affect the quality of analysis being conducted. Out of these factors, the major ones are used to determine the quality of water and include the biological, chemical, and physical properties.
2. Quality: Quality is very crucial in determining the properties of water. It involves the analysis of the chemical properties of water, including the oxygen dissolved in water, its pH level, and, most importantly, its oxidation reduction potential. The pH level helps in determining the alkalinity or acidity of the water. A pH level from six to nine is taken as normal in wireless distribution system (WDS). Oxygen dissolved in water should be about 0.5 mg per liter for the WDS. Furthermore, the oxidation reduction potential determines the ability of water to remove its own impurities. A high value of oxidation reduction potential represents good water quality.
3. There are various physical factors that also help in determining the quality of water, including its conductivity, temperature, and turbidity. Because of high temperatures, the quality of water gets reduced, as temperature directly affects the oxygen dissolved in the water. Turbidity of water represents the opacity in the water, which is mainly caused by the dissolution of microscopic materials. The turbidity of water should be greater than 1 NTU.
4. As far as the biological factors are considered, algae, bacteria, pesticides, etc., also affect the quality of the water. These are some of the basic components that are taken by WDS to determine the quality of water. Although in absence

of any standard method regarding the parameter's selection for the water quality level, there can be much confusion while determining the water quality.

5. Quantity: During the collection of the data for WDS, there are several quantity factors that should be considered, including the level of water and its velocity and flow pressure. Velocity and water flow measurements can be taken with the help of flow sensors and ultrasound sensors. However, the act of collection and management of sensor data in practicality is very complicated, as it is handled with the distributed approach. Efficient compression technology, along with the use of the hybrid approach, would help in the better management of real-time data. In addition, this would also help in the reduction of the transmission cost required for the network layer.
6. Technological factor: The smart water system is supported with an efficient and self-adapting network technology. The crucial factors required for gathering the information regarding water quality include proper selection of data fusion, station, and forecasting. This helps in efficient and quick decision making for the crucial situations that require quick determination of water quality. The system would provide more accurate results when used with the network layer architecture.
7. Topological factor: This includes both the active, as well as the passive, elements involved in the network. Active elements of the system include the pumps, turbines, and valves through which the pressure and flow of water can be controlled efficiently. On the other hand, passive elements include the reservoirs and pipes that help to support the active elements. The crucial factors that majorly impact the smart water system involve the operational cost and the real-time measurement and control of different sources of water. The smart water system can be made more efficient with the help of newly developed topology methods, such as clustering or convex programming.

12.7 PROPOSED SYSTEM

The IoT-based smart water supply system is considered an IoT system that monitors and controls the data, such as the flow of water, distance, and other types of sensor information being integrated into the system for the supply of water in home water tanks or commercial water tanks. The system uses the "plug and sense" methodology, which is considered smart sensing, as the system collects real-time data and operations and is displayed over smart devices, such as Android phones and tablets. The sensors generate data and transmute the connected device of the user for remote access via cloud computing and technology integrated into the application.

In the field, WSM system sensors are implemented, such as flow sensors, moister sensors, and distance sensors. The data received data from sensors are stored in databases and processed by the Android device. The system is controlled and monitored using the mobile application, which consists of ON/OFF buttons and livestock data output. The system automatically sends the notification to the user when the water comes into the pipeline and the flow meter increases speed to the rotator. The system also collects the data from distance sensors attached at the top of the water tank to measure water levels in the water tank.

This system can be enhanced by using smart valves in place of flow sensors and implemented in a city on a large scale, which can be useful for water supply chain management, water supply mapping, and faults in the water pipeline and water leakage in the pipelines in smart cities. The sensors and microcontrollers are connected to the interface for wireless communication between the system and application. Also, this system can be further modified by using machine-learning algorithms to study and define the procedure of WSM and to transform it into an automated system. This can also be used in water waste management in industrial companies, which increases the efficiency of water plants and fluid tanks. The system consists of the following:

- USB cable
- Arduino UNO
- Water flow sensor
- Bread board
- Ultrasonic sensor – HC-SR04
- Jumper wires
- Relay module
- Diode rectifier
- Motor (water motor)

The system consists of various components:

1. USB Cable A to B: This is used to connect two devices for connection, data sharing, data transfer, and power supply.
2. Arduino Uno: This supports the version 1.0 of Arduino IDE, which comes with a preprogrammed boot loader to upload code without an external hardware programmer.
3. Water Flow Sensor G1/2: This consists of a water rotor, hall-effect sensor, and valve body. It senses the flow of water through the rotor, which changes the speed at different flow rates of water.
4. Breadboard: This is used to hold connections, sensors, equipment, and modules for stable and well-defined connections between them without soldering.
5. Ultrasonic Sensor-HC-SR04: This is used for the accurate measurement of distance between sensor and target objects via sound waves and converts the reflected sound into electric signals.
6. Jumper wires: This is used in breadboards, antennas, and network cabling components.
7. Relay module: This is used to control the electronic device power supply provided to a machine or mechanical system from the main power supply.
8. 10 K Ohm resistor: This is used to make excellent pull-downs, pull-ups, and current limiters. It is used in breadboards, microcontrollers, electronic devices, etc.
9. Diode rectifier: This is an electrical device that converts AC periodically in the reverse direction to DC in only one direction.

12.8 DATA ANALYSIS

A significant task for this project was the gathering of information about the existing home automation devices in WSM and finding a suitable layout that could be built for older people and physically disabled people.

Implementation of our vision requires a clear understanding of home automation, working on Arduino, and working on the sensors to resolve the problem of the uncertain supply of water in cities to maintain the water storage and management.

Using Arduino in the project gives us many advantages, such as the following:

1. Larger community support to develop and implement code for IoT-based devices and applications
2. Easy to use and development of the code, system, and device gets easier.
3. Open-source hardware, in which the code can be modified according to user needs based on a specific device or a personal kit
4. Customization
5. Cost-efficient
6. Availability of components
7. High connectivity also supports the database for data collection and data transfer for operations
8. Cross-platform means the system runs on Mac OSX, Windows, and Linux operating systems, but microcontroller systems are highly supportive of Windows
9. Recyclable and reprogrammable means it the device can be reset as it connects to the system or desktop via a USB cable. Connections held between microcontrollers and sensors are pin ports that can be modified and changed based on needs.

Gathering information, such as how would people who are not able to understand technology would be able to use automation at their homes, was crucial, as well as how other people could benefit from a device like the one in the project

12.9 ARCHITECTURE

The IoT-based Android application is used to control and maintain the supply of water in households. Figure 12.1 shows the data flow in the system in IoT-based application development.

The architecture has three layers:

1. Perception layer: It has sensors for sensing and collecting data.
2. Network layer: It is responsible for connectivity to the network.
3. Application layer: It is responsible for the service provided via an application.

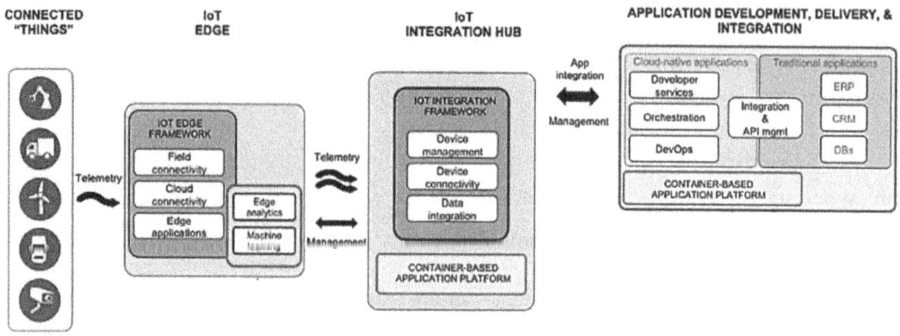

FIGURE 12.1 Architecture of IoT Based Application.

The main components of the architecture are shown in Figure 12.1:

1. IoT sensors and devices
2. Wi-Fi/Internet connectivity
3. IoT framework
4. Power/electricity supply
5. Integrating hub
6. Mobile application

12.10 CASE STUDY

In 2013, in Shirpur Warwade, Maharashtra, India, the city council of Shirpur launched a scheme and started a project to maintain and control the 24/7 supply of water for the citizens of Shirpur in the Dhule District. Earlier, citizens of the village acquired an irregular supply of water, which caused wastage of water, and there was no mechanism for tracking water consumption. The water supply company said Shirpur was under pressure to build a sustainable supply and demand management for the water supply to increase revenue. In the year 2013, the city council started a project to develop the continuous supply of water and tracking of water consumption for the entire town. This led them to smart water management for the standard evaluation of consumption and the discontinuous water supply. They selected "the ultrasonic metering technology" and "automatic meter reading system" offered by Kamstrup, which solved the problem of the water supply. Smart water management system increased the tracking of water, efficiency, accuracy, reliability, and life span of the system; managed revenues; and deceased the manual effort, error, cost of production, and other expenditures.

12.11 CHALLENGES IN IMPLEMENTING THE SMART WATER MANAGEMENT SYSTEM

1. Climate: Depleting rainfall and high temperatures are very common in the city.
2. Leakage in pipeline: Because of the unauthorized distribution connection by citizens, the pipeline was damaged and dripping.

3. Adaptation: The consumers were new to the technology and resistant to smart water consumption monitoring.
4. Accountability: Earlier, there was no accountability for the consumers to pay for the water they were consuming, which resulted in excess usage, as well as wastage of water.
5. Unauthorized connections: In the absence of an adequate monitoring system, people were stealing water through unauthorized connections.
6. Pipeline network: In the absence of timely information regarding the damages of and repairs to pipes, the entire network became vulnerable to leakages.
7. Repair
8. Delay in reading collection and billing

12.12 USAGE OF THE HOLISTIC APPROACH TO ADDRESS THE CHALLENGES

The municipal council of Shirpur wanted to increase the measure efficiency, 24/7 water supply, and monitoring rate; manage consumption; and establish fair billings system for the consumers. It is the first-ever smart water management project at the city level in India. The municipal focused on durability and quality of smart reading meters with long-term maintenance and operations. The solution consisted of the following:

- USB reading with antenna
- Third-party integration with billing software
- FlowIQ 2101 ultrasonic smart water meter
- Meter reading software
- Ten-year operation and management with extended warranty

The municipal provided 135 LPCD and installation and of a smart water metering system to every household. In addition, the smart alert system helps to detect any situation, such as leaky pipes, dry pipes, and burst pipes. The installation of the system was started in 2014. During installation, the issue of narrow lines and dense infrastructure occurred, which caused the inability to install smart meters. This problem was solved by the Kamstrup ultrasonic meters, which work perfectly in any direction.

The operation and maintenance started in 2017, and data was collected from the smart meters. The council took steps to verify consumer data by creating district metered areas for structured data for efficient reading and operational purposes.

The outcome of the smart water management system analysis is as follows:

1. Increases in data collection and comparative analysis of water supply patterns and consumption post meter installation
2. Difference in water consumption of commercial and residential connections
3. Consumer distribution per water consumption
4. New billing policy formed by the council based on Telescopic rates

12.13 CONCLUSION AND FUTURE SCOPE

The ever-evolving and technologically advanced nature of mobile technology has opened new opportunities for Android-based applications. People are getting more dependent on smartphones for the majority of their work, be it communication, calculation, or other crucial work. Hereby we are introducing "WSM Android Application," the Android application that would use its sensors to collect data regarding water levels and water flow in water tanks and water supply pipelines, respectively. The sensors would collect the real-time data of water levels and water flow and send it to the application with the help of Arduino. Users of the application can access this information from their user interfaces as based on their needs for water. They can also maintain or control the data provided by the sensors. The major advantages of this application are its user-friendliness, cost-effectiveness, and applicability in daily life for working people. This application would further provide several other benefits, including a reduction in wastage of water, less electricity consumption, ease of access, and control of the water supply. Further, a case study was done on citizens of Shirpur in Dhule District, Maharashtra, India. This study focused on smart water management for the standard evaluation of consumption and managed the discontinuous water supply.

Currently, the application is in limited scope to be used for household purposes. In the future, several extra features will be added to the device that will help in identifying faulty pipelines in a particular area or locality, which will decrease the cost of maintenance of the pipelines. Besides, an additional feature will be added to the application in regard to the billing system. This billing system will be connected to the government water supply board. In addition, a mobile application and query platform will be added to the application that will help address users' queries 24/7. Users will be able to use these features extensively and avail the available benefits, along with the standard bills.

REFERENCES

Alhamoud, A., Ruettiger, F., Reinhardt, A., Englert, F., Burgstahler, D., Bohnstedt, D., Gottron, C., & Steinmetz, R. (2014). SMARTENERGY.KOM: An intelligent system for energy saving in smart home, *39th Annual IEEE Conference on Local Computer Networks Workshops*, Edmonton, AB, Canada.

Anjana, S., Sahana, M. N., Ankith, S., Natarajan, K., Shobha, K. R., & Paventhan, A. (2015). An IoT based 6lowpan enabled experiment for water management, *Proceedings of IEEE International Conference on Advanced Networks and Telecommuncations Systems (ANTS)*, pp. 1–6, Kolkata, India.

Babovic, V., Drécourt, J. P., Keijzer, M., & Friss Hansen, P. (2002). A data mining approach to modelling of water supply assets, *Urban Water*, 4(4), 401–414.

Butler, D., Farmani, R., Fu, G., Ward, S., Diao, K., & Astaraie-Imani, M. (2014). A new approach to urban water management: Safe and sure. *Procedia Engineering*, 89, 347–354.

George, S., Bezdedeanu, L., Vasilescu, A., & Suciu. V. (2017). Unified intelligent water management using cyber infrastructures based on cloud computing and IoT, *Proceedings of 21st International Conference on Control Systems and Computer Science*, pp. 606–611, Bucharest, Romania.

Günther, M. Camhy, D., Steffelbauer, D., Neumayer, M., & Fuchs-Hanusch, D. (2015). Showcasing a smart water network based on an experimental water distribution system, *Procedia Engineering*, 119, 450–457.

Gupta, K., Kulkarni, M., Magdum, M., Baldawa, Y., & Patil, S. (2018). Smart water management in housing societies using IoT, *Second International Conference on Inventive Communication and Computational Technologies*, pp. 1609–1613, Coimbatore.

Hadipour, M., Derakhshandeh, J. F., & Shiran, M. A. (2019). An experimental setup of multi-intelligent control system (MICS) of water management using the Internet of Things (IoT), *ISA Transactions*, 96, 309–326.

Joseph, J., Manju, K. M., Sajith, M. R., Nair, S., Viay, V. P., & Krishnan, S. (2018). Water management system using IoT, *International Research Journal of Engineering and Technology*, 5(4), 1887–1890.

Kamienski, C., Soininen, J.-P., Taumberger, M., Dantas, R., Toscano, A., Cinotti, T. S., Maia, R. F., & Neto, A. T. (2019). Smart water management platform: IoT-based precision irrigation for agriculture, *Sensors*, 19(2), 276–286.

Lee, S. W., Sarp, S., Jeon, D. J., & Kim, J. H. (2015). Smart water grid: The future water management platform, *Desalination and Water Treatment*, 55, 339–346.

Lin, J., Yu, W., Zhang, N., Yang, X., Zhang, H., & Zhao, W. (2017). A survey on Internet of Things: Architecture enabling technologies security and privacy and applications, *IEEE Internet of Things Journal*, 4(5), 1125–1142.

Machado, M. R., Júnior, T. R., Silva, M. R., & Martins, J. B. (2019). Smart water management system using the Microcontroller ZR16S08 as IoT Solution, *IEEE 10th Latin American Symposium on Circuits & Systems (LASCAS)*, pp. 169–172, Armenia, Colombia.

Martinez, J. M. P., Llavori, R. B., Cabo, M. J. A., & Pedersen, T. B. (2007). Integrating data warehouses with web data: A survey, IEEE Trans. *Knowledge and Data Engineering*, 20(7), 940–955.

Misiunas, D., Vitkovsky, J., Olsson, G., Simpson, A., & Lambert, M. (2005). Pipeline break detection using pressure transient monitoring, *Journal of Water Resources Planning and Management*, 131(4), 316–325.

Mo, D., Ying, Z., & Chen, S. (2012). Automatic measurement and reporting system of water quality based on GSM, *Proceedings of Second International Conference on Intelligent System Design and Engineering Application*, pp. 1007–1010, Sanya, Hainan China.

Mounce, S., Machell, J., & Boxall, J. (2012). Water quality event detection and customer complaint clustering analysis in distribution systems, *Water Supply*, 12(5), 580–587.

Mukta, M., Islam, S., Barman, D., Reza, A. W., & Khan, M. S. H. (2019). IoT based smart water quality monitoring system, *Proceedings of 4th International Conference on Computer and Communication Systems*, pp. 669–673, Singapore.

Nguyen, K. A., Stewart, R. A., Zhang, H., Sahin, O., & Siriwardene, N. (2018). Re-engineering traditional urban water management practices with smart metering and informatics. *Environmental Modelling & Software*, 101, 256–267.

Parameswari, M., & Moses, M. B. (2018). Online measurement of water quality and reporting system using prominent rule controller based on aqua care–IOT, *Design Automation for Embedded Systems*, 22(1–2), 25–44.

Pradnya, A. H., & Ghewari, P. B. (2017). Review paper on IOT based technology, *International Research Journal of Engineering and Technology*, 4(1), 1580–1582.

Rajurkar, C., Prabaharan, S. R. S., & Muthulakshmi, S. (2017). IoT based water management, *International Conference on Nextgen Electronic Technologies: Silicon to Software*, pp. 255–259, Chennai, India.

Sharma, V., & Tiwari, R. (2016). A review paper on "IOT" & its smart applications, *International Journal of Science, Engineering and Technology Research*, 5(2), 472–476.

Sonaje, N. P., & Joshi, M. G. (2015). A review of modeling an application of water distribution networks (WDN) software. *International Journal of Technical Research and Applications*, 3(5), 174–178.

Tao, T., Li, J., Xin, K., Liu, P., & Xiong, X. (2016). Division method for water distribution networks in hilly areas, *Water Supply*, 16(3), 727–736.

Trina, M., Mohring, K., & Andersen, T. (2017). Semantic IoT: Intelligent water management for efficient urban outdoor water conservation, *Proceedings of Joint International Semantic Technology Conference*, pp. 304–317. Springer, Cham.

Wadekar, S., Vakare, V., Prajapati, R., Yadav, S., & Yadav V. (2016). Smart water management using IOT, *Proceedings of 5th International Conference on Wireless Networks & Embedded Systems*, pp. 1–4, Rajpura, India.

Xiaocong, M., Jiao, Q. X., & Shaohong, S. (2015). An IoT-based system for water resources monitoring and management, *7th International Conference on Intelligent Human-Machine Systems and Cybernetics*, pp. 365–368, Hangzhou.

Ye, Y., Liang, L., Zhao, H., & Jiang, Y. (2016). The system architecture of smart water grid for water security. *Procedia Engineering*, 154, 361–368.

Zanella, A., Bui, N., Castellani, A., Vangelista, L., & Zorzi, M. (2014). Internet of Things for smart cities, *IEEE Internet of Things Journal*, 1(1), 22–32.

Zhang, M., Sun, F., & Cheng, X. (2012). Architecture of Internet of Things and its key technology integration based-on RFID, *Proceedings of IEEE Fifth International Symposium on Computational Intelligence and Design*, 1(1), 294–297.

Zidan, N., Maree, M., & Samhan, S. (2018). An IoT based monitoring and controlling system for water chlorination treatment, *Proceedings of the 2nd International Conference on Future Networks and Distributed Systems*, pp. 31–36, Amman, Jordan.

13 A Hybrid Approach for Video Indexing Using Computer Vision and Speech Recognition

Saksham Jain
Amity School of Engineering and Technology
Delhi Campus, Noida, India

Akshit Pradhan
Kalinga Institute of Industrial Technology
Bhubaneswar, Odisha, India

Vijay Kumar
Department of Mathematics, AIAS, Amity University
Uttar Pradesh, Noida, India

CONTENTS

13.1 INTRODUCTION

Video has become a well-known storage medium of data over the web because of the fast advancement in recording devices and optimized video compression strategies, along with an increase in the network speed over the most recent couple of years.

Thus there has been an immense increase in the amount of video data on the web. Now for an end user, it might be very difficult to track down particular content inside a video without peeking inside the video file, and when the user has found out the related video information, it would be troublesome to decide whether a video is valuable by just looking at the title and other related metadata, which is most of the time short and describes the video on a higher level. In addition, the provided metadata might wind up in just a couple of moments; the user may, consequently, need to find out the part he or she is particularly in, and for that, the person will be required to view the whole video. The issue turns out to be how to recover the required part in a long video file more productively. The vast majority of the video search engines – for example, YouTube searches, rely on accessible metadata attached to a video (e.g., title, genre, and brief depiction of the video). This sort of metadata must be written by a human to guarantee a higher indexing in search engines, and the creation of this metadata is somewhat time and cost consuming. Besides that, the given metadata is normally short and a high-level description.

In some videos, the content may be hidden by the person or because camera qualities might not be visible; hence, optical character recognition (OCR) [1] can fail to detect this content. For these cases, we are incorporating speech recognition, along with a single OCR to provide better results to the end user.

Text is a high-level semantic feature that has regularly been used for content-based data retrieval. Say in lecture videos, notes from lecture slides fill in as a layout for the presentation and are significant for extraction. We can segment the video into smaller key frames and then remove the duplicate frames; thus, this video segmentation methodology will be executed on each key frame, and the separated content articles will be used in content acknowledgments and slide structure examination. Particularly, the separated basic metadata can empower progressively adaptable video search capacities.

Speech is also one of the most significant features in videos; therefore, it can be used for automatic video indexing, as the content will be spoken and can be detected by speech recognition frameworks. In our research, we planned to use automatic speech recognition (ASR) [2] to improve our model accuracy by combining the extracted text from speech with the key frame extracted from the video. In this manner, we built a more robust strategy so as to fill the gap of some of the bottlenecks discussed earlier.

A lot of text will be extracted by using OCR and ASR methods, which opens up the context of the videos. For content-based video searches, the search files are made from various data features, including manual metadata, OCR, and ASR catchphrases. In this manner, we propose a strategy for ranking keywords collected from different data sources and creating a term frequency inverse document frequency (TF-IDF) score [3]. The positioned keywords from both speech and text can be used straightforwardly for video content searching. Moreover, the video comparability can be determined by using the Cosine similarity measure [4] based on extracted keywords.

This chapter is organized in the following manner: Section 13.2 reviews related work in OCR and speech recognition regarding searching for the content of videos. Section 13.3 describes the overall architecture of the model that we have designed. Section 13.3.1 reviews in detail breaking the video into smaller frames and then extracting the key frames out of them for textual features and later processing them in the model, whereas Section 13.3.2 deals with the speech part of the model and how we detect breaks in the sentences and how they are mapped with the text to create a more robust model. Section 13.4 deals with the performance and accuracy our model has achieved, and, finally, Section 13.5 provides the conclusion and discussion of future works.

13.2 RELATED WORK

Speech recognition systems can be traced back to the 1950s. Researchers of one of the famous companies, AT&T Bell Labs, extracted data related to changes of vowel formant frequency in the voice and developed the world's first system for identifying about ten English digital pronunciations and thus was considered the foundation of one of the major domains in the tech industry. Since the starting of this new technology, several researchers have done remarkable work in this area to improve the quality of the work. Paliwal et al. [5] used dynamic spectral sub-band centroids, which proved that the new dynamic spectral separation coefficients are more resilient to noise than the mel-frequency cepstral coefficients (MFCC) features. Zavarehei et al. [6] used the concept of sequence modeling, along with joint semantic-lexical modeling, through which they concluded there was an increase in semantic information used and a tightness of integration between lexical and semantic items. And work done by Patel et al. [7] related to resolution decomposition with separating frequency as the mapping function shows an improvement in the quality metrics of speech recognition with respect to computational time and learning accuracy of the speech recognition system. Work done by Sharma et al. [8] used different types of filters for denoising of voices, which had a significant advantage over other similar methods. Singh et al. [9] used the Hidden Markov model for developing a voice-based user machine interface system, which is used by the user to interact with the system. And the work by Pramkeaw et al. [10] for improving the recognition rates of spoken words was done using MFCC-based speech classification with a finite impulse response (FIR) filter, which is a lot better than the previous approaches used for the task.

The computer vision domain came into existence in the late 1960s, having the main goal of mimicking the human visual system. It works on the principle of analyzing the different components of an image and has use cases from as small as edge detection in an image to as complex as object recognition in a video sequence, 3D pose estimation, and many more. Computer vision is also referred to as the computer's eye, as it is defined as the science and technology of machines that enables them to see and gain information from images, as well as multidimensional data. In a recent study, researchers worked on getting to know the image passed to the model by asking questions related to the image using recurrent neural networks [11]. For finding the similarities between the input video, researchers have used neural sentence embeddings [12]. And a great study was done by Aggrawal et al. [13] related to object recognition systems, visual-odometry, and key-point matching systems. And Wang et al. [14] worked on sparse coding models made for high resolution as a neural network. Boundaries can be predicted by using object-level features of a pre-trained object classification networks [15]. A work by Doersch et al. [16] included implementing unsupervised object detection models with very high accuracy. ConvNet architecture is used by researchers to implement the pose detection model for a human body [17]. Many researchers have worked on predicting the probable future motion of an image using optical flow based on a pixel value present in the image under consideration [18]. Some of the great researchers argued that the great convergence of ConvNets is making the field less diverse, but in contrast, it is making the techniques easier to implement on similar use cases, thus making it easier for new researchers to start working in the field.

13.3 ARCHITECTURE

The overall architecture is depicted in Figure 13.1. A multimedia file is taken as an input for indexing and searching; in the next stage, the video file is broken into small corresponding image parts using FFMPEG [19]. After this, there will be many similar images that need to be removed to remove the duplicates and build a perfect indexing system. We will remove the visually similar duplicate frames from the image sequence by comparing their pixels positions. Now, these unique images sequences will be ingested into our video indexing algorithm in Figure 13.2.

The unique image sequences are then treated as a single source file, which has audio and video as its components. We process the video by extracting the key frames. Textual data from the key frames of the video are extracted using OCR. Later, we perform the text normalization, removal of punctuation, and standardization. TF-IDF is applied to the OCR extracted text. The audio segment is bifurcated using our pitch detection algorithm, and then using speech to text [20], the textual data is extracted, and then on that we apply the TF-IDF to normalize the data for easier indexing. Now these two TF-IDF matrices are processed, and the vectors are matched and indexed. Using these matched keywords, we search from the given query from the user and output to the desired time stamp of that particular sequence.

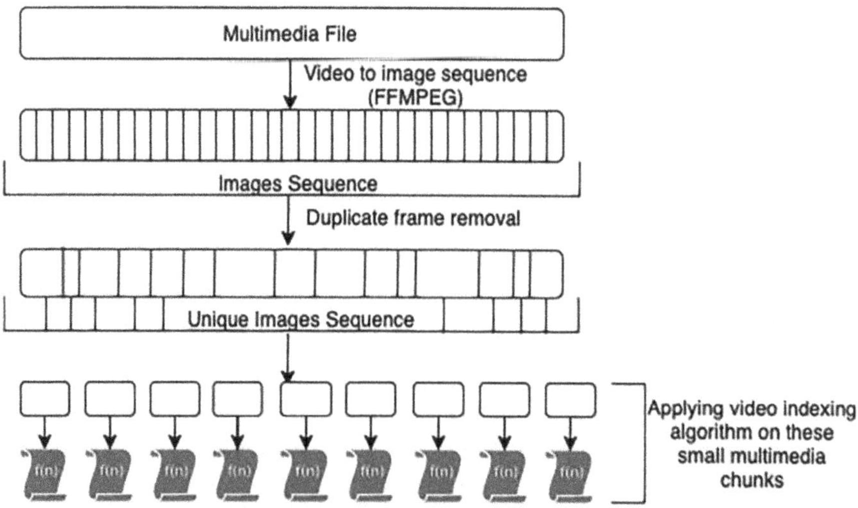

FIGURE 13.1 Complete Architecture of Proposed Approach.

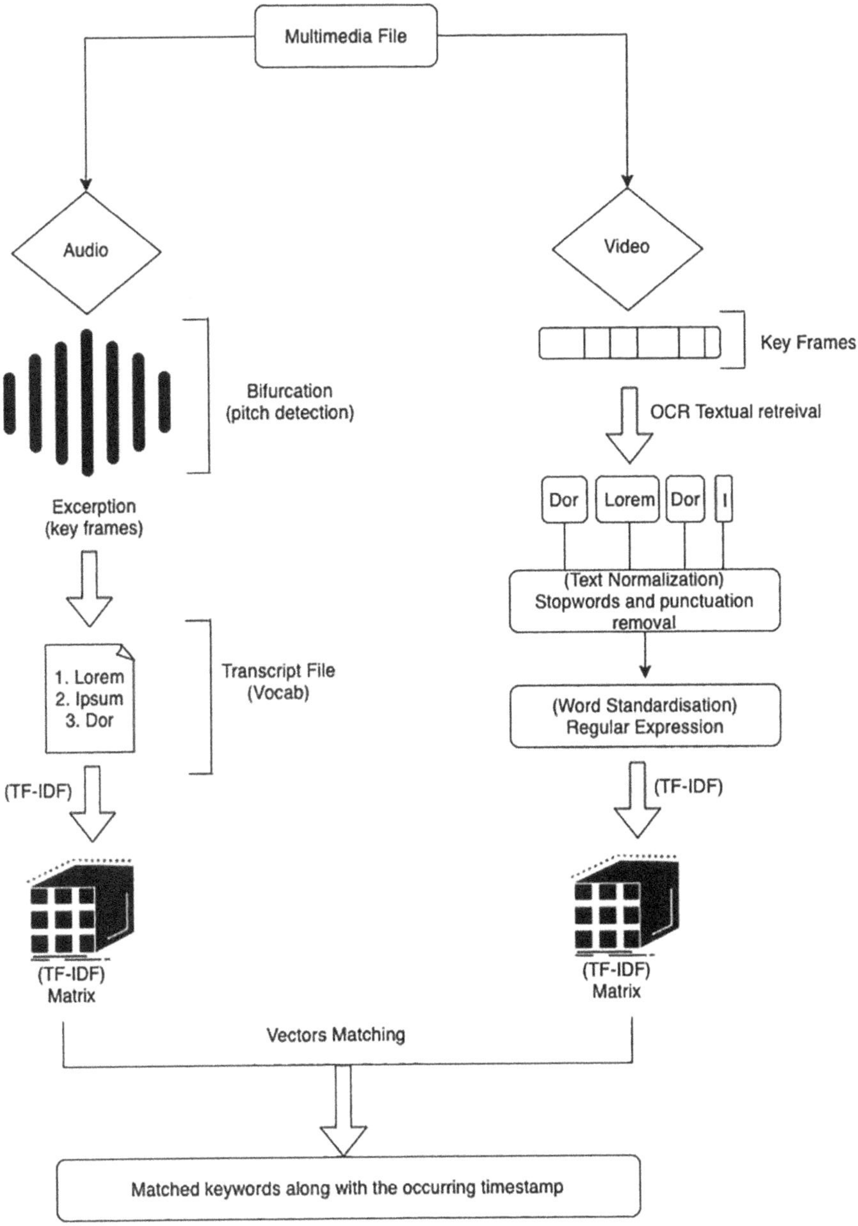

FIGURE 13.2 Architecture in Detail.

13.3.1 Key Frame Detection

In this section, we will introduce four investigation forms for recovering important metadata from the two fundamental pieces of the addressed video, in particular the video and audio. From the video, we right off recognize the slide changes and concentrate every interesting slide outline with its fleeting degree considered the video fragment. At that point, the video OCR investigation is performed for recovering literary metadata from slide outlines. In view of the OCR results, we propose a novel answer for address layout extraction by using stroke width and geometric data of recognized content lines. In discourse to content examination, we applied the open-source ASR programming CMU Sphinx 4 [21]. To construct the acoustic and language model, we gathered discourse preparing information from open-source corpora and our talk recordings. In this way, we built an answer to fill this hole and made it accessible for further research use.

Video perusing can be accomplished by dividing the video into delegate key casings. The chosen key casings can give a visual rule to route in the talk video entrance. In addition, video division and key-outline choice are regularly embraced as preprocessing for different investigation assignments – for example, video OCR and visual idea identification. Picking an adequate division technique depends on the meaning of "video fragment" and for the most part relies on the class of the video. In the video area, the video succession of an individual talk point or subtopic is regularly considered a video portion. This can be generally dictated by dissecting the transient extent of talk slides. Our segmentation technique comprises two stages: In the initial step, the whole slide video is investigated. We attempt to catch each information change between adjoining outlines, for which we set up investigation time frame seconds by taking both precision and productivity into account. This implies that fragments with a span smaller than three seconds might be disposed of in our framework. Since there are very few point fragments shorter than three seconds, this setting is in this manner not basic. At that point, we make watchful edge maps for neighboring casings and assemble the pixel differential picture from the edge maps [22]. The outcome along these lines is an unreasonable excess for video ordering. Subsequently, the process proceeds with the subsequent division step in view of the edges in the initial step.

In the subsequent division step, the genuine slide advances will be caught. The title and substance areas of a slide outline are first characterized. We set up the substance appropriation of normally used slide styles by breaking down a lot of talk recordings in our database. Hence any little changes inside the title area may cause slide progress. For example, two slides frequently contrast from one another in an arbitrary section number.

Since the proposed technique is intended for sectioning slide recordings, it may be not appropriate when recordings with differing types have been implanted into the slides and are played during the introduction. To take care of this difficulty, we have broadened the first calculation by using a support vector machine (SVM) [23] classifier and picture power histogram highlights. To make a correlation, we additionally applied the histogram of oriented gradients (HOG) included in the examination. We adjusted the HOG with eight inclination bearings, while the nearby locale size was set to 64 × 64. In addition, in using this methodology, our video division strategy is additionally reasonable for handling such one-screen address recordings with a continuous switch among slide-and-speaker scenes.

The next step is the vectorization of the transcript of the file as a TF-IDF, which is defined as converting a collection of raw documents to a matrix of TF-IDF features.

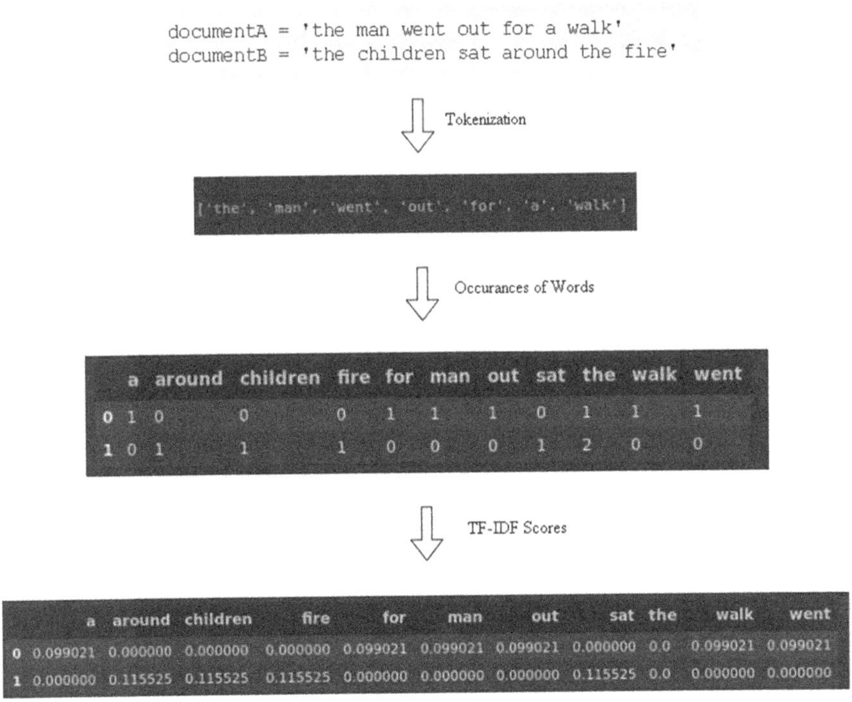

FIGURE 13.3 TF-IDF Architecture.

The process of vectorization is combined with the bag-of-words methodology so that the process of vector matching can be done effectively. TF-IDF can be defined as the process of extracting important keywords from a document to get a sense of what characterizes a document. As shown in Figure 13.3, vectorization has three steps: tokenization, occurrence of words in the document, and, lastly, computing TF-IDF scores.

13.3.2 Speech Recognition

Speech recognition is a kind of technology that includes a device to transform the voice signal to an associated text. Speech recognition, most commonly referred to as automated speech recognition (ASR) [25] system, is similar to a pattern recognition system, which includes three subunits: extraction of features, pattern matching, and referencing the model library. The input speech is converted to electrical signals. Then those electrical signals are analyzed, and the desired features are extracted, which commonly includes removing the noise present in the input to give the output text.

ASR systems are classified by different standards. One way is based on the relevance of the speaker with the system under consideration; it is divided into three subcategories: (1) speaker-dependent ASR, works only with a particular person's voice; (2) speaker-independent ASR, not based on a particular individual but may need a large database for getting better accuracy; and (3) group specific ASR for a group of peoples can identify a group of human voices, only needed to be trained on the database containing the data related to the peoples who are a part of the group under consideration.

An ASR system can also be divided based on the size of the vocabulary of the system under consideration into three subcategories: (1) small vocabulary ASR, having dozens of words; (2) medium-sized vocabulary system, having between 100–1,000 words; and (3) large vocabulary systems, include more than 1,000 words.

One can say that the accuracy of such speech systems has improved a lot because of recent advancements that occurred in the machine-learning domain, neural networks to be precise, and the size of databases that such systems are used to train on. Speech recognition systems have use cases from as small as a virtual personal assistant to as complex as giving an artificial voice to a person who can't speak, often referred to as "mute." And speech recognition systems had a wide variety of use cases from automobiles, toys, telephone communication, education, household systems, etc.

Components of a speech recognition system are as follows:

1. Voice Input: With the help of a microphone, audio is input into the system or an audio clip can be used.
2. Digitalization: The process of converting the analog signal into digital form is known as digitalization.
3. Acoustic Model: The acoustic model is created by taking audio recordings of speech and their text transcriptions and using software to create statistical representations of the sounds that make each word.
4. Language Model: It is used in natural language processing, which has applications, such as speech recognition. Speech recognition tries to capture the properties of a language and to predict the next word in the speech sequence.
5. Speech Engine: It converts the input audio into text.

For the scope of this chapter, we have used speech recognition for detecting pitch in the input video, which is converted to an audio sample, as shown in Figure 13.2. After getting the audio file of the input file, that audio file is subjected to a pitch detection module so that the relative time (time stamp) of the word that occurred in the input file can be generated, which we will be saving for generating the final results. After the relative time of word that occurred in the video is saved in a file, then that input file is subjected to a speech recognition system for transcripting the input file. And for the sake of getting higher accuracy, we have a pretrained model, developed by Google.

The next step is vectorization of the transcript of the file as a TF-IDF, which is defined as converting a collection of raw documents to a matrix of TF-IDF features, as described in Section 13.4.

13.3.3 Combining Audio and Video Parts

For the concluding part, we need to combine both sections (i.e., video perusing part and audio extraction part). Input for our system is the multimedia file upon which indexing needs to be done, which is broken down further after it is computed separately by the subsections. In the video subsystem, duplicate frames are removed by comparing the subsequent frames in the video, which are considered a single source file. Processing of video frames is done by using the OCR system followed by text normalization, some cleaning, and standardization. And for integrating the system with the audio system, the data is converted using TF-IDF. For the audio part, the single source file is then subjected to pitch detection, which helps in dividing the audio part by the breaks present in the source file. Dividing the file using the speech detected by the speech detection system is followed by computing TF-IDF values of the detected text. After getting the TF-IDF values of the video part, as well as the audio part, values of TF-IDF are compared to get the idea about the correlation of both parts, thus enacting the sequence of the single source file joined to make a complete multimedia file, thus getting to know the approximate indexing of the words present in the video.

13.4 PERFORMANCE

To assess our slide-video division technique, we have assembled a video test data set, which consists of 20 arbitrarily chosen videos from various YouTube channels with different formats and textual styles. Every remarkable slide with complete content is viewed as a right match. The general number of identified sections is 530, of which 506 are correct. The accomplished division recall and accuracy are 96 and 94 percent for the test data set, respectively. To assess the slide-picture classifier, 2,597 slide frames and 5,224 non-slide outlines have been gathered for the preparation. All slide outlines are gathered from our talk video database. To make a non-slide outline set with fluctuating picture classes, we have gathered an extra 3,000 pictures from Google. Moreover, around 2,000 non-slide video outlines have been gathered from the video database. The test data set comprises 240 slide outlines and 233 non-slide outlines, which vary from the preparation set. In the classifier, preparing the SVM-boundaries was dictated by using the framework search work.

13.4.1 Video OCR for Lecture Videos

Text in the video files was firmly identified with the speech content and would thus be able to give significant data to the recovery task. In our system, we built a novel video OCR framework for the extraction of the video text. For text recognition, we built another localization verification process. In the discovery stage, an edge-based multiscale text identifier is used to rapidly confine competitor text areas with a low dismissal rate. For the resulting text region confirmation, a picture entropy-based versatile refinement calculation not only serves to dismiss bogus positives that uncover low edge thickness but also further parts the most content and non-text areas into independent squares. At that point, stroke width transform (SWT) [26] based checking methodologies are applied to expel the non-text squares. Since the SWT verifier can't effectively distinguish exceptional non-text examples, we embraced an extra SVM classifier to sift through these non-text designs so as to further improve the location exactness. For text division and acknowledgment, we built a novel binarization approach in which we use a picture skeleton and edge maps to distinguish the content pixels. The proposed strategy comprises three primary advances, text slope heading investigation, seed pixel choice, and seed-area developing. After the seed-locale developing procedure, the video text pictures were changed over into an appropriate configuration for the standard OCR model. The ensuing spell-looking procedure will additionally sort inaccurate words from the acknowledgment results. Our video OCR framework has been assessed by applying a few test data set indexes. Our content identification strategy accomplished the subsequent spot, and our content binarization and word acknowledgment technique accomplished the primary spot in the related positioning rundown on the data set. An inside and out conversation of the created video OCR approach and the point-by-point test results can be found. By applying the open-source print OCR model tesseract OCR, we accomplished acknowledgment of 92 percent accuracy and 85 percent of all words accurately for video slides. The gathered test data set included 180 videos.

13.4.2 Slide Structure Analysis

By and large, in the talk slide, the substance of the title, caption, and key points have more criticalness than the typical slide text, as they sum up each slide. Because of this reality, we characterize the sort of text lines perceived from slide outlines by using geometrical data and stroke width highlights. The advantages of the structure examination strategy can be summed up as follows: The talk blueprint can be extricated using grouped text lines. It can give a quick outline of a talk video, and each frame with the time stamp can thus be received for video perusing. The ease of use of the layout included has been assessed by a client study given in the structure of text lines that can mirror their extraordinary hugeness. This data is significant for a search/ordering motor. Like web search motors that used the expressly precharacterized HTML structure (HTML-labels) for computing the weight of writings on site pages, our strategy further opens up the video content and empowers the pursuit motor to give increasingly precise and adaptable hunt results dependent on the organized video text. The procedure starts with a title line distinguishing proof strategy. A book line will be considered as an up-and-comer title line at the point when it confines the upper third piece of the casing, it has multiple characters, it is one of the three most elevated content lines, and it has the highest vertical position. At that point, the relating text line items will be named as the title items, and we will rehash the procedure on the rest of the content questions in a similar way. The further recognized title lines must have a comparative stature (the resistance is up to 10 px) and stroke width esteem (the resistance is up to 5 px) as the first one. For our motivations, we permit up to three title lines to be identified for each slide outline. All non-title text line objects are additionally arranged into three classes: (1) content, (2) key point and, what's more, (3) foot line indicates the most extreme vertical situation of a book line object (begins from the upper left corner of the picture). To additionally remove the video diagram, we right off the bat apply an illuminate checker to sort text line objects, which don't fulfill the accompanying conditions: a substantial book line object must have more than three characters; a substantial book line object must contain at any rate one thing; the printed character check of a legitimate book line object must be in excess of 50 percent of the whole string length.

The rest of the items will be marked as the framework object. In this way, we consolidate all title lines inside the same slide per their position. Other content line objects from this slide will be considered as the subitem of the title line. At that point, we combine text line objects from nearby edges with the comparable title content (with 90 percent content cover). The closeness is estimated by figuring the sum of the same characters and same words. After a blending procedure, all copied text lines will be expelled. At long last, the address layout is made by allocating all legitimate blueprint objects into a back-to-back structure per their events.

13.4.3 Experimental Results

We assessed the talk layout extraction technique by choosing 180 portioned exceptional slide outlines from the test recordings used in our test, both diagram type characterization precision and word acknowledgment exactness were measured. Thus the precision of diagram extraction depends on the OCR acknowledgment result. We have characterized the review and accuracy metric as follows: i) review is 25% of effectively recovered framework words of all layout words in grounded truth and ii) accuracy is 25% of accurately recovered framework words of recovered layout words, as shown in Figures 13.4 and 13.5.

$$\text{Precision} = \frac{tp}{tp + fp}$$

FIGURE 13.4 Equation of Precision.

$$\text{Recall} = \frac{tp}{tp + fn}$$

FIGURE 13.5 Equation of Recall.

The figures show the test results for the title and key point extraction, separately. Despite the fact that the extraction pace of the key point despite everything has a certain improvement space, the separated titles are now appropriate for the programmed video ordering. Besides, the plot extraction exactness can be additionally improved by giving better OCR results.

13.5 CONCLUSION

In this chapter, we introduced a methodology for indexing content-based video ordering and retrieving the content in a large video repository. To evaluate the performance, we used visual content, as well as audio content, from the video for separating content-based metadata consequently. Many novel indexing algorithms and features have been used in the video retrieval process by using the metadata extracted.

For future work, the ease of use and utility study for the indexing algorithm in the video repository will be done. Automatic annotation process, which is using OCR, is the key point for retrieving the data correctly and is the most important process. Therefore, more efficient search and recommendation methods could be developed for the video repositories.

REFERENCES

1. Smith, R. (2007). An overview of the tesseract OCR engine, *Ninth International Conference on Document Analysis and Recognition (ICDAR 2007), Parana*, pp. 629–633. doi:10.1109/ICDAR.2007.4376991.
2. Chiu, C.-C., Sainath, T., Wu, Y., Prabhavalkar, R., Nguyen, P, Chen, Z., Kannan, A., Weiss, R., Rao, K., Gonina, E., Jaitly, N., Li, B., Chorowski, J., & Bacchiani, M. (2018). State-of-the-art speech recognition with sequence-to-sequence models. 4774–4778. doi:10.1109/ICASSP.2018.8462105.
3. Tokunaga, T., & Makoto, I. (1994). Text categorization based on weighted inverse document frequency. In *Special Interest Groups and Information Process Society of Japan (SIG-IPSJ)*.
4. Rahutomo, F., Kitasuka, T., & Aritsugi, M. (2012, October). Semantic cosine similarity. In *The 7th International Student Conference on Advanced Science and Technology ICAST* (Vol. 4, No. 1).
5. Chen, J., Huang, Y., Li, Q., & Paliwal, K. K. (2004). Recognition of noisy speech using dynamic spectral subband centroids. *IEEE Signal Processing Letters*, 11(2), 258–261.
6. Zavarehei, E., Vaseghi, S., & Van, Q. (2005). Speech enhancement using Kalman filters for restoration of short-time DFT trajectories. In *IEEE Workshop on Automatic Speech Recognition and Understanding* (pp. 313–318). San Juan, Puerto Rico: IEEE.
7. Patel, I., & Rao, Y. S. (2010). Speech recognition using HMM with MFCC-An analysis using frequency specral decomposion technique. *Signal & Image Processing: An International Journal (SIPIJ)*, 1(2), 101–110.
8. Sharma, K., & Haksar, P. (2012). Speech denoising using different types of filters. *International Journal of Engineering Research and Applications*, 2(1), 809–811.
9. Singh, B., Kapur, N., & Kaur, P. (2012). Speech recognition with hidden Markov model: A review. *International Journal of Advanced Research in Computer Science and Software Engineering*, 2(3), 400–403.
10. Ittichaichareon, C., & Pramkeaw, P. (2012). Improving MFCC-based speech classification with FIR filter. In *International Conference on Computer Graphics, Simulation and Modelling (ICGSM 2012)* July (pp. 28–29).
11. Malinowski, M., Rohrbach, M., & Fritz, M. (2015). Ask your neurons: A neural-based approach to answering questions about images. In *Proceedings of the IEEE International Conference on Computer Vision* (pp. 1–9).
12. Zhu, Y., Kiros, R., Zemel, R., Salakhutdinov, R., Urtasun, R., Torralba, A., & Fidler, S. (2015). Aligning books and movies: Towards story-like visual explanations by watching movies and reading books. In *Proceedings of the IEEE International Conference on Computer Vision* (pp. 19–27).
13. Agrawal, P., Carreira, J., & Malik, J. (2015). Learning to see by moving. In *Proceedings of the IEEE International Conference on Computer Vision* (pp. 37–45).
14. Wang, Z., Liu, D., Yang, J., Han, W., & Huang, T. (2015). Deep networks for image super-resolution with sparse prior. In *Proceedings of the IEEE International Conference on Computer Vision* (pp. 370–378).

15. Bertasius, G., Shi, J., & Torresani, L. (2015). High-for-low and low-for-high: Efficient boundary detection from deep object features and its applications to high-level vision. In *Proceedings of the IEEE International Conference on Computer Vision* (pp. 504–512).
16. Doersch, C., Gupta, A., & Efros, A. A. (2015). Unsupervised visual representation learning by context prediction. In *Proceedings of the IEEE International Conference on Computer Vision* (pp. 1422–1430).
17. Pfister, T., Charles, J., & Zisserman, A. (2015). Flowing convnets for human pose estimation in videos. In *Proceedings of the IEEE International Conference on Computer Vision* (pp. 1913–1921).
18. Walker, J., Gupta, A., & Hebert, M. (2015). Dense optical flow prediction from a static image. In *Proceedings of the IEEE International Conference on Computer Vision* (pp. 2443–2451).
19. FFmpeg Developers. (2016). ffmpeg tool (Version be1d324) [Software].
20. National Academy of Sciences. (1994). *State of the art in continuous speech recognition. Voice Communication Between Humans and Machines.* Washington, DC: The National Academies Press. doi:10.17226/2308.
21. Lamere, P., Kwok, P., Gouvea, E., Raj, B., Singh, R., Walker, W., ... & Wolf, P. (2003, April). The CMU SPHINX-4 speech recognition system. In *IEEE International Conference on Acoustics, Speech and Signal Processing (ICASSP 2003)*, Hong Kong (Vol. 1, pp. 2–5).
22. Ansari, M., Kurchaniya, D., & Dixit, M. (2017). A comprehensive analysis of image edge detection techniques. *International Journal of Multimedia and Ubiquitous Engineering*, 12, 1–12. doi:10.14257/ijmue.2017.12.11.01.
23. Hearst, M. A. (1998). Support vector machines. *IEEE Intelligent Systems*, 13(4), 18–28. doi:10.1109/5254.708428.
25. Lee, K. F. (1988). *Automatic Speech Recognition: The Development of the SPHINX System* (Vol. 62). Springer Science & Business Media.
26. Epshtein, B., Ofek, E., & Wexler, Y. (2010, June). Detecting text in natural scenes with stroke width transform. In *2010 IEEE Computer Society Conference on Computer Vision and Pattern Recognition* (pp. 2963–2970). IEEE.

14 Statistical Methodology for Software Reliability with Environmental Factors

R. Jeromia Muthuraj
Manonmaniam Sundaranar University
Tirunelveli, Tamil Nadu, India

A. Mohamed Ashik
Merit Arts and Science College
Tirunelveli, Tamil Nadu, India

A. Loganathan
Manonmaniam Sundaranar University
Tirunelveli, Tamil Nadu, India

CONTENTS

14.1 INTRODUCTION

Software plays an important role in every moment of life among all humans. Software is the inevitable thing at the moment because all technological activities depend on software. The fast-growing technological world expects reliable software. In a software users' point of view, software plays an important role in both safety-critical and civil applications. From the software developers' point of view, the is difficulty in the delivery of high-quality software on time and within a required budget. Present-day end users of software are expecting failure-free and reliable software. There are many analytical models, which could be found in the literature addressing the problem of software reliability measurements. These models are developed mainly based on the failure record of software, and they can be classified according to the failure process. An attempt made by Jelinski and Moranda (1972) may be considered the pioneering work on software reliability. Later, many models were devolved to study the reliability of the software through statistical methodologies. Shooman (1972), Littlewood and Verrall (1973); Schik and Wolverton (1973); Goel and Okumoto (1978, 1979), Goel (1985); Littlewood (1980); Yamada and Osaki (1984, 1985); Yamada et al. (1993); Kumar et al. (2016); Pham and Pham (2019); Kumar and Sahni (2020); and many models were devolved. All the statistical software reliability models that were developed are based the information on the software development process or the information (data) that was collected while testing the software – that is, software failure data.

Zhang and Pham (2000) introduced factors related to the development of the software that are also important in deciding the reliability of software. Factors such as complexity and category of the programs that built the software; design of the software; skill and domain knowledge of the personnel who develop the programs; devices such as processors and storage devices; and system software that are related to the hardware environment are also responsible for the reliability of the software. Since these kinds of factors are related to the environment in which the software is developed, they are called environmental factors (EFs). There are 32 EFs proposed and categorized into five phases and discussed the impact of these EFs on software reliability assessment. Patwa and Malviya (2014) proposed two more impartment EFs – namely, complexity in logic and random access memory. In this work, it was decided that all 34 EFs can be a part of deciding the reliability of software. A survey was conducted on the study of the significance level of all 34 EFs in view of the technical experts who are involved in the development and maintenance of software. A questionnaire was developed to collect the opinions of the managers, system engineers, software testers, and programmers. The questionnaire was sent to software development experts working in various software development organizations. The respondents communicated through email to which the questionnaire was attached. As pointed out in Zhang and Pham (2000) and Zhu et al. (2015), the determination of software reliability modeling using EFs requires independent explanatory variables. Although the EFs were found to be correlated for the present data, this property may be expected for other data also. It is difficult to generate independent explanatory variables from these EFs. However, it is possible to extract uncorrelated explanatory variables from these EFs when conducting principal components analysis (PCA).

A new statistical methodology evolved for the determination of principal components such that they explain a threshold of 90% of variation from the original information. Logistic regression models were fitted for studying software reliability using the principal components extracted from three different procedures. The model fit the principal components as predictors. Validity of the model is discussed with respect to relevant criteria in each case.

14.2 SOFTWARE RELIABILITY MODELS

The software reliability models were proposed mainly based on the failure history of software. The models are classified into four types based on the nature of the failure process.

14.2.1 Times Between Failures Models

In this model, the process under study is the time between failures. The approach is to guess that the time between, say, the $(i - I)^{st}$ and the ith failures follow a distribution whose parameter depends on the number of fault left over in the program during this period.

14.2.2 Failure Count Models

The importance of this type of model is the number of fault or failure in a particular time interval rather than time between failure. The failure counts are assumed to follow a known random process with a time-dependent discrete or continuous failure rate.

14.2.3 Fault-Seeding Models

The main approach of this model is to "seed" an identified number of fault in a program that is assumed to have an unknown number of native fault. The program is tested and observed, and the number of seeded and indigenous faults are counted.

14.2.4 Input Domain Based Models

Another model approach taken here is to produce a set of test cases from an input distribution that is assumed to be representative of the operational usage of the program. Since there is difficulty obtaining this distribution, the input domain is partitioned into a set of equivalence classes, each of which is usually associated with a program path.

14.3 EFS AND THEIR SIGNIFICANCE

As mentioned in the introduction, Zhang and Pham (2000) proposed a set of 32 EFs. They insisted on the need for including information about the EFs in software reliability modeling and analysis. They grouped the EFs into five phases under the heads of general (7), analysis and design (7), coding (6), testing (7) and hardware systems (5). The numerals within the parentheses represent the number of EFs that are grouped under the respective categories.

Patwa and Malviya (2014) also discussed a set of 26 potential EFs. However, except for two, the other 24 were already available in the list of EFs proposed by Zhang and Pham (2000). Among the two EFs, one belongs to a "general" (F08, complexity in logic) category and the other belongs to "hardware system" (F34, RAM). Thus now there are eight EFs under "general" category and six EFs under the "hardware systems" category. All 34 EFs are presented in Table 14.1 under the five categories.

A survey was conducted to study the significance level of all 34 EFs regarding the reliability of software in view of the technical experts who are involved in the development and maintenance of software. A questionnaire was developed to collect the opinions of the managers, system engineers, software testers, and programmers. The was questionnaire divided into two components – viz., demographic information and professional experience of the respondents and the opinions of the respondents on each EF. The pie-chart graph for the respondents in the study is shown in Figure 14.1. The respondents were given eight options to express their opinions about the significance of each EF on determining software reliability. The opinions were scaled in an eight-point scale, as listed in Table 14.2. If the respondent felt that the significance level of an EF depended on the software, his or her opinion about that particular EF would be "may or may not be significant," and the corresponding score would be 3.

The questionnaire was sent to a randomly selected 50 persons who were working in 12 software development organizations. The questionnaires were sent to the respondents along with software for collecting their opinions on the significance of

TABLE 14.1
EFs

Factor Number	Category	EF
F01	**General**	Program Complexity
F02		Program Categories
F03		Difficulty of Programming
F04		Amount of Programming Effort
F05		Level of Programming Technologies
F06		Percentage of Reused Modules
F07		Programming Language
F08		Complexity in Logic
F09	**Analysis and Design**	Frequency of Program Specification Change
F10		Volume of Program Design Documents
F11		Design Methodology
F12		Requirements Analysis
F13		Relationship of Detailed Design to Requirement
F14		Work Standards
F15		Development Management
F16	**Coding**	Programmer Skill
F17		Programmer Organization
F18		Development Team Size
F19		Program Workload
F20		Domain Knowledge
F21		Human Nature
F22	**Testing**	Testing Environment
F23		Testing Effort
F24		Testing Resource Allocation
F25		Testing Methodologies
F26		Testing Coverage
F27		Testing Tools
F28		Documentation
F29	**Hardware systems**	Processors
F30		Storage Devices
F31		Input/Output Devices
F32		Telecommunication Devices
F33		System Software
F34		Random Access Memory

each EF with the aim of studying the reliability of the software. The total number of respondents was 50. Among them, 8% (4) of them were managers, 24% (12) of them were system engineers, another 24% (12) of them were testers, and the remaining 44% (22) were programmers. The background information about the respondents is summarized in Table 14.3.

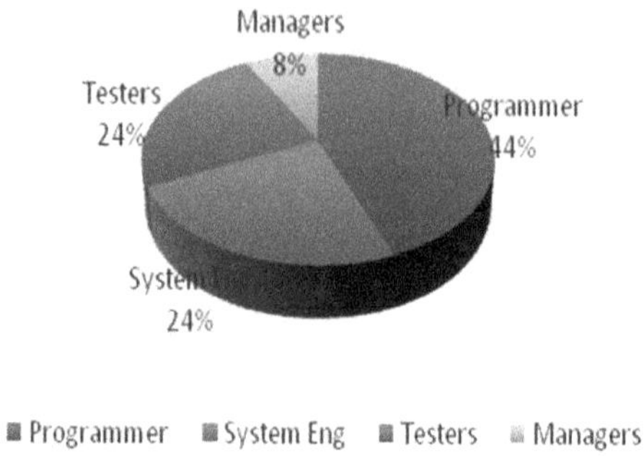

FIGURE 14.1 Respondents of Study.

TABLE 14.2
Opinion of Respondents and Scores

S. No.	Opinion	Score
1	Extremely Significant	7
2	More Significant	6
3	Moderately Significant	5
4	Significant	4
5	May or May Not Be Significant	3
6	Less Significant	2
7	Moderately Insignificant	1
8	Not Significant	0

TABLE 14.3
Background Information about the Respondents

S. No.	Demographic Character	Data
1	No. of respondents	50
2	Gender (male: female)	46:4
3	Designations of the respondents	Managers (4) System engineers (12) Testers (12) Programmers (22)
4	Professional experience (in years)	30 (more than 10 years) 20 (less than or equal to 10 years)
5	Average working experience	8.9 years
6	No. of software organizations	12
7	Kinds of software developed by the respondents	Commercial – 35 User-oriented applications – 15

14.4 SIGNIFICANCE LEVEL OF EF

The opinions of the respondents on the impact level of each of the 34 EFs are consolidated and presented in Table 14.4. Each entry in the table represents the number of respondents and who expressed their opinions about the significance level of each EF in determining the reliability of the software. All the respondents considered

TABLE 14.4
EFs and Scores

EFs	Scores							
	0	**1**	**2**	**3**	**4**	**5**	**6**	**7**
F01	0	0	0	0	0	4	32	14
F02	0	0	0	3	20	20	3	4
F03	0	0	0	3	10	26	4	7
F04	0	0	0	0	10	21	15	4
F05	0	0	0	4	13	14	16	3
F06	0	0	0	0	0	13	34	3
F07	0	0	0	0	6	34	10	0
F08	0	0	0	0	4	26	20	0
F09	0	0	0	0	3	10	21	16
F10	0	0	0	3	4	27	16	0
F11	0	0	0	3	4	34	9	0
F12	0	0	0	0	0	8	33	9
F13	0	0	0	6	4	21	19	0
F14	0	0	0	0	31	19	0	0
F15	0	0	0	0	22	14	11	3
F16	0	0	0	0	0	3	10	37
F17	0	0	0	0	3	18	22	7
F18	0	0	0	0	16	14	20	0
F19	0	0	0	0	23	24	0	3
F20	0	0	0	0	6	23	21	0
F21	0	0	0	0	6	27	17	0
F22	0	0	0	0	0	7	19	24
F23	0	0	0	0	0	6	9	35
F24	0	0	0	3	3	7	34	3
F25	0	0	0	0	3	6	21	20
F26	0	0	0	0	6	0	6	38
F27	0	0	0	0	6	8	22	14
F28	0	0	0	15	23	8	4	0
F29	0	0	13	34	3	0	0	0
F30	0	3	21	23	3	0	0	0
F31	0	15	25	4	6	0	0	0
F32	6	13	20	7	4	0	0	0
F33	0	0	4	26	26	0	0	0
F34	0	6	23	21	0	0	0	0

"program complexity" as one of the significant EFs. Sixty-four percent (32) of them mentioned its impact on software reliability as the most significant EF, and 28% (14) of them pointed out it as extremely significant.

It may be observed from the reading of the data presented in Table 14.4 that the respondents have considered the impact of the factor F01 through F28 upon determining software reliability as significant but at varied levels. Similarly, they have rated the significance of factors F29 through F34 at different levels of significance. Also, the number of respondents who ranked different EFs at the same level is not uniform. Thus the weights of the EFs on determination reliability of software vary. It may be required that we find the weights of the EFs. Since the respondents hold different positions in various types of organizations, the scores allocated based on the opinions of the respondents should be normalized. The following relative weights method has been discussed in Zhang and Pham (2000), Pham (2006) and Zhu et al. (2015) to find relative weight. Let s_{ij}, $(i = 1,2,\ldots,34; j = 1,2,\ldots,50)$ be the score allocated to i_{th} EF based on the opinions presented by the j_{th} respondent. The scores may be normalized first by applying

$$w_{ij} = \frac{s_{ij}}{\sum_{i=1}^{34} s_{ij}}, i = 1,2,\ldots,34 \text{ and } j = 1,2,\ldots,50$$

That is, the scores are normalized with respect to EFs for each respondent. As a result of this computation, the total weight score under each respondent will be 1.

$$i.e., \sum_{i=1}^{34} w_{ij} = \sum_{i=1}^{34} \frac{s_{ij}}{\sum_{i=1}^{34} s_{ij}} = 1$$

Now the EFs may be compared to understand their impact. But the normalized scores for each EF, w_{ij}, varies over the respondents. To facilitate the comparison of EFs, the average normalized score may be computed for each EF using

$$w_i = \frac{1}{50} \sum_{j=1}^{50} w_{ij}, i = 1,2,\ldots,34.$$

These average relative weights may be used to compare the significant levels of the 34 EFs and arrange them in an order with reference to the average relative weights. The EF with largest average relative weight can be considered the factor with the highest priority. The priority/significance levels of the EFs decline, as their average relative weights are in the decreasing order (Figure 14.5).

The average relative weights of the EFs are calculated based on the opinions of the respondents. The calculated values are presented in Table 14.5. It may be noted that the EFs are arranged according to their priority level. The proportion of the

TABLE 14.5
Average Relative Weights of EF

Priority value	Factor Number	EF	Average Relative Weight
1	F16	Programmer skill	0.042605
2	F9	Frequency of program specification change	0.042002
3	F23	Testing effort	0.041551
4	F22	Testing environment	0.040552
5	F1	Program complexity	0.039476
6	F25	Testing methodologies	0.038989
7	F6	Percentage of reused modules	0.038619
8	F26	Testing coverage	0.037325
9	F27	Testing tools	0.036769
10	F12	Requirements analysis	0.036574
11	F17	Programmer organization	0.035895
12	F24	Testing resource allocation	0.034364
13	F4	Amount of programming effort	0.0330002
14	F11	Design methodology	0.032899
15	F20	Domain knowledge	0.032685
16	F7	Programming language	0.031156
17	F3	Difficulty of programming	0.031105
18	F5	Level of programming technologies	0.030483
19	F10	Volume of program design documents	0.030157
20	F2	Program categories	0.029933
21	F18	Development team size	0.029749
22	F15	Development management	0.02923
23	F13	Relationship of detailed design to requirement	0.027203
24	F14	Work standards	0.024024
25	F19	Program workload (stress)	0.023285
26	F08	Complexity in logic	0.022551
27	F28	Documentation	0.021772
28	F34	Random access memory	0.021184
29	F21	Human nature	0.02004
30	F29	Processors	0.014883
31	F30	Storage devices	0.013609
32	F31	Input/output devices	0.012697
33	F32	Telecommunication devices	0.012369
34	F33	System software	0.011265

combinations of each EF can be calculated by multiplying its respective average relative weight by 100. Table 14.5 shows that the contribution of all the EFs ranges between 4.26% and 1.13%. "Programmer skill" and "frequency of program specification change," respectively, are the largest proportion of the contribution (i.e., 4.26% and 4.20%, respectively). The contribution level of "system software" is the lowest at 1.12%. In general, differences in the proportion of the contribution of the EFs are not very significant. Some of the EFs have relatively equal significance/priority levels.

14.5 RELATIONSHIP AMONG EFs

All the EFs are grouped under five categories (Table 14.1) and, hence, there may be an inherent relationship among them because of grouping. Also, a relationship may arise among the EFs across the categories. It is important to note that the application of Pearson's measure requires the fulfillment of assumptions regarding linear relationships and distributional properties. But the data on EFs in the present form cannot be expected to satisfy such assumptions. However, there are other formulas that compute the correlation coefficient, and some of them are more suitable to this kind of data. It should be recalled that the data on each EF are ordinal measurements. Kendall (1970) proposed a formula, which is popularly known as the *tau* coefficient, to ascertain the strength of the relationship between pairs of variables using the data observed as scores.

$$\tau = \frac{n_c - n_d}{\sqrt{\frac{n(n-1)}{2} - T_x}\sqrt{\frac{n(n-1)}{2} - T_y}},$$

where (X, Y) is a pair of variables, n- represents the number of observations made on (X, Y), n_c- represents the number of concordant pairs, n_d- represents the number of discordant pairs, $T_X = \frac{1}{2}\Sigma t_X(t_X - 1)$, $T_Y = \frac{1}{2}\Sigma t_Y(t_Y - 1)$, t_X- represents number of tied observations of X with the same score, and t_Y- represents number of tied observations of Y with the same score. Kendall's τ value ranges from -1 to $+1$; the sign of τ indicates the direction of the relationship between X and Y. The magnitude of τ indicates the strength of association between X and Y. The coefficient of correlation is calculated using Kendall's τ coefficient to each pair of EFs. The values are presented in Table 14.6. The following specific observations are made from the upper diagonal part of Table 14.6.

14.6 UNCORRELATED LINEAR COMBINATION OF EF

The exploratory analysis of information about the impact level of the EFs revealed the existence of correlated structures among the EFs. Some of the EFs were found to have high correlations among themselves and some other have moderate levels of correlation. As pointed out in Pham (2006) and Zhu et al. (2015), determination of software reliability modeling using EFs requires independent explanatory variables. Although the EFs are found to be correlated for the present data, this property may be expected for other data also. It is difficult to generate independent explanatory variables from these EFs. However, it is possible to extract uncorrelated explanatory variables from these EFs conducting PCA. The PCA is one of the classical methods of multivariate statistical data analysis. PCA is a dimensionality reduction technique, which is used to transform a high-dimensional data set into a smaller-dimensional subspace. As mentioned in Dunteman (1989), this technique was first introduced by Pearson (1901) and developed independently by Hotelling (1933). Like many other

TABLE 14.6
Coefficient of Correlation

EFs	F01	F02	F03	F04	F05	F06	F07	F08	F09	F10	F11	F12	F13	F14	F15	F16	F17
F01	1.00	0.22	.316	0.01	0.23	−0.26	−0.05	.390	−0.02	.275	−0.26	−0.07	−0.21	−0.09	−0.02	0.23	0.14
F02	0.22	1.00	.580	.330	−0.10	0.06	0.08	.313	0.06	0.04	−0.09	−0.16	.260	0.09	−0.15	.354	.425
F03	.316	.580	1.00	.443	−0.04	0.09	.499	0.24	−0.03	−0.01	0.03	−.307	.249	0.26	0.18	−0.05	0.15
F04	0.01	.330	.443	1.00	0.18	0.22	0.22	.343	−0.12	−.319	−.255	−.481	−0.19	0.09	−0.10	−0.03	0.16
F05	0.23	−0.10	−0.04	0.18	1.00	.446	.274	.350	0.18	0.22	0.18	−0.12	−.643	0.09	−0.07	0.12	0.20
F06	−0.26	0.06	0.09	0.22	.446	1.00	0.22	0.02	−0.10	−0.07	0.21	−.369	−0.24	0.03	−0.18	−0.15	0.01
F07	−0.05	0.08	.499	0.22	.274	0.22	1.00	0.17	.257	0.10	.389	−0.23	0.14	.610	.398	−0.20	0.13
F08	.390	.313	0.24	.343	.350	0.02	0.17	1.00	.369	−0.18	−.264	−.297	−.374	0.03	0.12	.527	0.08
F09	−0.02	0.06	−0.03	−0.12	0.18	−0.10	.257	.369	1.00	.412	0.10	0.08	0.08	.413	−0.13	.325	0.15
F10	.275	0.04	−0.01	−.319	0.22	−0.07	0.10	−0.18	.412	1.00	.464	.446	0.08	.326	−.374	0.00	.369
F11	−0.26	−0.09	0.03	−.255	0.18	0.21	.389	−.264	0.10	.464	1.00	.556	0.05	.364	0.11	−0.24	.307
F12	−0.07	−0.16	−.307	−.481	−0.12	−.369	−0.23	−.297	0.08	.446	.556	1.00	0.05	0.11	−0.09	0.02	0.18
F13	−0.21	.260	.249	−0.19	−.643	−0.24	0.14	−.374	0.08	0.08	0.05	0.05	1.00	0.16	0.12	−0.05	0.15
F14	−0.09	0.09	0.26	0.09	0.09	0.03	.610	0.03	.413	.326	.364	0.11	0.16	1.00	0.03	−.392	.287
F15	−0.02	−0.15	0.18	−0.10	−0.07	−0.18	.398	0.12	−0.13	−.374	0.11	−0.09	0.12	0.03	1.00	−0.12	−0.13
F16	0.23	.354	−0.05	−0.03	0.12	−0.15	−0.20	.527	.325	0.00	−0.24	0.02	−0.05	−.392	−0.12	1.00	0.22
F17	0.14	.425	0.15	0.16	0.20	0.01	0.13	0.08	0.15	.369	.307	0.18	0.15	.287	−0.13	0.22	1.00
F18	−0.07	.449	0.08	.324	−0.16	−0.24	−0.20	0.00	−0.05	0.07	−0.09	−0.04	0.10	−.284	−.302	.305	.256
F19	0.11	0.18	0.21	0.18	.430	.713	0.22	−0.13	−0.16	0.22	0.18	−.323	−0.17	.280	−.278	−.308	.316
F20	−.287	−.288	−0.11	−0.23	−.334	0.22	−.357	−.418	−0.22	−0.18	0.02	−0.05	0.15	−.493	−0.02	−0.15	−.333
F21	0.07	−0.12	−0.11	−0.22	.406	0.12	.278	0.02	−0.16	−0.12	0.11	−0.06	−0.16	−0.07	.517	0.13	0.08
F22	−0.02	−0.12	−.290	−0.17	−.335	−.450	−.485	0.05	−0.08	−.312	−.511	0.05	−0.05	−.563	0.11	0.25	−.570
F23	0.25	−0.22	0.08	−0.19	−0.25	−.269	0.02	0.19	.304	−0.13	−.396	−.332	0.23	−0.04	.300	0.02	−.257
F24	0.16	−.464	−.254	−0.16	−0.05	−.382	−0.21	0.00	0.19	−0.06	−.418	0.03	−0.16	−0.24	0.05	0.12	−.431
F25	0.21	−0.07	0.14	0.04	0.14	0.00	0.16	.405	.582	0.23	−0.01	−0.15	0.09	0.02	−0.06	0.25	0.06

(Continued)

TABLE 14.6 *(Continued)*

EFs	F01	F02	F03	F04	F05	F06	F07	F08	F09	F10	F11	F12	F13	F14	F15	F16	F17
F26	0.22	−0.18	−0.10	0.10	0.13	−0.22	0.02	.326	0.13	−0.19	−.424	−.366	0.01	−.446	0.21	.401	−0.07
F27	.288	0.00	0.07	0.14	−0.15	−.487	−.279	0.01	0.06	0.14	−0.16	0.11	0.14	−.371	−0.07	0.21	0.17
F28	0.12	0.24	.332	0.17	−.332	−0.18	0.02	0.00	0.13	−0.02	−.380	−.398	.522	−0.23	0.02	0.13	0.00
F29	−0.05	−0.11	0.09	0.07	−.275	0.05	−0.23	−0.22	−.441	−.410	−.366	−0.12	−0.08	−.288	−0.10	−0.18	−.706
F30	0.07	−0.03	.301	.249	−0.10	−0.06	−0.02	−.406	−.421	−0.06	0.02	0.12	−0.01	0.06	0.02	−.401	−0.04
F31	−0.09	0.02	.402	0.04	−0.12	.393	.309	−.325	−0.04	0.04	0.03	−.263	.257	.305	0.02	−.501	−0.23
F32	0.17	0.07	.349	−0.02	−.279	−0.05	0.03	−.410	−.544	−0.18	−0.18	−0.12	.274	−0.21	0.18	−0.23	−0.22
F33	.390	0.01	−0.05	−.466	−.312	−.623	−.429	−.324	−0.17	.360	0.12	.668	0.23	−0.23	0.00	0.06	0.14
F34	0.25	.548	.369	0.21	.335	.368	0.15	0.11	−0.22	0.08	0.09	−.272	−0.03	−0.23	−0.02	.335	.466
EFs	**F18**	**F19**	**F20**	**F21**	**F22**	**F23**	**F24**	**F25**	**F26**	**F27**	**F28**	**F29**	**F30**	**F31**	**F32**	**F33**	**F34**
F01	−0.07	0.11	−.287	0.07	−0.02	0.25	0.16	0.21	0.22	.288*	0.12	−0.05	0.07	−0.09	0.17	.390	0.25
F02	.449	0.18	−.288	−0.12	−0.12	−0.22	−.464	−0.07	−0.18	0.00	0.24	−0.11	−0.03	0.02	0.07	0.01	.548
F03	0.08	0.21	−0.11	−0.11	−.290	0.08	−.254	0.14	−0.10	0.07	.332	0.09	.301	.402	349	−0.05	.369
F04	.324	0.18	−0.23	−0.22	−0.17	−0.19	−0.16	0.04	0.10	0.14	0.17	0.07	.249*	0.04	−0.02	−.466	0.21
F05	−0.16	.430	−.334	.406	−.335	−0.25	−0.05	0.14	0.13	−0.15	−.332	−.275	−0.10	−0.12	−.279	−.312	.335
F06	−0.24	.713	0.22	0.12	−.450	−.269	−.382	0.00	−0.22	−.487	−0.18	0.05	−0.06	.393	−0.05	−.623	.368
F07	−0.20	0.22	−.357	.278	−.485	0.02	−0.21	0.16	0.02	−.279	0.02	−0.23	−0.02	.309	0.03	−.429	0.15
F08	0.00	−0.13	−.418	0.02	0.05	0.19	0.00	.405	.326	0.01	0.00	−0.22	−.406	−.325	−.410	−.324	0.11
F09	−0.05	−0.16	−0.22	−0.16	−0.08	.304*	0.19	.582	0.13	0.06	0.13	−.441	−.421	−0.04	−.544	−0.17	−0.22
F10	0.07	0.22	−0.18	−0.12	−.312	−0.13	−0.06	0.23	−0.19	0.14	−0.02	−.410	−0.06	0.04	−0.18	.360	0.08
F11	−0.09	0.18	0.02	0.11	−.511	−.396	−.418	−0.01	−.424	−0.16	−.380	−.366	0.02	0.03	−0.18	0.12	0.09
F12	−0.04	−.323	−0.05	−0.06	0.05	−.332	0.03	−0.15	−.366	0.11	−.398	−0.12	0.12	−.263	−0.12	.668	−.272
F13	0.10	−0.17	0.15	−0.16	−0.05	0.23	−0.16	0.09	0.01	0.14	.522	−0.08	−0.01	.257	.274	0.23	−0.03
F14	−.284	.280	−.493	−0.07	−.563	−0.04	−0.24	0.02	−.446	−.371	−0.23	−.288	0.06	.305	−0.21	−0.23	−0.23
F15	−.302	−.278	−0.02	.517	0.11	.300	0.05	−0.06	0.21	−0.07	0.02	−0.10	0.02	0.02	0.18	0.00	−0.02
F16	.305	−.308	−0.15	0.13	0.25	0.02	0.12	0.25	.401	0.21	0.13	−0.18	−.401	−.501	−0.23	0.06	.335
F17	.256	.316	−.333	0.08	−.570	−.257	−.431	0.06	−0.07	0.17	0.00	−.706	−0.04	−0.23	−0.22	0.14	.466

F18	1.00	−0.12	−0.06	−.302	0.13	−.334	−0.22	−0.06	0.09	.361	.282	−0.13	−0.06	−.375	−0.06	0.12	.334
F19	−0.12	1.00	0.01	0.13	−.730	−0.23	−.43	−0.14	−.355	−.392	−0.16	−0.11	0.14	.414	0.07	−.342	.505
F20	−0.06	0.01	1.00	−0.23	0.11	.274	0.16	0.08	−0.01	0.15	0.22	.314	0.16	.313	0.22	0.07	−0.04
F21	−.302	0.13	−0.23	1.00	−0.05	−0.08	0.06	−.339	0.17	−.316	−0.25	−0.13	0.00	−0.07	0.16	−0.08	.382
F22	0.13	−.730	0.11	−0.05	1.00	0.18	.486	−0.10	.295	0.21	0.22	.394	−0.09	−0.23	0.10	0.25	−.328
F23	−.334	−0.23	.274	−0.08	0.18	1.00	.505	.522	.489	0.23	.479	−0.05	−0.15	0.22	0.01	0.00	−.315
F24	−0.22	−.438	0.16	0.06	.486	.505	1.00	0.12	.381	.313	0.15	.32	0.24	0.06	0.15	0.17	−.404
F25	−0.06	−0.14	0.08	−.339	−0.10	.522	0.12	1.00	.524	.360	.409	−.394	−.475	−0.07	−.391	−0.11	−0.15
F26	0.09	−.355	−0.01	0.17	.295	.489	.381	.524	1.00	.461	.506	−0.22	−.369	−.335	−0.07	−0.08	0.02
F27	.361	−.392	0.15	−.316	0.21	0.23	.313	.360	.461	1.00	.501	−0.13	0.19	−.253	0.01	.452	−0.03
F28	.282	−0.16	0.22	−0.25	0.22	.479	0.15	.409	.506	.501	1.00	−0.03	−0.03	.273	.259	0.03	0.10
F29	−0.13	−0.11	.314	−0.13	.394	−0.05	.322	−.394	−0.22	−0.13	−0.03	1.00	.487	.393	.581	0.02	−0.19
F30	−0.06	0.14	0.16	0.00	−0.09	−0.15	0.24	−.475	−.369	0.19	−0.03	.487	1.00	.444	.588	.263	0.05
F31	−.375	.414	.313	−0.07	−0.23	0.22	0.06	−0.07	−.335	−.253	.273	.393	.444	1.00	.457	−0.19	0.00
F32	−0.06	0.07	0.22	0.16	0.10	0.01	0.15	−.391	−0.07	0.01	.259	.581	.588	.457	1.00	.275	0.20
F33	0.12	−.342	0.07	−0.08	0.25	0.00	0.17	−0.11	−0.08	.452	0.03	0.02	.263	−0.19	.275	1.00	−0.08
F34	.334	.505	−0.04	.382	−.328	−.315	−.404	−0.15	0.02	−0.03	0.10	−0.19	0.05	0.00	0.20	−0.08	1.00

multivariate statistical methods, it was not widely used until the advent of electronic computers. But it is now well entrenched in every statistical computer package. The central idea of PCA is to reduce the dimensionality of a data set, which consists of a large number of correlated variables, retaining as much as possible to explain the larger preparation of variation present in the data set. This reduction is achieved by transforming to a new set of variables, the principal components, which are uncorrelated and ordered so that the first few explain most of the variation present in all of the original variables.

14.7 PRINCIPAL COMPONENTS EXTRACTION FROM CLUSTERS OF EF

14.7.1 Formation of Clusters

Cluster analysis is the process of grouping objects into subsets such that similar objects are put into the same groups and dissimilar objects are put into different groups. The final set of groups will be mutually exclusive and exhaustive. The objects are thereby organized into an efficient representation that characterizes the population being sampled. Unlike classification, clustering does not rely on predefined classes. Cluster analysis subdivides a given set of individuals/objects into a number of disjointed subsets called clusters such that individuals/objects in the same cluster are more similar than individuals/objects in different clusters according to the same acceptable criterion of similarity. Clustering methods can be broadly categorized into two – *viz.*, hierarchical methods and nonhierarchical methods. In nonhierarchical clustering methods, the individuals/objects are grouped into a single classification of "k" clusters, where k is either specified "a prior" or is determined as part of the clustering method. Then cluster memberships are altered to obtain a better set of clusters (Anderberg, 1973). Hierarchical clustering method means a method of arraigning individuals/objects in a graded series according to the same acceptable notion of "grading." These methods are broadly classified under the heads – *viz.*, agglomerative hierarchical and divisive hierarchical. If the data consists of a measurement on "n" individuals/objects, the agglomerative method begins with a set of n single element clusters. These clusters are fused to obtain bigger clusters according to the same acceptable criteria. This process is continued until a required number of clusters are obtained or all the items are grouped into a single cluster. The divisive method is a reverse process of the agglomerative method. The results of both the methods may be represented by a two-dimensional diagram known as "dendrogram." The fusion or division of clusters is performed based on the values calculated for the same similarity measures, such as Euclidean distance, Minlouski distance, and Mahalanobis distance. The correlation coefficient is also used as a measure of similarity for fusion or division of clusters (Anderberg, 1973).

The 34 EFs disregarding their category are considered for clustering. The single linkage method is the simplest among hierarchical methods. This method is applied to the data on the scores of the EFs, which have been assigned based on the options of respondents. Euclidean distance is calculated as the measure of similarity. At each stage of the hierarchical procedure, Euclidean distance is calculated for each pair of clusters. Two clusters are merged/agglomerated when the distance is small. The agglomeration procedure is continued until all the EFs are put in a single cluster. The process of fusion is described in the dendrogram, which is displayed in Figure 14.2.

A set of six clusters was selected, and the details of the cluster members are presented in Table 14.7. Opinions of the respondents on the significance of the EFs formed a new grouping of EFs. It may be noted that each cluster consists of EFs from

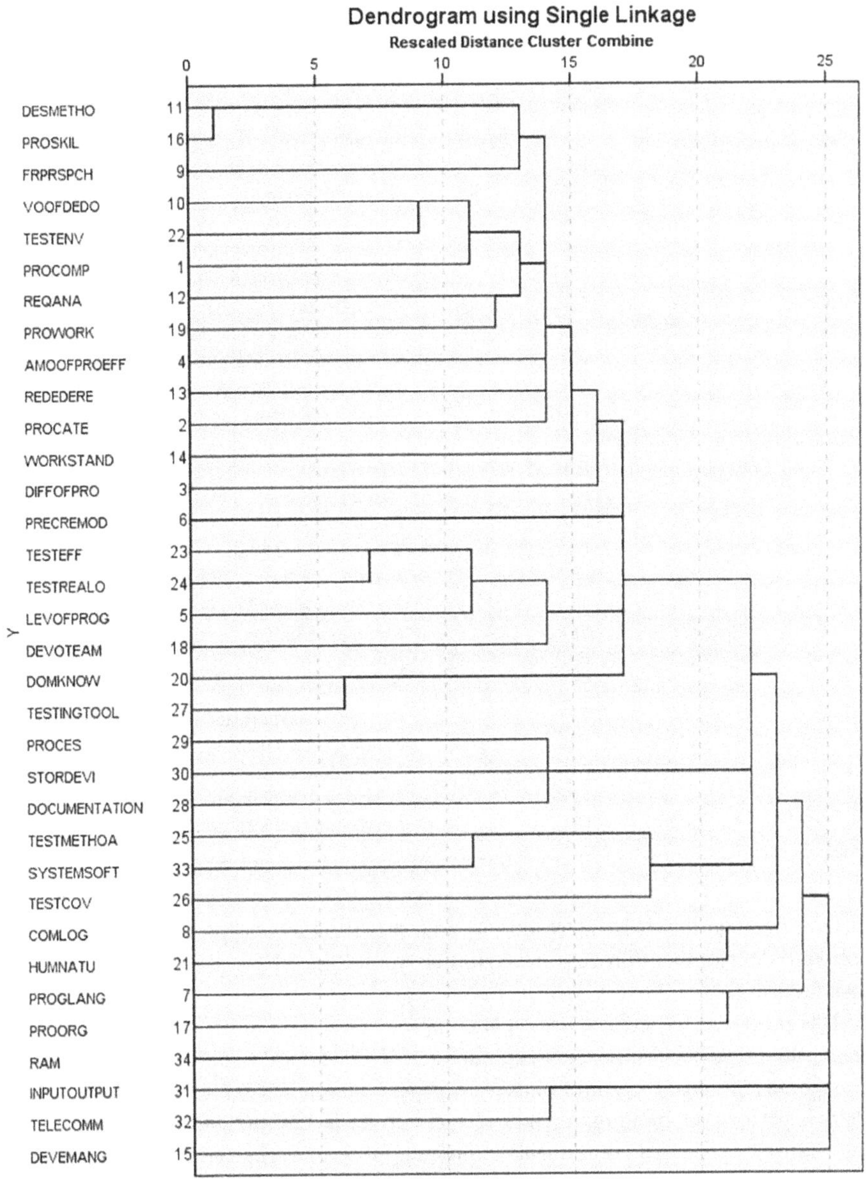

FIGURE 14.2 Dendrogram of Cluster Formation.

TABLE 14.7
Cluster of EF

Cluster	No. of EFs	Factor Number	Factors
Cluster 1	8	F11	Design methodology
		F16	Programmer skill
		F9	Frequency of program specification change
		F10	Volume of program design documents
		F22	Testing environment
		F1	Program complexity
		F12	Requirements analysis
		F19	Program workload (stress)
Cluster 2	6	F4	Amount of programming effort
		F13	Relationship of detailed design to requirement
		F2	program categories
		F14	Work standards
		F3	Difficulty of programming
		F6	Percentage of reused modules
Cluster 3	5	F23	Testing effort
		F24	Testing resource allocation
		F5	Level of programming technologies
		F18	Development team size
		F20	Domain knowledge
Cluster 4	6	F27	Testing tools
		F29	Storage devices
		F30	Human nature
		F28	Documentation
		F25	Testing methodologies
		F33	System software
Cluster 5	4	F26	Testing coverage
		F8	Complexity in logic
		F21	processors
		F7	Programming language
Cluster 6	5	F17	Programmer organization
		F34	Random access memory
		F31	Input/output devices
		F32	Telecommunication devices
		F15	Development management

different categories. It is interesting to note that the EFs in different phases of the software development process have similarity, and the EFs in the same phase have dissimilarity.

It is proposed to perform principal component analysis in two stages. In the first stage, the principal components explaining 90% of variation among the EFs in each cluster are extracted (i.e., intra-cluster principal components extraction). Then, again, PCA is performed, treating the intra-clusters principal components as variables. The inter-clusters principal components explaining 90% of variation of the intra-cluster principal components are extracted and such principal components will be considered for further analysis.

14.7.2 Intra-cluster Principal Components Extraction

14.7.2.1 Cluster I

According to Table 14.7, the first cluster of EFs, consist of eight EFs – *viz.*, "design methodology," "programmer skill," "frequency of program specification change," "volume of program design documents," "testing environment," "program complexity," "requirements analysis," and "program workload." Applying PCA, eight principal components are extracted. The eigenvalues and the proportion of total variance explained by each of these principal components are presented in Table 14.8, along with cumulative values. The eigenvalue corresponding to each principal component is displayed in Figure 14.3. The principal component numbers are located along the x-axis and the eigenvalues are displayed in Y-axis.

14.7.2.2 Cluster II

According to Table 14.7, the Cluster III of EFs consist of six EFs – *viz.*, "amount of programming effort," "relationship of detailed design to requirement," "program categories," "work standards," "difficulty of programming," and "percentage of reused modules." Applying PCA, six principal components are extracted. The eigenvalue and the proportion of total variance explained by each of these principal components are presented in Table 14.9, along with cumulative values. The eigenvalue corresponding to each principal component is displayed in Figure 14.4.

14.7.2.3 Cluster III

According to Table 14.7, the Cluster III of EFs consist of five EFs – *viz.*, "testing effort," "testing resource allocation," "level of programming technologies," "development team size," and "domain knowledge." Applying PCA, five principal components are extracted. The eigenvalue and the proportion of total variance explained by each of these principal components are presented in Table 14.10, along with cumulative values. The eigenvalue corresponding to each principal component is displayed in Figure 14.5.

TABLE 14.8
Proportion of Total Variance by Principal Components for Cluster I

Principal Components	Eigenvalues	% of Variance	Cumulative % of Variance
1	2.726	34.071	34.071
2	2.097	26.208	60.279
3	1.322	16.524	76.803
4	.946	11.827	88.630
5	.431	5.391	94.021
6	.258	3.219	97.240
7	.149	1.864	99.104
8	.072	.896	100.000

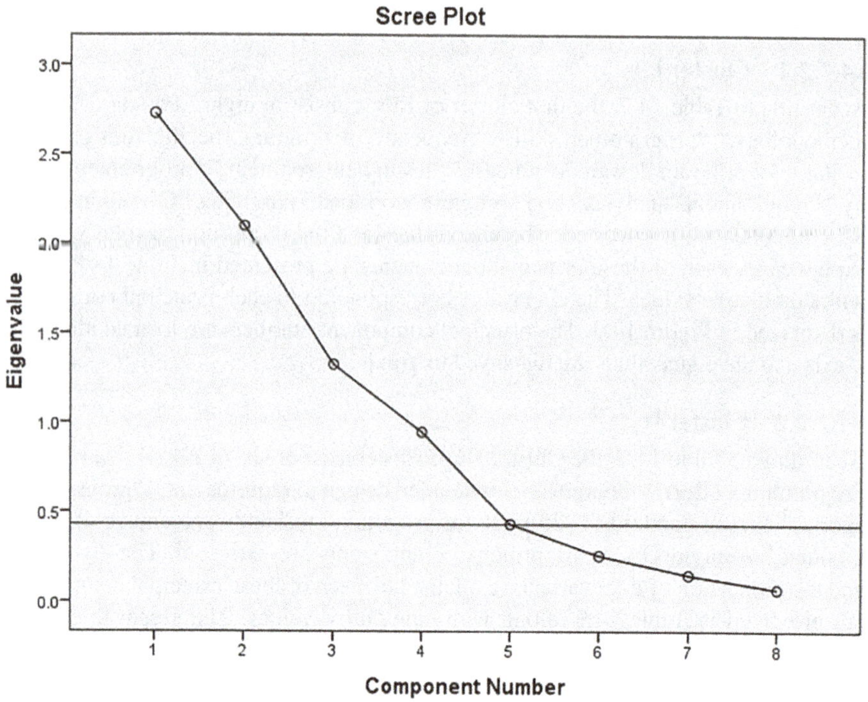

FIGURE 14.3 Scree Plot for Cluster I.

TABLE 14.9
Proportion of Total Variance by Principal Components for Cluster II

Principal Components	Eigenvalues	% of Variance	Cumulative % of Variance
1	2.413	40.217	40.217
2	1.461	24.351	64.568
3	.905	15.077	79.644
4	.696	11.606	91.250
5	.278	4.626	95.876
6	.247	4.124	100.000

14.7.2.4 Cluster IV

According to Table 14.7, the Cluster IV of EFs consist of six EFs – *viz.*, "testing tools," "storage devices," "human nature," "documentation," "testing methodologies," and "system software." Applying PCA, six principal components are extracted. The eigenvalue and the proportion of total variance explained by each of these principal components are presented in Table 14.11, along with cumulative values. The eigenvalue corresponding to each principal component is displayed in Figure 14.6.

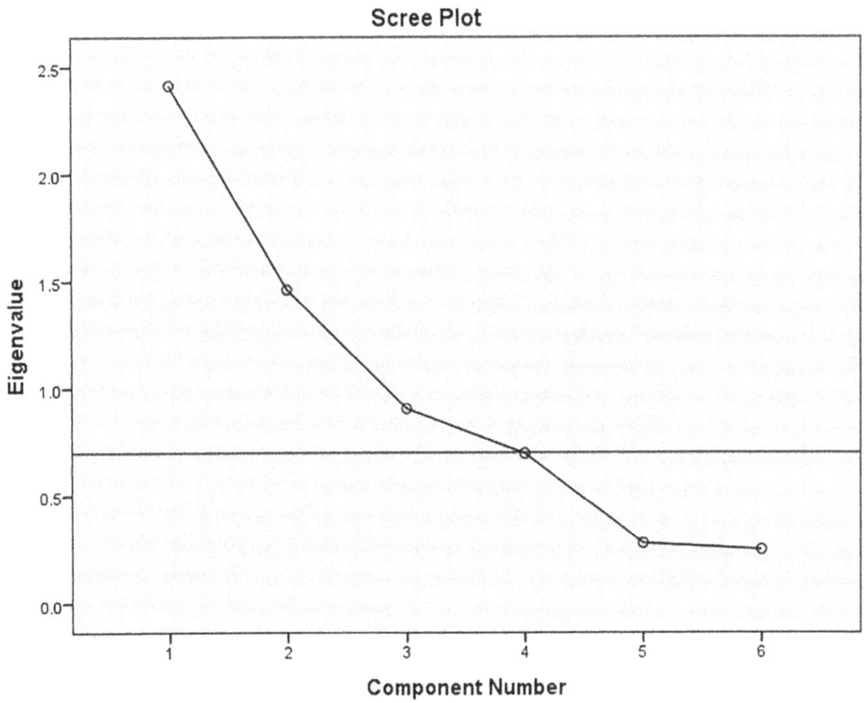

FIGURE 14.4 Scree Plot for Cluster II.

TABLE 14.10
Proportion of Total Variance by Principal Components for Cluster III

Principal Components	Eigenvalues	% of Variance	Cumulative % of Variance
1	2.584	51.687	51.687
2	1.179	23.588	75.276
3	.831	16.620	91.895
4	.405	8.105	100.000
5	1.427E-016	2.855E-015	100.000

14.7.2.5 Cluster V

According to Table 14.7, the Cluster V of EFs consist of four EFs – *viz.*, "testing coverage," "complexity in logic," "processors," and "programming language." Applying PCA, four principal components are extracted. The eigenvalue and the proportion of total variance explained by each of these principal components are presented in Table 14.12, along with cumulative values. The eigenvalue corresponding to each principal component is displayed in Figure 14.7

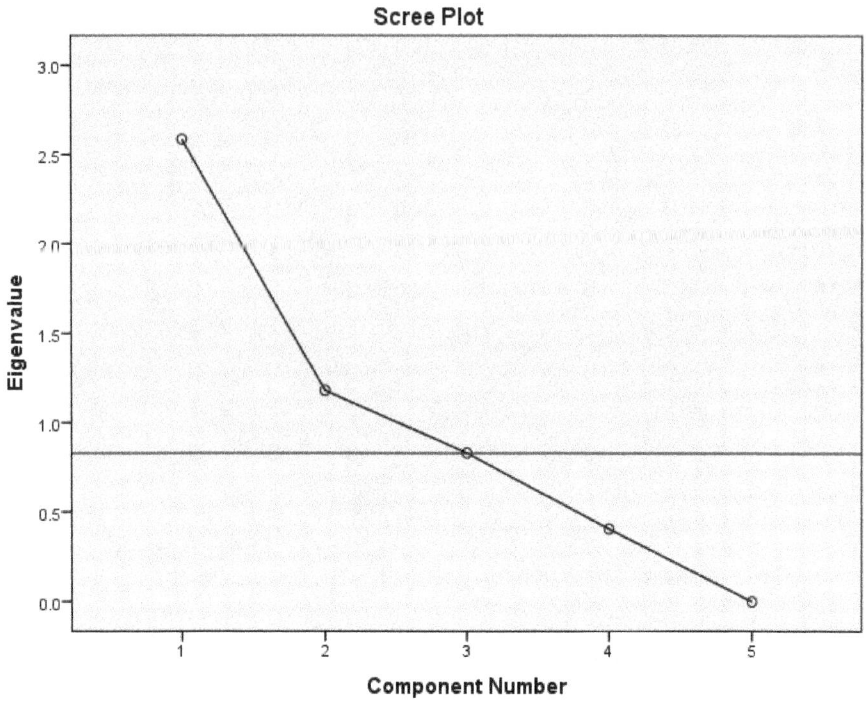

FIGURE 14.5 Scree Plot for Cluster III.

TABLE 14.11
Proportion of Total Variance by Principal Components for Cluster IV

Principal Components	Eigenvalues	% of Variance	Cumulative % of Variance
1	2.357	39.290	39.290
2	1.573	26.214	65.504
3	.820	13.664	79.168
4	.772	12.863	92.031
5	.290	4.826	96.856
6	.189	3.144	100.000

14.7.2.6 Cluster VI

According to Table 14.7, the Cluster IV of EFs consist of five EFs – *viz.*, "testing coverage," "complexity in logic," "processors," and "programming language." Applying PCA, five principal components are extracted. The eigenvalue and the proportion of total variance explained by each of these principal components are presented in Table 14.13, along with cumulative values. The eigenvalue corresponding to each principal component is displayed in Figure 14.8.

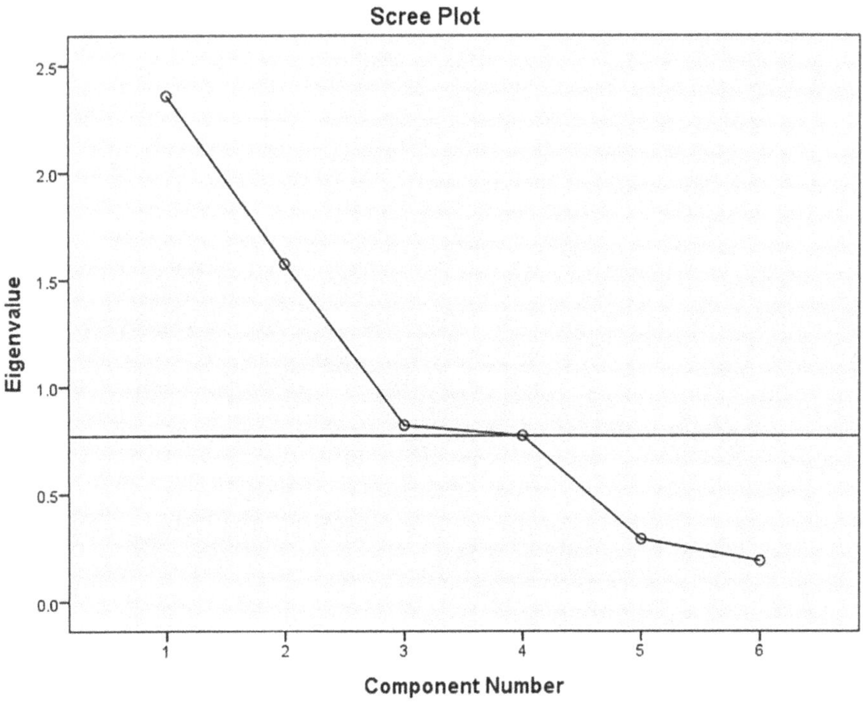

FIGURE 14.6 Scree Plot for Cluster IV.

TABLE 14.12
Proportion of Total Variance by Principal Components for Cluster V

Principal Components	Eigenvalues	% of Variance	Cumulative (%) of Variance
1	2.288	57.188	57.188
2	0.814	20.348	77.536
3	0.609	15.228	92.764
4	0.289	7.236	100.000

14.7.3 Inter-cluster Principal Components Extraction

The number of intra-cluster principal components extracted from the EFs are clusters according to procedure explained in Section 8.1. The details are presented in Table 14.14. However, there is a common characteristic among the extracted principal components. Each of them describe about 90% of total variance of the EFs in the respective categories.

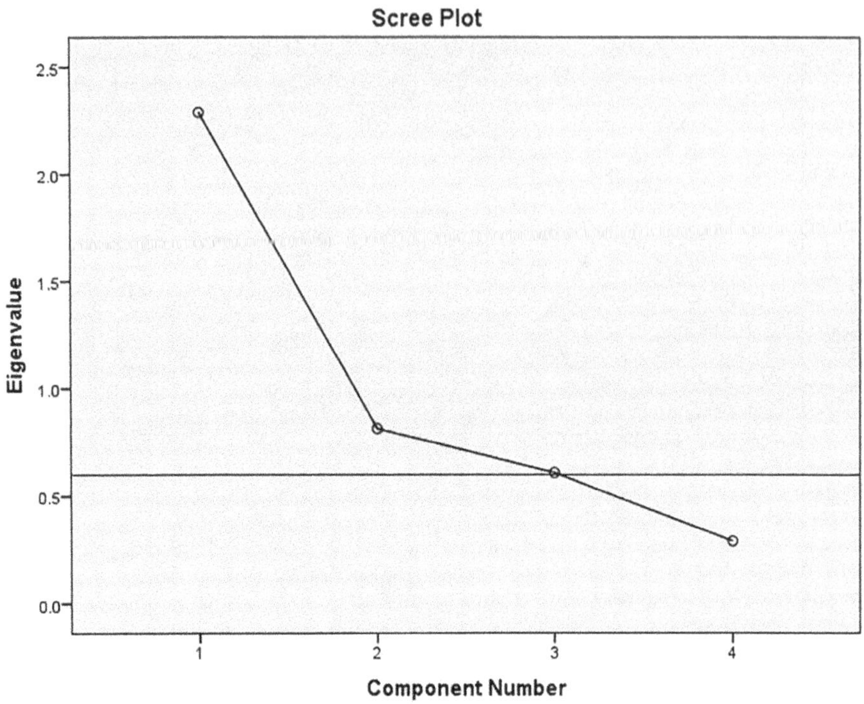

FIGURE 14.7 Scree Plot for Cluster V.

TABLE 14.13
Proportion of Total Variance by Principal Components for Cluster VI

Principal Components	Eigenvalues	% of Variance	Cumulative (%) of Variance
1	2.819	56.387	56.387
2	1.062	21.237	77.624
3	0.763	15.256	92.881
4	0.356	7.119	100.000
5	1.677E-016	3.354E-015	100.000

It should be noted that the PCA conducted in Section 8.2 has extracted the principal components that are uncorrelated within the respective clusters only. Moreover, the principal components are a linear combination of the EFs in the respective groups. Under this circumstance, we should conduct PCA considering these 22 principal as variables. As in Section 8.2, the principal components are extracted, explaining 90% of the variation of these 22 variates.

The eigenvalues and proportion of variation explained by principal components, along with their cumulative values, are presented in Table 14.15. The principal

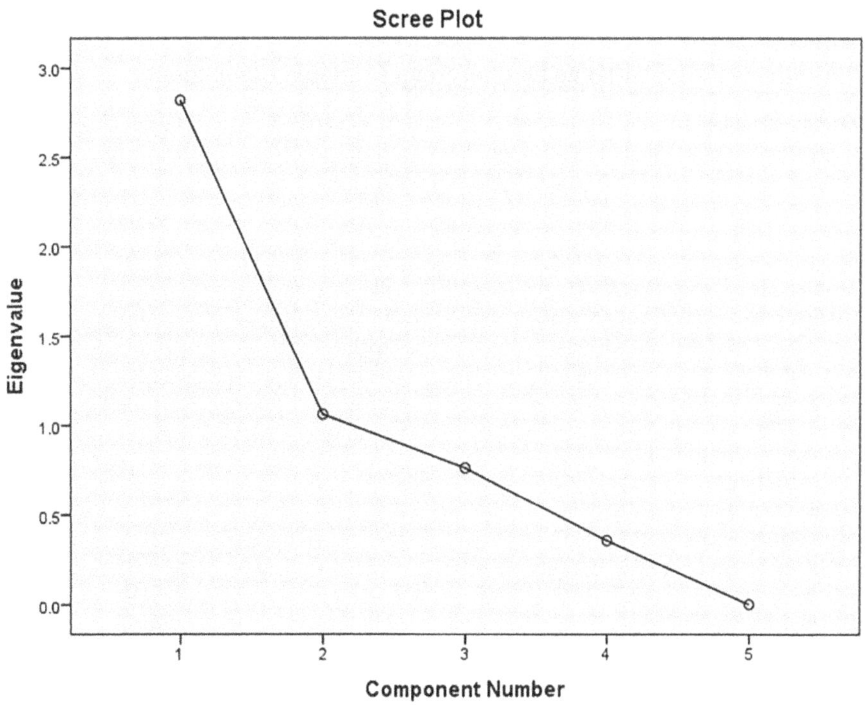

FIGURE 14.8 Scree Plot for Cluster VI.

TABLE 14.14
No. of Intra-cluster Principal Components

Cluster of EFs	No. of Principal Components
Cluster I	5
Cluster II	4
Cluster III	3
Cluster IV	4
Cluster V	3
Cluster VI	3
Total	**22**

component number and the corresponding eigenvalues are displayed in Figure 14.9. A perusal of Table 14.15 shows that the first 11 principal components explain about 91.732% of the total variance of the 22 intra-cluster principal components. Among them, the first two intergroup principal components explain, respectively, 26.186% and 14.530 of total variance. Also, the first six principal components explain, in total, 73.759% of total variance. Inclusion of the principal components from seven to 11 has added 17.973% of total variance.

TABLE 14.15
Proportion of Total Variance by Intergroup Principal Components

Principal Components	Eigenvalues	% of Variance	Cumulative % of Variance
1	5.761	26.186	26.186
2	3.197	14.530	40.715
3	2.360	10.727	51.443
4	2.221	10.093	61.536
5	1.464	6.653	68.189
6	1.225	5.570	73.759
7	.999	4.539	78.299
8	.902	4.101	82.400
9	.868	3.945	86.345
10	.609	2.770	89.115
11	.576	2.616	91.732
12	.513	2.333	94.065
13	.352	1.599	95.664
14	.296	1.345	97.009
15	.227	1.034	98.043
16	.167	.758	98.800
17	.117	.530	99.330
18	.068	.308	99.638
19	.045	.203	99.841
20	.021	.095	99.936
21	.010	.047	99.983
22	.004	.017	100.000

14.7.4 Intra-cluster Principal Components Extraction

14.7.4.1 Cluster I

According to Table 14.7, the first cluster of EFs, consist of eight EFs – *viz.*, "design methodology," "programmer skill," "frequency of program specification change," "volume of program design documents," "testing environment," "program complexity," "requirements analysis," and "program workload." Applying PCA, eight principal components are extracted. The eigenvalues and the proportion of total variance explained by each of these principal components are presented in Table 14.8, along with cumulative values. The eigenvalue corresponding to each principal component is displayed in Figure 14.3. The principal component numbers are located along the x-axis and the eigenvalues are displayed in Y-axis.

14.7.4.2 Cluster II

According to Table 14.7, the Cluster III of EFs consist six EFs – *viz.*, "amount of programming effort," "relationship of detailed design to requirement," "program categories," "work standards," "difficulty of programming," and "percentage of reused

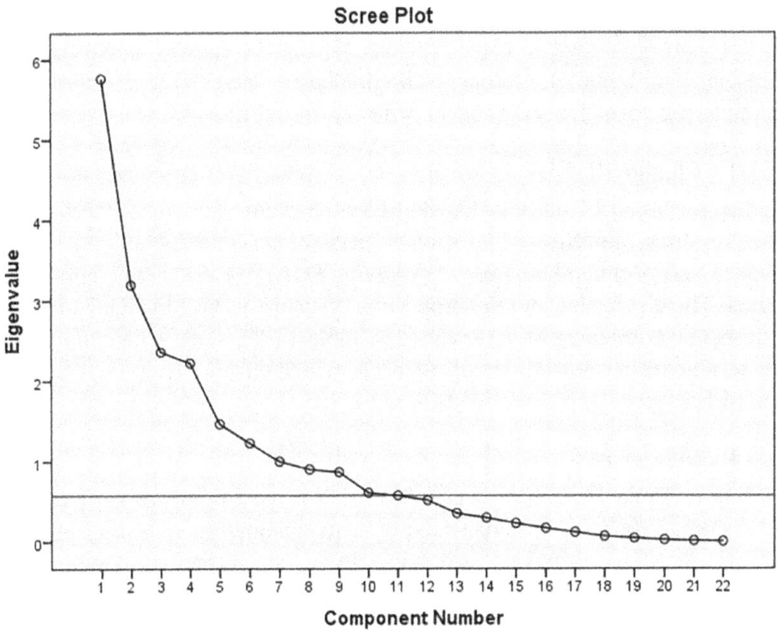

FIGURE 14.9 Scree Plot for Inter Cluster PCA.

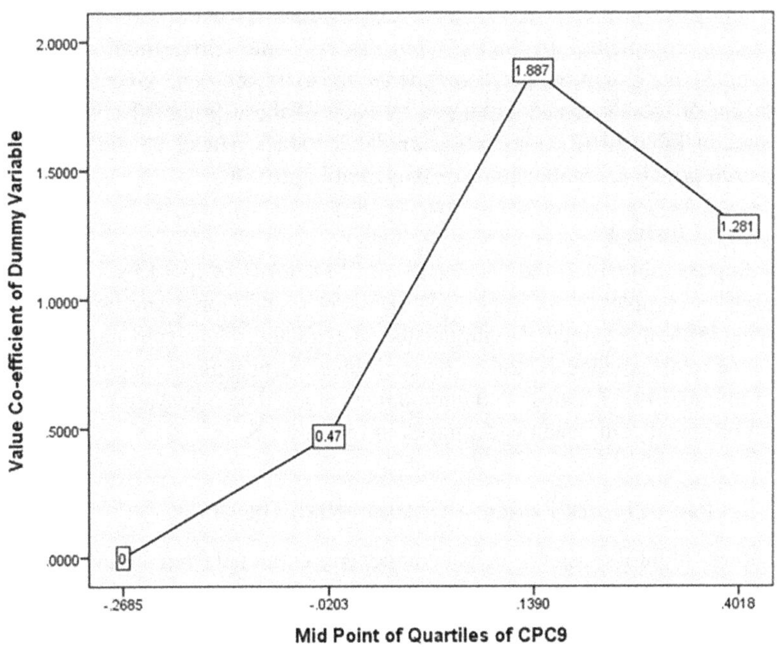

FIGURE 14.10 Plot of Coefficient Quartile Dummy Variables against Midpoints of CPC_9.

modules." Applying PCA, six principal components are extracted. The eigenvalue and the proportion of total variance explained by each of these principal components are presented in Table 14.9, along with cumulative values. The eigenvalue corresponding to each principal component is displayed in Figure 14.4.

14.7.4.3 Cluster III

According to Table 14.7, the Cluster III of EFs consist five EFs – *viz.*, "testing effort," "testing resource allocation," "level of programming technologies," "development team size," and "domain knowledge." Applying PCA, five principal components are extracted. The eigenvalue and the proportion of total variance explained by each of these principal components are presented in Table 14.10, along with cumulative values. The eigenvalue corresponding to each principal component is displayed in Figure 14.5.

14.7.4.4 Cluster IV

According to Table 14.7, the Cluster IV of EFs consist of six EFs – *viz.*, "testing tools," "storage devices," "human nature," "documentation," "testing methodologies," and "system software." Applying PCA, six principal components are extracted. The eigenvalue and the proportion of total variance explained by each of these principal components are presented in Table 14.11, along with cumulative values. The eigenvalue corresponding to each principal component is displayed in Figure 14.6.

14.7.4.5 Cluster V

According to Table 14.7, the Cluster V of EFs consist of four EFs – *viz.*, "testing coverage," "complexity in logic," "processors," and "programming language." Applying PCA, four principal components are extracted. The eigenvalue and the proportion of total variance explained by each of these principal components are presented in Table 14.12, along with cumulative values. The eigenvalue corresponding to each principal component is displayed in Figure 14.7.

14.7.4.6 Cluster VI

According to Table 14.7, the Cluster IV of EFs consist of five EFs – *viz.*, "testing coverage," "complexity in logic," "processors," and "programming language." Applying PCA, five principal components are extracted. The eigenvalue and the proportion of total variance explained by each of these principal components are presented in Table 14.13, along with cumulative values. The eigenvalue corresponding to each principal component is displayed in Figure 14.8.

14.7.5 Inter-cluster Principal Components Extraction

The number of intra-cluster principal components extracted from the EFs are clusters according to procedure explained in Section 8.1. The details are presented in Table 14.14. However, there is a common characteristic among the extracted principal components. Each of them describe about 90% of total variance of the EFs in the respective categories.

It should be noted that the PCA conducted in Section 8.2 has extracted the principal components that are uncorrelated within the respective clusters only. Moreover, the principal components are a linear combination of the EFs in the respective groups. Under this circumstance, we should conduct PCA considering these 22 principals as variables. As in Section 8.2, the principal components are extracted, explaining 90% of the variation of these 22 variates.

The eigenvalues and proportion of variation explained by principal components, along with their cumulative values, are presented in Table 14.15. The principal component number and the corresponding eigenvalues are displayed in Figure 14.9. A perusal of Table 14.15 shows that the first 11 principal components explain about 91.732% of the total variance of the 22 intra-cluster principal components. Among them, the first two intergroup principal components explain, respectively, 26.186% and 14.530 of total variance. Also, the first six principal components explain, in total, 73.759% of total variance. Inclusion of the principal components from seven to 11 has added 17.973% of total variance.

14.8 LOGISTIC REGRESSION MODEL

A multiple logistic regression model is fitted using the 11 principal components as the covariates extracted from based inter-cluster principal components. The maximum likelihood estimates of the coefficients, their standard errors, and the value of Wald statistics (w_j) of their significant p-values and 95% Wald statistic based confidence limits for the model coefficients are calculated. The results are presented in Table 14.16.

The sampling distribution of G under H_0 is χ^2 with 11 *d. f.* and the significant *p*-value can be computed as $p = p(\chi_{11}^2 > 26.57) = 0.0053$. Since this *p*-value is smaller, it is decided that the data have sufficient evidence for the rejection of H_0. Hence, the estimated logistic regression model can be considered as significant at 5% level of

TABLE 14.16
Fitted Logistic Regression Model (PCs from Clusters)

Variable	Coefficient	SE	w_j	*p*	[95% Confidence Limits]	
cpc1	3.662417	2.558648	1.43	0.152	−1.35244	8.677274
cpc2	4.727238	2.625481	1.8	0.072	−0.41861	9.873086
cpc3	−3.41584	2.159374	−1.58	0.114	−7.64814	0.816454
cpc4	2.118547	1.890878	1.12	0.263	−1.58751	5.824599
cpc5	−2.78825	1.888645	−1.48	0.14	−6.48992	0.91343
cpc6	4.313636	2.312824	1.87	0.062	−0.21942	8.846688
cpc7	6.463114	3.108014	2.08	0.038	0.371518	12.55471
cpc8	2.39743	1.936911	1.24	0.216	−1.39885	6.193707
cpc9	−10.3935	3.960206	−2.62	0.009	−18.1553	−2.63161
cpc10	−4.61897	1.982148	−2.33	0.02	−8.50391	−0.73403
cpc11	1.086495	1.536359	0.71	0.479	−1.92471	4.097704
constant	0.814845	2.641073	0.31	0.758	−4.36156	5.991253

significance. Significance of the constant term and each of the 11 CPCs in the fitted model may be investigated by testing the following sets of hypotheses.

$$H_{0j} : \beta_j = 0, j = 0,1,2,\ldots,11$$

Against $H_{1j} : \beta_j \neq 0, j = 1, 2, \ldots 11$

The values of Wald statistics calculated for the testing the above sets of hypotheses are presented in the fourth column of Table 14.16. The p-value calculated under the respective H_{0j} is shown in the fifth column of Table 14.16. It may be noted that the p-value of CPC_7, CPC_9, and CPC_{10} are less than 0.05. Therefore, the hypothesis H_{07}, H_{09}, and H_{10} are rejected at 5% of the level of significance. The data does not provide sufficient evidence to reject the remaining nine hypotheses. Thus it is inferred that the covariates CPC_7, CPC_9, and CPC_{10} are found significant in determining the software reliability. The constant term and predictors CPC_1, CPC_2, CPC_3, CPC_4, CPC_5, CPC_6, CPC_8, and CPC_{11} are not significant in the model.

14.8.1 Logistic Regression Model (Reduced Model I)

Since the main aim is of fitting the best model by minimizing the number of parameters (Hosmer et al., 2013), it is proposed to fit a reduced model with the principal components CPC_7, CPC_9, and CPC_{10} as the covariates. The results of fitting the reduced model are presented in Table 14.17.

The sampling distribution of G under H_0 is χ^2 with $3d.f.$, and the significant p-value can be computed as $p = p(\chi_3{}^2 > 11.214) = 0.0015$. Hence the estimated logistic regression model can be considered as significant at 5% of the level of significance.

Significance of the constant term and each of the three CPCs in the estimated model may be investigated by testing the following sets of hypotheses. $H_{0j} : \beta_j = 0$, $j = 1, 2, 3$. Against $H_{1j} : \beta_j \neq 0, j = 1, 2, 3$. The values of Wald statistics calculated for testing the previous sets of hypotheses are presented in the fourth column of Table 14.17. The p-values calculated under the respective H_{0j} are shown in the fifth column of Table 14.17. It may be noted that the p-value of CPC_9 only is less than 0.05. Therefore, the hypotheses H_{02} is rejected. Also, the data does not provide sufficient evidence to reject the remaining two hypothesis H_{01} and H_{03}. Thus it is inferred that the covariate CPC_9 is found significant in determining the software reliability. The constant term and the predictors CPC_7 and CPC_{10} are not significant in the model.

TABLE 14.17
Reduced Multiple Logistic Regression Model Summary I

Variable	Coefficient	SE	w_j	p	[95% Confidence Limits]	
cpc7	2.018	1.567	1.29	0.198	– 1.054121	5.091444
cpc9	–4.834	1.763	– 2.74	0.006	– 8.291518	– 1.377699
cpc10	– 2.080	1.252	– 1.66	0.097	– 4.534515	0.3740401
constant	3.660	1.402	2.61	0.009	0.9109543	6.410091

14.8.2 Simple Logistic Regression Model (Reduced Model II)

It is proposed to fit a reduced simple logistic regression model with the principal component CPC_9 as the covariate. Results of fitting the reduced model are presented in Table 14.18.

The estimated simple logistic regression model $\hat{p}(x) = \frac{e^{\hat{g}(x)}}{1+e^{\hat{g}(x)}}$ and the estimated logit is *g(x) = 3.199 – 4.178CPC9*. The log-likelihood statistic for the fitted simple logistic regression model is calculated as –25.7365. Therefore, the value of the likelihood ratio test statistic for testing the hypotheses is calculated from the log-likelihood values of the estimated model and the null model as

$$H_0 : \beta_j = 0, j = 1.$$

Against $H_1 : \beta_j \neq 0, j = 1.$

G= −2[−25.736−(22.8667)}; G= 5.7396 $p = p(\chi_2^2 > 5.7396) = 0.0019$. The sampling distribution of G under H_0 is χ^2 with 1 degree of freedom. The corresponding significant *p*-value is 0.0019. Therefore, the data has strong evidence to reject H_0 at 1 level of significance. The significance of the constant term and the covariates term CPC_9 in the model may be investigated by testing the following sets of hypotheses.

$$H_{0j} : \beta_j = 0, j = 0 \text{ and } 1.$$

against $H_{1j} : \beta_j \neq 0, j = 0$ and 1

The values of the Wald statistic calculated for testing the above sets of hypotheses are presented in the fourth column of Table 14.18. The *p*-values calculated under the respective H_{0j} are shown in the fifth column of Table 14.18. It may be noted that the *p*-value is calculated for testing the constant and covariate terms are less than 0.01. Thus it is inferred that both the constant and covariate CPC_9 are found significant in determining software reliability. The reduced model II provides a better fit to the data for assessing the reliability of the software.

Now the estimated simple logistic regression model is obtained, assuming that the covariates CPC_9 has a linear relationship with log (odds) or logit function. The linear relationship is verified graphically as follows: The quartiles of CPC_9 are −0.122, 0.0815, and 0.1965. A multiple logistic regression model is fitted for y with the three quartiles of CPC_9 as dummy variables. The estimates of the coefficients of the dummy variables and the midpoints of the quantile groups of CPC_9 are as follows (Table 14.10): The quartiles of CPC_9 are −0.122, 0.0815, and 0.1965. A multiple

TABLE 14.18
Reduced Simple Logistic Regression Model with CPC_9 as the Covariate

Variable	Coefficient	SE	w_j	*p*	[95% Confidence Limits]	
cpc9	−4.178	1.537	− 2.72	0.007	− 7.191	− 1.165407
constant	3.199	1.0005	3.20	0.001	1.238	5.160467

logistic regression model is fitted for y with the three quartiles of CPC_9 as dummy variables. The estimates of the coefficients of the dummy variables and the midpoints of the quantile groups of CPC_9 are as given in Table 14.19.

The plot of the midpoints of the four quantile groups and the estimates of the coefficient of the dummy variables are displayed in Figure 14.4. The line joining the plotted points exhibits the approximately linear relationship between the covariate CPC_9 and the outcome variable y.

14.8.2.1 Assessing the Fitness of the Model

The goodness of the fitted simple logistic regression model has to be analyzed before applying it for estimation of logistic probability. Hosmer-Lemeshow test (Table 14.20), classification table, and the area under the receiver operating characteristic curve are applied to assess the goodness of the fitted model.

Since the total number of observations is 50, and the deciles are considered total frequency if each decile is 5. Thus the sum of the expected frequencies corresponding to the two values of the outcome variable, Y = 1 (software is reliable) and Y = 0 (software is nor reliable) should be 5. This is possible only when all the expected frequencies are less than 5 provided none of them in each case is zero. Because of

TABLE 14.19
Estimated Coefficient and Midpoint of Quantile Group

Quantile Group	Estimated Coefficient	Midpoint of Quantile Group
< −0.122	0.000	−0.2685
(−0.122, 0.0815)	0.470	−0.0203
(0.0815, 0.1965)	1.887	0.139
> 0.1965	1.281	0.4018

TABLE 14.20
Contingency Table for Hosmer-Lemeshow Test

Decile Group	Y = 0		Y = 1		Total
	Observed	Expected	Observed	Expected	
1	4	3.213	1	1.787	5
2	1	2.715	4	2.285	5
3	3	2.320	2	2.680	5
4	2	1.913	3	3.087	5
5	2	1.592	3	3.408	5
6	1	1.193	4	3.807	5
7	1	0.819	4	4.181	5
8	0	0.535	5	4.465	5
9	1	0.405	4	4.595	5
10	0	0.294	5	4.706	5

these reasons, the expected frequencies in all 20 cells of Table 14.19 are less than 5. However, a careful reading of the difference between the observed and the expected frequencies in each group are moderate and small. The largest difference should be noticed as 1.715 in the second group. In all other groups, a closed assessment between the observed and expected frequencies can be noticed. The value of the Hosmer-Lemeshow test statistic is calculated as $\theta = 5.392$. The sampling distribution of the test statistic under H'_0 is χ^2 with 8 *d.f.* and the significant *p*-value is 0.715. It points out what is very evident in the data for rejection of H'_0. It shows that the simple logistic regression model seems to fit well to the data. Assessment of the validity of the fitted logistic regression model can also be done by analyzing its ability to correctly classifying the observations. The 50 respondents have already classified the commercial billing software into reliable (35 respondents) and not reliable (15 respondents). But the fitted logistic regression model classifies the software based on the estimated probability into reliable (39 respondents) and not reliable (11 respondents). The classification done by the estimated model agrees with the observed data in 36 cases and does not agree with the remaining 14 cases. Details of the cross-classification are presented in Table 14.21.

The overall rate of correct classification is, hence, the $\frac{30}{50} \times 100 = 72.0\%$ and incorrect classification is 28.0%. Sensitivity is a measure used in health-care sciences to calculate the ability in terms of rate or correctly identifying the presence of disease. Specificity is another measure that is used to calculate the ability in terms of the rate of correctly identifying the absence of disease. Here in software reliability analysis, reliable and unreliable status are considered equivalent to the presence or absence of disease. However, the values of these two measures can be used to assess the classification ability of the estimated model. In the present case, the values of these two measures are

$$Sensitivity = \frac{30}{35} \times 100 = 85.71\%; Specificity = \frac{6}{15} \times 100 = 40.00\%$$

As mentioned in Hosmer et al. (2013), sensitivity and specificity are calculated based on a single cutpoint (ie., 0.5). But these two measures can be calculated by changing the cutpoint value. Plotting the pairs (1-specificity, sensitivity) calculated for various cutpoints drawing a curve in two-dimensional plane yields the receiver

TABLE 14.21
Classification Table Based on Estimated Simple Logistic Regression Model with the Cutpoint of 0.5

		Observed		Total
		Reliable	**Not Reliable**	
Classified	Reliable	30	9	39
	Not Reliable	5	6	11
	Total	35	15	50

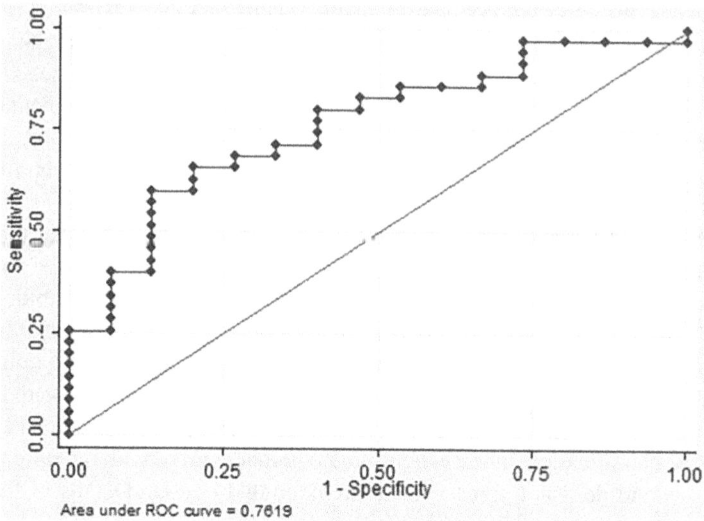

FIGURE 14.11 Receiver Operating Characteristic Curve.

operating characteristic curve (ROC). The ROC curve drawn for the software reliability data is displayed in Figure 14.11. The area under the curve is calculated as 0.7619. This value is also a measure of the classification accuracy of the fitted model. According to Hosmer et al. (2013), the classification accuracy of the fitted simple logistic regression model can be categorized as belonging to the status of "acceptable discrimination."

14.9 CONCLUSION

This chapter provided a clear idea about EFs and their significance to the study of the reliability of software using a statistical methodology. This study attempted to examine the most significant EFs using the relative average (relative weight) method. It provides the significance of the factors with their ranks. Kendall's tau was used to study the relationship between the variables (EFs). That is how the factors were correlated with other factors. There are 34 EFs, which were proposed to study the reliability of software. The factors were grouped into homogeneous clusters. To reduce the dimensionality, the PCA was used in two stages. In stage one, the components were extracted from the clusters that were intra-cluster PCs. The second stage, the components were extracted from between clusters – that is, inter-cluster PCs. The PCs that were extracted from inter-cluster were taken into consideration for further analysis. Here the response variable is dichotomous; hence, logistic regression was fitted to the data that was extracted from inter-cluster PCs. The validity of the model was justified with the Hosmer-Lemeshow Test and ROC curve. It can be decided that the estimated logistic regression model enables us to classify the opinions on the reliability of the given software with an acceptable level of accuracy.

REFERENCES

Anderberg, M. R. (1973). *Cluster Analysis for Applications.* Academic Press, New York.

Dunteman, G. H. (1989). *Principal Components Analysis*, QASS series. Sage Publications, London.

Goel, A. L. (1985). Software Reliability Models: Assumptions, Limitations, and Applicability, *IEEE Transactions on Software Engineering*, vol. SE-11, no. 12, pp. 1411–1423.

Goel, A. L., & Okumoto, K. (1978). An Analysis of Recurrent Software Failures in a Real – Time Control System, *In Proceedings of ACM Annual Technology Conference*, ACM, Washington, DC, pp. 496–500.

Goel, A. L., & Okumoto, K. (1979). Time Dependent Error Detection Rate Model for Software Reliability and Other Performance Measures, *IEEE Transaction on Reliability*, vol. R-28, no. 3, pp. 206–211.

Hosmer, D. W., Lemeshow, S., & Sturdivant, R. X. (2013). *Applied Logistic Regression.* Wiley-Inter-Science Publication. John Wiley & Sons, INC, Hoboken, NJ.

Jelinski, Z., & Moranda, P. (1972). Software Reliability Research. *In Statistical Computer Performance Evaluation,* ed. W. Freiberger, Academic, New York, pp. 465–484.

Kumar, V., Mathur, P., Sahni, R., & Anand, M. (2016). Two-Dimensional Multi-Release Software Reliability Modeling for Fault Detection and Fault Correction Processes, *International Journal of Reliability, Quality and Safety Engineering*, vol. 23, no. 3, p. 1640002.

Kumar, V., & Sahni, R. (2020). Dynamic Testing Resource Allocation Modeling for Multi-Release Software using Optimal Control Theory and Genetic Algorithm, *International Journal of Quality & Reliability Management* (Ahead-of Print).

Littlewood, B. (1980). The Littlewood – Verrall Model for Software Reliability Compared with Some Rivals, *The Journal of Systems and Software*, vol. 1, pp. 251–258.

Littlewood, B., & Verrall, J. L. (1973). A Bayesian Reliability Growth Model for Computer Software, *Applied Statistics*, vol. 22, no. 3, pp. 332–346.

Patwa, S., & Malviya, A. K. (2014). A Survey on Factors Affecting Testing Techniques in Object Oriented Software, *International Journal of Applied Research on Information Technology and Computing*, vol. 5, no. 1, pp. 78–85.

Pham, H. (2006). *System Software Reliability, Springer Series in Reliability Engineering*, Springer-verlag, London.

Pham, T., & Pham, H. (2019). A Generalized Software Reliability Model with Stochastic Fault-Detection Rate, *Annals of Operations Research*, vol. 277, no. 1, pp. 83–93.

Yamada, S., Hishitani, J., & Osaki, S. (1993). Software Reliability Growth Model with Weibull Testing – Effort: A model and Application, *IEEE Transaction on Reliability*, vol. R-42, pp. 100–106.

Yamada, S., & Osaki, S. (1984). *Non-homogeneous Error Detection Rate for Software Reliability Growth, Stochastic Models in Reliability Theory*, Springer-Verlag, New York.

Yamada, S., & Osaki, S. (1985). Software Reliability Growth Modeling: Models and Applications, *IEEE Transactions on Software Engineering*, vol. SE-11, no. 12, pp. 1431–1437.

Zhang, X., & Pham, H. (2000). An Analysis of Factors Affecting Software Reliability. *Journal of Systems and Software*, vol. 50, no. 1, pp. 43–56.

Zhu, M., Zhang, X., & Pham, H. (2015). A Comparison Analysis of Environmental Factors Affecting Software Reliability, *Journal of Systems and Software*, vol. 109, pp. 150–160.

15 Maintenance Data-Trends Based Reliability Availability and Maintainability (RAM) Assessment of a Steam Boiler

Suyog S. Patil
Zeal College of Engineering
Savitribai Phule Pune University
Pune, Maharashtra, India

Sharad Institute of Technology College of Engineering
Yadrav, Maharashtra, India.

Anand K. Bewoor
Cummins College of Engineering
Pune, Maharashtra, India

Rajkumar Bhimgonda Patil
Center for Advanced Life Cycle Engineering (CALCE),
University of Maryland
College Park, MD, USA

Annasaheb Dange College of Engineering & Technology
Ashta, India

Mohamed Arezki Mellal
M'Hamed Bougara University
Boumerdes, Algeria

Center for Advanced Life Cycle Engineering (CALCE)
University of Maryland
College Park, MD, USA

CONTENTS

15.1 INTRODUCTION

The process industries are the industries that work either in batches or continuous. Such industries exist in every corner of the world and have a considerable share in the country's economy. In many of the process industries, such as food, beverages, chemicals, pharmaceuticals, petroleum, ceramics, base metals, coal, plastics, rubber, textiles, wood and wood products, paper, and paper products, steam is used as a heat transfer medium. The steam is a vaporized water, part of the gas and part of the liquid water and continuously generated in the boiler in a controlled manner. As an essential element of its operations, many process industries rely on industrial steam boilers. The boiler is a complex system that uses process integration, modern technology, software interfaces, and interdisciplinary tasks. Developing new products now involves increased organizational requirements, increased complexity, and lower costs for product development [1]. Reliability, availability, maintainability (RAM), and safety include the integrity of the system features and their components [2–6]. The availability study helps to classify the boiler system, subsystems, and components based on their reliability and maintainability that contribute to system failure and safety-related issues. Therefore, reliability, maintainability, and optimization of product life-cycle costs have now become the focus of interest [7, 8]. The availability of a plant can be enhanced by improving uptime and downtime of boiler subsystems and components. It is, therefore, decided to carry out the RAM analysis of the components of typical steam boilers used in process industries.

The primary objective of this study is to review the present studies on the reliability and maintainability of the steam boilers, classify them into different subsystems and components, and identify the critical components and failure modes. Very few studies have been conducted on the equipment failure of the boiler. The risk of boiler tube failure is higher because the boiler tubes are continuously in contact with water. The failure analysis of the boiler tubes is carried out by various techniques, such as visual inspection, microstructural, hardness measurements, and residual stress measurements by means of x-ray diffraction methods [9], scanning electron microscopy, optical microscopy, energy dispersive spectroscopy methods [10], failure modes and effects analysis (FMEA), and stochastic techniques [11], and it is expressed that the salt deposits, fly ash, and quick chemical responses among water and tube material [12] stress corrosion cracks, pits on boiler tube surfaces, creep, and gradual destruction [13] are the main reasons for boiler tube failure. Microstructure assessment and thermo-mechanical analysis of the boiler welded joints have been conducted [14] using the finite element method and the electron backscattered diffraction analytic technique.

In the RAM studies of the boiler, the researchers, and practitioners have also concentrated on the failure analysis of the combustion, ignition, or fuel supply systems. Vandermeer et al. [15] recorded the possible reasons for ignition flame failure of the boiler, and a flame loss detector is proposed to avoid this type of failure. The wear and tear of the rotors, end of the life of the mill, utilization of faulty hammers, and issues in the coal bunkers are stated as some of the possible reasons for coal-crusher failures [16]. An alternative solution to fuel-feeding system failure is provided in [17] using FMEA, Taguchi method, and cause and effect diagram. Based on plant maintenance records, Barry and Hudson [18] suggested a probabilistic feedback scheme to plan the maintenance schedule of the fuel-feeding system.

The availability of the existing plant can be enhanced by assessing the reliability and maintenance processes. Few researchers identified and analyzed some critical components of the boiler system, such as induced draft (ID) fan, forced draft (FD) fan, and boiler water drum. Parthiban et al. [16] examined various issues related to boiler fans and identified inappropriate inlet air motion as a major cause for FD fan failure. The conditions that need to be considered while choosing fans are suggested in [19]. The use of a coordinated condition monitoring approach for accurate maintenance of the ID fan used in a thermal power plant is recommended by Jagtap et al. [20]. Rajkumar et al. [21] expressed that low water levels in the boiler drum may cause the detonation and high-water levels will cause the presence of water particles in the steam as some of the issues related to the water level in the boiler. Jagtap and Bewoor [22] prioritized the equipment of a thermal power plant using the analytical hierarchy process. Carazas et al. [23] proposed a reliability and availability assessment method using FMEA.

To optimize different maintenance actions [24], to decide maintenance strategies, and to minimize the operational costs and cleaning costs of the biomass boiler [25], a dynamic programming algorithm is developed. Kiran et al. [26] proposed and applied a model to form an optimum maintenance schedule for improving plant availability in the process plant. Jagtap and Bewoor [27] presented the Markov probabilistic approach-based availability simulation modeling approach to evaluate the performance of the subsystem of the thermal power plant. Marquez and Iung [28] used Monte Carlo simulation to evaluate the availability of a cogenerating plant considering system complexity, failure modes, and environmental factors affecting system performance. Arjunwadkar et al. [29] investigated several major operational and performance-related problems, such as agglomeration, gas refluxation, back-sifting, emission control, and bed temperature control of the circulating fluidized bed boiler components. Woo [30] identified key plant failure parameters and improved plant efficiency by altering working procedures to ensure minimal plant outage. Patil and Bewoor [31] analyzed a boiler system's components by using the best-fit failure model method and the expert-judgment method model and identified critical components from a reliability perspective.

It is observed from the literature review that, until now, there has been a scarce study on the failure analysis of the few boiler components. A few researchers have tried to reduce the boiler failure rate and decided on optimum maintenance activities in the process industry. The present studies have shown that the RAM assessment of the steam boilers need to be performed. This chapter examines a case study conducted on a typical steam boiler in the process industry to estimate the preventive maintenance plan based on reliability and to improve capacity utilization using RAM analyses. The remainder of the chapter is organized as follows: Section 15.2 presents the methodology used for RAM analysis of steam boilers used in process industries. A case study illustrating the reliability maintainability and availability characteristics of an industrial steam boiler at a different time interval is discussed in Section 15.3. Finally, the discussion on the results of RAM analysis of the steam boiler and conclusions are presented in Section 15.4.

15.2 RAM ANALYSIS METHODOLOGY

To choose time-to-failure (TTF) and time-to-repair (TTR) models, Barabady and Kumar [32] presented a framework based on various data trends. This approach is robust and simplified and can be used to analyze the maintenance data. On that basis, a new framework that includes a number of measures to choose models is proposed. The framework presented in Figure 15.1 is complete and easier to use for industrial applications in the model selection process. It shows a comprehensive flow diagram used here to identify and analyze the faults.

The analysis begins with the categorizing the steam boiler system into different levels, including assembly, subassembly, and components when selecting a system for analysis. RAM analysis begins with data collection from a variety of sources, such as maintenance history cards, registers, opinions of maintenance analysts, and experts. The Bayesian approach can be used when there is not enough data available. If adequate information is available, then it can be parametric or nonparametrically analyzed. Many system failures are often seen as small in number; therefore, the Pareto chart analysis technique is essential for the identification of critical components. The next step is to analyze data trends through some graphical and analytical approaches after critical analysis. The existence of a trend in the failure data can be checked by some graphical techniques such as cumulative failure versus time plot, timeline plot, and serial co-relation plot and by analytical techniques, such as the Mann test and military handbook tests.

Afterward, estimate the goodness of fit of the failure data, and, finally, provide the time to failure model. The data can be fitted to various distributions, such as Weibull, normal, log-normal, and exponential. The most probable distribution is then evaluated based on the Chi-square, the classic *p*-value test, or the Kolmogorov-Smirnov (K-S) test and estimate distribution parameters. Finally, reliability and maintainability characteristics are evaluated. When the reliability features are discovered, the notion of prominence measurement can be used to detect the criticality of each element and the partial failure subsystem. It defines the weakest regions of the system and explains the changes that will enhance system reliability.

15.3 RAM STUDY OF AN INDUSTRIAL STEAM BOILER – A CASE STUDY

For the assessment of reliability data, different models are available. The required system analysis model must be chosen based on TTF and TTR data. Various methods for modelling the reliability of repairable and nonrepairable systems have been

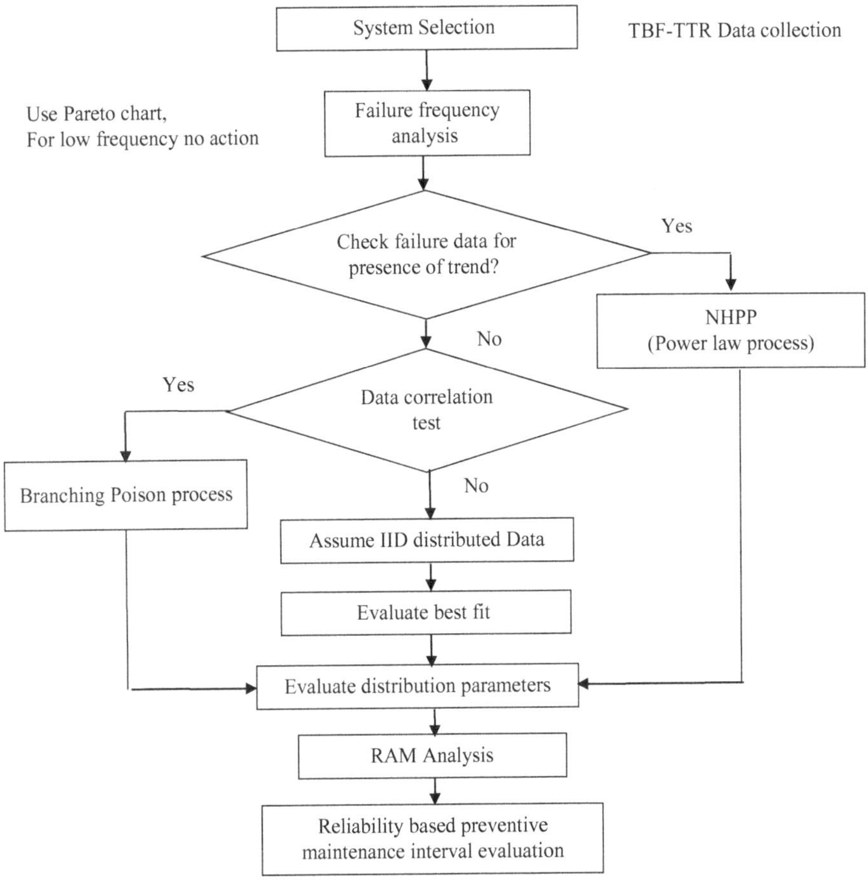

FIGURE 15.1 TTF-TTR Model Selection Framework (Adapted from [31] and Modified).

presented in the literature. A large number of studies have to be carried out in many processes to determine the trends in the data and to check the goodness of fit. The framework must, therefore, be simplified and flexible so that it can be used appropriately assessment and the duration of the analysis can be reduced.

15.3.1 System Selection

Mechanical steam boilers mainly power many of the process industries. Any component, assembly, or subsystem failure may cause the entire boiler system to fail and may cause the entire system to stop working. To increase the availability of the entire plant, the reliability and performance of the boiler must be improved. For analytical purposes, the entire boiler system is divided into nine different subsystems: combustion and ignition system (CIS), feed-water supply system (FWS), blow-down system (BDS), emission control system (ECS), fuel supply system (FSS), steam circulation system (SCS), control system (CS), electrical system (ES), and other systems (OS). There is a large number of components in each subsystem. Further assessment considers that there is only adequate failure and repair information for the critical elements from each subsystem.

15.3.2 Data Collection

The failure and repair data are collected from maintenance history sheets, as well as opinions from industry experts based on their experience. More accurate analytical results can be obtained if sufficient data is available. In most real cases, the number of failure data points available is not sufficient to apply the general parameter estimation techniques. Therefore, in such cases, failure data can be collected by the expert-judgment technique. The idea behind the expert-judgment data collection is using the knowledge and experience of the maintenance personnel. The maintenance people have sufficient knowledge of the machines and their failures. Such specialists have several years of data on the failure and repair of boilers. The knowledge of the industrial experts for a random variable t (i.e., TTF-TTR data) is collected for the reliability studies.

After the collection of maintenance data, a frequency failure and criticality analysis must be performed to identify critical subsystems. Pareto chart analysis can be used to identify the weaker subsystems [33]. The failure frequency analysis of boiler subsystems was carried out using the Pareto analysis method for the identification of critical subsystems and results are shown in Figure 15.2. The failure count of the FWS is nearly 78 (which contributes nearly 27.18% of the total failures), SCS contributes nearly 17.08%, and BDS contributes nearly 13.24% of the total failures. Failure rates of the aforementioned subsystems have been reported to be comparatively higher than in other subsystems.

15.3.3 Trend Testing

The existence of the trend in the failure data must be determined before the analysis model is selected. The key stage in the analysis of failure data is to ensure that the data is distributed independently and identically. Certain graphical and analytical methods, such as cumulative failure versus time, time line plot, and scatter plot have been suggested by Louit et al. [34] to check the presence of a trend in the data. Trend testing is conducted at various subsystem levels in the data. Further, the scatter plot of the service life of $(i)^{th}$ failure versus $(i - 1)^{th}$ failure is plotted for the failure data of the boiler system. Cumulative failure versus cumulative time and scatter plots have been plotted to check the trend in the failure data. Figure 15.3 shows the results of

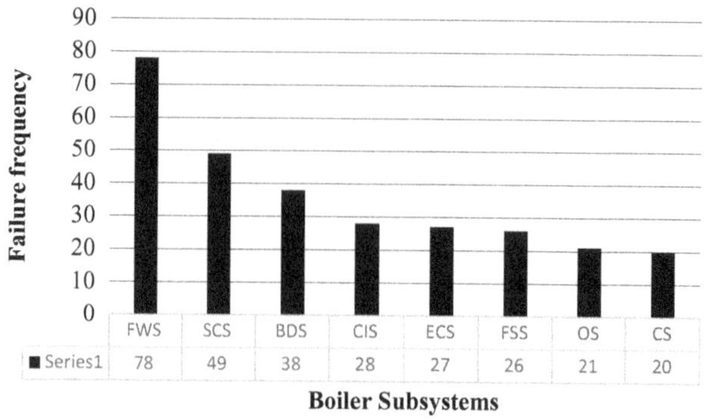

FIGURE 15.2 Frequency Failure Analysis.

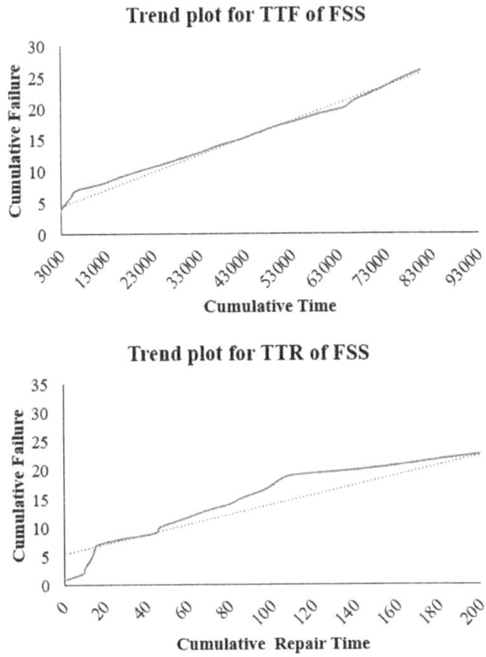

FIGURE 15.3 Cumulative Failure Trend Test Graphs for FSS.

cumulative failure versus cumulative time plot for TTF and TTR data of the FSS, and it is observed as a linear plot with the least error. Therefore, it is concluded that there is no trend, and it is shown in Figure 15.4, and a single cluster with few anomalies is observed, which shows no trend in the data. From these graphical tests, it is concluded that there is no trend in the maintenance data and hence data is independent and identically distributed. Similarly, trend testing and serial correlation tests are also carried out for other subsystems, and it is observed that there is no trend in the data.

15.3.4 Goodness-of-Fit Test

From trend testing plots, it is observed that there is no trend in the failure data so that trend-free data are further evaluated to establish the accuracy of the failure distributions for the assessment of the reliability of the boiler system. The K-S test is carried out by ReliaSoft Weibull++10 software to find the goodness of fit, and the results are shown in Table 15.1. Weibull-3P is the best distribution of TTF and TTR data in most boiler system components.

15.3.5 RAM Analysis

The boiler is a complex system, and it consists of parallel or series configuration of components or subsystems. All boiler subsystems are connected to a sequence that ensures that the entire boiler operates in a satisfactory condition when all subsystems work. The reliability analysis is carried out in accordance with the framework

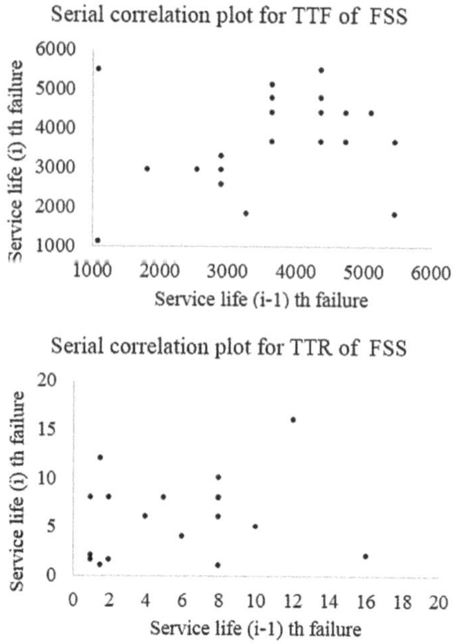

FIGURE 15.4 Serial Correlation Test Plots for FSS.

specified in Section 15.3. The theoretical reliability of all subsystems with the best-fit distribution has been estimated by using ReliaSoft's Weibull++10 Software package at different intervals. The reliability at the end of 1,460 days, 2,920 days, 4,380 days, and 5,840 days and reliable life at 99%, 90%, 75%, and 50% reliability levels has been estimated and presented in Table 15.2.

TABLE 15.1
Best-Fit Distribution Results for Boiler Subsystems

Sr. No.	Subsystems	TBF Data Analysis		TTR Data Analysis	
		Best Distribution	Distribution Parameters	Best Distribution	Distribution Parameters
1	CIS	Weibull 3P	$\beta = 1.569$ $\theta = 1402$ $\gamma = 2762$	Weibull 3P	$\beta = 1.57$, $\theta = 36.89$, $\gamma = 1.58$
2	FWS	Weibull 3P	$\beta = 1.913$ $\theta = 3816$ $\gamma = 418.9$	Weibull 3P	$\beta = 1.57$, $\theta = 36.89$, $\gamma = 1.58$
3	BDS	Weibull 3P	$\beta = 0.885$ $\theta = 2788$, $\gamma = 37.00$	Weibull 3P	$\beta = 0.703$, $\theta = 2.007$, $\gamma = 0.88$
4	ECS	Normal	$\mu = 3582$, $\sigma = 1115$	Log- normal	$\mu' = 1.894$, $\sigma' = 0.34$
5	FSS	Normal	$\mu = 3095$, $\sigma = 1621$	Weibull 3P	$\beta = 0.78$, $\theta = 9.86$, $\gamma = 0.678$
6	SCS	Normal	$\mu = 3754.3$, $\sigma = 961.85$	Weibull 3P	$\beta = 0.6$, $\theta = 4.38$, $\gamma = 0.98$
7	CS	Normal	$\mu = 3060$, $\sigma = 1879$	Log- normal	$\mu' = 0.748$, $\sigma' = 0.29$
8	OS	Log- normal	$\mu' = 8.42$, $\sigma' = 0.12$	Weibull 3P	$\beta = 1.98$, $\theta = 1.72$, $\gamma = 1.245$

TABLE 15.2
Reliability Results of Boiler Subsystems

Sub-system	Design life (days) for a given reliability				Reliability at a given time				MTBF (days)
	0.99	**0.90**	**0.75**	**0.50**	**1460**	**2920**	**4380**	**5840**	
CIS	26.13	273.9	747.9	1802.19	0.570	0325	0.185	0.106	2600.0
FWS	0.00	757.9	1570.69	2731.84	0.7727	0.4609	0.2122	0.0760	2966.68
BDS	52.37	255.98	718.70	1879.3	0.5760	0.3569	0.2276	0.1477	2999.11
ECS	987.48	2152.9	2830.04	3582.40	0.9714	0.7236	0.2372	0.0214	3582.40
FSS	0.00	1018.0	2001.95	3095.19	0.8434	0.5430	0.2139	0.0451	3095.19
SCS	1516.67	2521.6	3105.52	3754.28	0.9914	0.8071	0.2576	0.015	3754.28
CS	0.00	652.40	1792.86	3060.00	0.8028	0.5297	0.2411	0.0694	3060.00
OS	3416.89	3868.1	4157.15	4503.71	1.00	0.9998	0.5927	0.0143	4535.55

TABLE 15.3
Maintainability and Availability Analysis of Boiler Subsystems

Sr. No.	Subassembly	MTBF (days)	MTTR (Hrs)	Availability
01	CIS	2600.0	31.5578	0.988
02	FWS	2966.68	17.7900	0.994
03	BDS	2999.11	3.4108	0.998
04	ECS	3582.40	7.0315	0.998
05	FSS	3095.19	12.0161	0.996
06	SCS	3754.28	7.5143	0.998
07	CS	3060.00	2.2001	0.999
08	OS	4535.55	2.7684	0.999
Boiler System Availability =				**0.9703**

Availability analysis is assessed using the corresponding most likelihood distribution and parameters based on the reliability and maintainability of the boiler subsystems. Likewise, subsystems operational availability can be calculated as follows:

$$\text{Aop} = \frac{\text{MTBF}}{MTBF + MTTR} \tag{15.1}$$

By putting these values of mean time between failure (MTBF) and mean time to repair (MTTR) of the FSS in Eq. (15.1), we get availability as follows:

$$\text{Aop} = \frac{74285}{74285 + 12.0161} = 0.9998 \tag{15.2}$$

Similarly, the availability analysis of all subsystems is carried out and presented in Table 15.3. The findings of the availability analysis show that all boiler subsystems are over 98% available, and the combustion system is less readily available. The RAM analysis shows that the combustion system is the critical subsystem from the reliability and availability perspective. Moreover, it is seen that the reliability of the steam circulation system is less, and it is 0.015 after 16 years of operation, and it is higher for the BDS, and time required to repair the combustion system is more with MTTR of nearly 32 hours. It is observed that the reliability of the boiler is decreasing as the operating time increases.

15.4 CONCLUSIONS

The study presented the qualitative and quantitative RAM analysis of the steam boiler system. A simplified framework for the analysis of the TTF and TTR data was developed and used for the analysis of the steam boiler under consideration. The analysis revealed that TTF and TTR data sets are independent and identically distributed and most of the subsystems of the steam boiler follow the Weibull distribution. Although the FWS and SCS frequencies are higher than other subsystems, the shape parameter of the Weibull distribution of FWS is greater than one, showing an increasing failure rate owing to the aging process. BDS follows a decreasing failure rate ($\beta < 1$) – i.e., burn-in period – therefore, corrective maintenance can be suggested for BDS.

The availability of the plant depends on the reliability and maintainability of the subsystems. The current RAM levels of the various subsystems of the boiler system have been calculated in this analysis. Based on the reliability analysis, a preventive maintenance schedule is determined for all subsystems to obtain the desired level of plant availability. The analysis reveals that the combustion system has the lowest MTBF, and MTTR was longer than other subsystems; therefore, it is a critical subsystem from the RAM perspective. In addition, the FWS and FSS have lower MTBF and availability and comparatively higher MTTR of other systems; thus, in order to improve the availability of the steam boiler, it is suggested that maintenance services should be allocated to these subsystems at the appropriate time.

REFERENCES

1. C. E. Ebeling, *An introduction to reliability and maintainability engineering*. New York, USA: McGraw Hill, 1997.
2. A. Birolini, *Reliability engineering: Theory and practice*, 6th ed. Berlin, Germany: Springer-Verlag Berlin Heidelberg, 2010.
3. M. A. Mellal and E. Zio, "System reliability-redundancy optimization with cold-standby strategy by an enhanced nest cuckoo optimization algorithm," *Reliab. Eng. Syst. Saf.*, vol. 201, pp. 106973, 2020.
4. M. A. Mellal, E. Zio, and E. J. Williams, "Cost minimization of repairable systems subject to availability constraints using efficient cuckoo optimization algorithm," *Qual. Reliab. Eng. Int.*, vol. 36, no. 3, pp. 1098–1110, Apr. 2020.
5. M. A. Mellal and E. Zio, "An adaptive cuckoo optimization algorithm for system design optimization under failure dependencies," *J. Risk Reliab.*, vol. 233, no. 6, pp. 1099–1105, 2019.
6. M. A. Mellal and E. Zio, "An adaptive particle swarm optimization method for multi-objective system reliability optimization," *Proc. Inst. Mech. Eng. Part O J. Risk Reliab.*, p. 1748006X1985281, vol. 233, no. 6, pp. 990–1001, Jun. 2019.

7. H. E. Ascher and H. Feingold, *Repairable system reliability: Modeling, interface, misconception and their causes*. New York, USA: Marcel Dekker, 1984.
8. M. A. Mellal and E. J. Williams, "Optimal replacement strategy of obsolete industrial components under fuzzy data," *Proc. Inst. Mech. Eng. Part I J. Syst. Control Eng.*, p. 095965181985348, vol. 234, no. 3, pp. 349–357, 2020.
9. C. A. Duarte, E. Espejo, and J. C. Martinez, "Failure analysis of the wall tubes of a water-tube boiler," *Eng. Fail. Anal.*, vol. 79, pp. 704–713, Sep. 2017.
10. S. W. Liu, W. Z. Wang, and C. J. Liu, "Failure analysis of the boiler water-wall tube," *Case Stud. Eng. Fail. Anal.*, vol. 9, pp. 35–39, Oct. 2017.
11. L. N. Moghanlou and M. Pourgol-Mohammad, "Assessment of the pitting corrosion degradation lifetime: A case study of boiler tubes," *ASCE-ASME J. Risk Uncert. Engrg. Sys. Part B Mech. Engrg.*, vol. 3, no. 4, Dec. 2017.
12. S. Chaudhuri and R. Singh, "High temperature boiler tube failures – Case studies," *Proceedings: COFA*, pp. 107–120, 1997.
13. S. K. Dhua, "Metallurgical investigation of failed boiler water-wall tubes received from a thermal power station," *Eng. Fail. Anal.*, vol. 17, no. 7–8, pp. 1572–1579, Oct. 2010.
14. H. Sang Lee, J. Sung Jung, D. Soo Kim, and K. Bong Yoo, "Failure analysis on welded joints of 347H austenitic boiler tubes," *Eng. Fail. Anal.*, vol. 57, pp. 413–422, 2015.
15. W. Vandermeer, "Flame safeguard control in multi-burner environment," pp. 1–33, 1998.
16. K. K. Parthiban, "Fans at work in boilers," *Venus Energy Audit System Report*, 2006.
17. A. Mariajayaprakash and T. Senthilvelan, "Failure detection and optimization of sugar mill boiler using FMEA and Taguchi method," *Eng. Fail. Anal.*, vol. 30, pp. 17–26, 2013.
18. D. M. Barry and M. W. Hudson, "Reliability Modelling for the Scheduling of Plant Work in Majority Vote Mode," *Int. J. Qual. Reliab. Manag.*, vol. 3, no. 2, pp. 12–20, Feb. 1986.
19. R. Thompson and D. Wong, "Boiler induced draft fan optimisation," in *Australian Society of Sugar Cane Technologists*, 2010.
20. H. P. Jagtap, A. K. Bewoor, and R. Kumar, "Failure analysis of induced draft fan used in a thermal power plant using coordinated condition monitoring approach: A case study," *Eng. Fail. Anal.*, 2020.
21. T. Rajkumar and V. M. Ramaa Priyaa, "Boiler drum level control by using wide open control with three element control system," *Int. J. Sci. Eng. Res.*, vol. 4, no. 5, pp. 204–2010, 2013.
22. H. P. Jagtap and A. K. Bewoor, "Use of Analytic Hierarchy Process Methodology for Criticality Analysis of Thermal Power Plant Equipments," *Materials Today Proc.*, vol. 4, no. 2, pp. 1927–1936, 2017.
23. F. J. G. Carazas, C. H. Salazar, and G. F. M. Souza, "Availability analysis of heat recovery steam generators used in thermal power plants," *Energy*, vol. 36, pp. 3855–3870, 2011.
24. S. Agarwal and A. Suhane, "Study of Boiler Maintenance for Enhanced Reliability of System A Review," *Materials Today: Proceedings*, pp. 1542–1549, 2017.
25. K. Macek, P. Endel, N. Cauchi, and A. Abate, "Long-term predictive maintenance: A study of optimal cleaning of biomass boilers," *Energy Build.*, vol. 150, pp. 111–117, Sep. 2017.
26. S. Kiran, K. P. Prajeeth Kumar, B. Sreejith, and M. Muralidharan, "Reliability evaluation and risk based maintenance in a process plant," *Procedia Technol.*, vol. 24, pp. 576–583, 2016.
27. H. P. Jagtap and A. K. Bewoor, "Markov probabilistic approach-based availability simulation modeling and performance evaluation of coal supply system of thermal power plant," *Lecture Notes Mech. Eng.*, pp. 813–824, 2020.

28. A. C. Marquez and B. Iung, "A structured approach for the assessment of system availability and reliability using Monte Carlo simulation," *J. Qual. Maint. Eng.*, vol. 13, no. 2, pp. 125–136, 2007.
29. A. Arjunwadkar, P. Basu, and B. Acharya, "A review of some operation and maintenance issues of CFBC boilers," *Applied Thermal Eng.*, vol. 102, pp. 672–694, 2016.
30. T. G. Woo, "Reliability analysis of a fluidized-bed boiler for a coal-fueled power plant," *IEEE Trans. Reliab.*, vol. R-29, no. 5, pp. 422–424, 1980.
31. S. S. Patil and A. K. Bewoor, "Reliability analysis of a steam boiler system by expert judgment method and best-fit failure model method: A new approach," *Int. J. Qual. Reliab. Manag.*, 2020.
32. J. Barabady and U. Kumar, "Reliability analysis of mining equipment: A case study of a crushing plant at Jajarm Bauxite Mine in Iran," *Reliab. Eng. Syst. Saf.*, vol. 93, no. 4, pp. 647–653, Apr. 2008.
33. R. B. Patil, B. S. Kothavale, and L. Y. Waghmode, "Selection of time-to-failure model for computerized numerical control turning center based on the assessment of trends in maintenance data," *Proc. Inst. Mech. Eng. Part O J. Risk Reliab.*, vol. 233, no. 2, pp. 105–117, Apr. 2019.
34. D. M. Louit, R. Pascual, and A. K. S. Jardine, "A practical procedure for the selection of time-to-failure models based on the assessment of trends in maintenance data," *Reliab. Eng. Syst. Saf.*, vol. 94, no. 10, pp. 1618–1628, Oct. 2009.

Index